Decision Making for a Sustainable Environment: A Systemic Approach

Social-Environmental Sustainability Series

Series Editor

Chris Maser

Published Titles

Decision Making for a Sustainable Environment: A Systemic Approach, **Chris Maser**

Economics and Ecology: United for a Sustainable World, **Charles R. Beaton and Chris Maser**

Resolving Environmental Conflicts, Second Edition, **Chris Maser and Carol A. Pollio**

Sustainable Development: Principles, Frameworks, and Case Studies, **Okechukwu Ukaga, Chris Maser, and Michael Reichenbach**

Social-Environmental Planning: The Design Interface Between Everyforest and Everycity, **Chris Maser**

Decision Making for a Sustainable Environment: A Systemic Approach

Chris Maser

CRC Press
Taylor & Francis Group
Boca Raton London New York

CRC Press is an imprint of the
Taylor & Francis Group, an **informa** business

CRC Press
Taylor & Francis Group
6000 Broken Sound Parkway NW, Suite 300
Boca Raton, FL 33487-2742

First issued in paperback 2019

ISBN-13: 978-1-4665-5216-6 (hbk)
ISBN-13: 978-0-367-86725-6 (pbk)

Library of Congress Cataloging-in-Publication Data

Maser, Chris.
Decision making for a sustainable environment : a systemic approach / Chris Maser.
p. cm. -- (Social-environmental sustainability)
Includes bibliographical references and index.
ISBN 978-1-4665-5216-6
1. Sustainable development--Decision making. 2. Environmental policy--Decision making. 3. Environmental protection--Decision making. I. Title.

HC79.E5M3594 2012
333.72--dc23 2012004999

Visit the Taylor & Francis Web site at
http://www.taylorandfrancis.com

and the CRC Press Web site at
http://www.crcpress.com

I lovingly dedicate this book to my three grandchildren Camden, Scott, and Evan Richards.

May our decisions begin NOW—and continue henceforth—to create a lasting legacy of social-environmental sustainability for them and all children, present and future.

Contents

Foreword ... xiii
Editor's Note ... xvii
Preface ... xix
Acknowledgments ... xxv

SECTION I Nature's Urgent Message to the World's Decision Makers

Chapter 1 The Principles of Biophysical Sustainability ... 3

The Waterbed Principle ... 4
The Law of Cosmic Unification ... 6
Nature's Inviolable Biophysical Principles ... 9
Principle 1—Everything Is a Relationship ... 9
Principle 2—All Relationships Are All Inclusive and Productive of an Outcome ... 10
Principle 3—The Only True Investment in Our Global Ecosystem Is Energy from Sunlight ... 11
Principle 4—All Systems Are Defined by Their Function ... 12
Principle 5—All Relationships Result in a Transfer of Energy ... 13
Principle 6—All Relationships Are Self-Reinforcing Feedback Loops ... 14
Principle 7—All Relationships Have One or More Trade-offs ... 17
Principle 8—Change Is a Process of Eternal Becoming ... 19
Immediate Change ... 19
Understanding Historical Change ... 19
Change as a Historical Continuum ... 20
Principle 9—All Relationships Are Irreversible ... 22
Principle 10—All Systems Are Based on Composition, Structure, and Function ... 22
Principle 11—All Systems Have Cumulative Effects, Lag Periods, and Thresholds ... 24
Principle 12—All Systems Are Cyclical, But None Are Perfect Circles ... 27
Principle 13—Systemic Change Is Based on Self-Organized Criticality ... 28
Principle 14—Dynamic Disequilibrium Rules All Systems ... 29

Summary.......30
Notes.......31

Chapter 2 Decision Making and Nature's Rules of Engagement.......37

Every Decision Has Its Antithesis.......37
Air: The Breath of Life—And of Death.......38
Water: A Captive of Gravity.......39
Soil: The Great Placenta.......40
Biodiversity: The Variety of Life.......40
Human Population: A Matter of Equality.......41
Sunlight: The Source of Global Energy.......42
Climate: The Global Arbiter.......43
Turning the Key.......43
The Inviolable Rules of Decision Making.......45
Rule 1—Everything Is a Relationship.......45
Intra-Personal.......45
Inter-Personal.......46
Between People and the Environment.......48
Between People in the Present and Those of the Future.......50
Rule 2—All Relationships Are All Inclusive and Productive of an Outcome.......51
Rule 3—The Only True Investment Is Energy from Sunlight.......55
Rule 4—All Relationships Involve a Transfer of Energy.......55
Rule 5—All Systems Are Based on Composition, Structure, and Function.......61
Rule 6—All Relationships Have One or More Trade-offs.......62
Rule 7—All Systems Have Cumulative Effects, Lag Periods, and Thresholds.......65
Rule 8—Change Is an Irreversible Process of Eternal Becoming.......66
Rule 9—Systemic Change Is Based on Self-Organized Criticality.......69
Rule 10—Dynamic Disequilibrium Rules All Systems.......70
Rule 11—Success or Failure Lies in the Interpretation of an Event.......71
Rule 12—People Must Be Equally Informed If They Are to Function as a Truly Democratic Society.......72
Rule 13—We Must Consciously Limit Our "Wants".......72
Rule 14—Simplicity Is the Key to Contentment, Adaptability, and Survival.......72
Rule 15—Nature, Environmental/Cultural Wisdom, and Human Well-Being Are Paramount.......73
Rule 16—Every Legal Citizen Deserves the Right to Vote.......75

Rule 17—This Present Moment, the Here and Now, Is All We Ever Have ... 75
Summary ... 76
Notes ... 76

SECTION II When the Decision Making Becomes Self-Serving

Chapter 3 The Stages of an Environmental Agency ... 87

Introduction ... 87
An Introduction to Command-and-Control Decision Making ... 88
The Inception of an Agency ... 89
We Are the Agency ... 90
Stages in the Cycle of an Agency ... 90
Stage One ... 90
Stage Two ... 92
Stage Three ... 92
Stage Four ... 92
When Dysfunctional Decision Making Creeps In ... 93
The Homeostatic Command-and-Control of Decision Making ... 96
The Mythology of Abnegating Personal Responsibility ... 99
Myth One—I Can't Change Because I'm Locked In ... 100
Myth Two—I Can't Commit Future Leaders to a Course of Action ... 101
Myth Three—It's Not My Responsibility Because It's Not in My Job Description ... 101
Myth Four—I Can't Make a Decision Because I Lack Definitive Data ... 101
Myth Five—It Can't Be True, So I Won't Believe It ... 102
Myth Six—Yes, But I Have to Face Reality ... 103
Myth Seven—What You Are Asking Can't Be Done—It's Impossible ... 104
The Unpredictability of Environmental Laws ... 105
Setting the Stage for the Engagement of Coping Mechanisms ... 108
How Coping Mechanisms Influence the Effectiveness of Decision Making ... 110
Anger and Aggression ... 110
Appraisal ... 111
Defensiveness ... 112
Denial ... 112
Displacement ... 114
Filters ... 116

Projection 116
Rationalization 117
Repression 118
Resistance 119
The Upshot 120
Summary 121
Notes 121

Chapter 4 Dissonance within Duplicity: Sustainability Programs at the Southern Nevada Water Authority, a Case Study 125

Jessica K. La Porte, PhD
Summary 137
Notes 137

Chapter 5 Breaking the Dysfunctional Cycle of Environmental Agency Decision Making 141

How Job Descriptions Limit Professional Decision Making 143
Leaders and Managers Make Fundamentally Different Kinds of Decisions 144
Categories of Value 145
Why a Vision Is Important 146
Worldview 147
Perception 148
Imagination 149
Understanding a Vision 152
Vision, Goals, and Objectives 152
The Negotiability of Constraints 153
Monitoring Tests a Vision's Effectiveness 154
Framing a Relevant Question 155
Seven Steps of Monitoring 155
Summary 158
Notes 159

SECTION III What Makes a Psychologically Mature Decision Maker?

Chapter 6 The Essence of a Responsible Decision Maker 163

Personal Values and Philosophy of Life 163
Characteristics of an Effective Decision Maker 165
Other Centered and Authentic 167
An Honorable Person 170

Balancing the Masculine and Feminine 171
A Therapeutic Person 173
Detachment and Equanimity 176
A Good Follower 177
Servant Leader 180
Shared Leadership 183
Willingness to Delegate Authority 184
Encourage Leadership in Others 186
Summary 187
Notes 188

Chapter 7 Coping with the Responsibilities and Pressures of Social-Environmental Decision Making 191

Introduction 191
Circumstances Faced by Decision Makers 191
Dealing with Anxiety 192
Use of Power 193
Criticism in the Form of Projection 194
Projection as a Means of Avoiding Personal Responsibility 194
Projection Can Be Either Negative or Positive 196
Criticism and Your Image 196
Responding to Negative Criticism 197
Being and Disclosing Yourself 200
Perfectionism and the Fear of Making Decisions 200
Honesty with Followers 201
Understanding Silence 203
Understanding the Need to Be Heard 204
Establishing Your Boundaries 205
Dealing with the Uncommitted 206
Accepting Slow or Delayed Results 206
Learning Your Limits 207
Avoiding Self-Deception 209
Overinvestment in Followers 209
The Value of Humor 210
Coping with Someone You Dislike 210
Identify Those Specific Behaviors that Trigger Your Own Emotions 211
Talk to the Individual Privately and Use the "Sandwich Technique" 211
Timing Is Critical 211
Remember Your Role as Decision Maker 211
A Leader Does Not Quit When Things Get Tough 211
If Nothing Else Works, Find a Facilitator 212
Imagine Yourself as Different People 212

Inspiring Performance ... 213
Nurturing Creativity ... 215
Making Do with What You Have ... 216
Establishing Realistic Objectives ... 219
The Need for Urgency ... 222
Give Counsel, Not Advice ... 223
The Questions We Ask ... 223
Maintaining Viability as a Leader ... 225
Causes of Burnout ... 226
Remedies for Burnout ... 226
Preventing Burnout ... 227
Social-Environmental Decision Making within Organizations ... 228
Summary ... 230
Notes ... 230

Chapter 8 Giving Children a Voice in Their Future through Shared Decision Making ... 233

The Great American Irony: Children Have No Voice in Their Future ... 236
First, I Asked the Children What They Wanted ... 237
And This Is What They Said ... 238
Then, I Asked Their Teacher What She Had Learned ... 242
Changing Our Adult Thinking ... 243
The Ultimate Democratic Revolution in Decision Making: Giving Children a Voice in Their Future ... 248
Address the School Board ... 249
Interviewing New Teachers ... 249
Creating a Community Vision ... 249
Resolving Conflicts ... 250
Healing the Land ... 251
Children's Advisory Councils ... 251
Summary ... 254
Notes ... 255

Afterword ... 257

Appendix: Common and Scientific Names of Plants and Animals ... 261

Index ... 263

Foreword

How often do you start to read a book on sustainability and get "bored?" Bored because you "get it"; bored because you are "burned out" out preaching and practicing sustainability in your workplace and community; bored because of despair over rates of consumption of natural resources and social systems that, ironically, "benefit" you but obviously do not benefit the vast majority of people on this earth at this moment and certainly not future generations; bored because you know that ecological and social systems are not infinitely resistant; bored because you know that many decision makers and leaders do not and will never get it?

And then maybe you think about your own children (in my case an eight-year-old daughter) or grandchildren or simply children and future children. And then you think about the people you wish would read the book you are reading—people whose decisions can turn a corporation or city or university etc. toward a sustainable path. How if they "got it," systematic change could and would happen. If you could compel a president of a country or a city council member or a university president or facilities managers or the vast majority of academics and conservation managers to read what you are reading, would you come out of your "boredom?" If you thought decision makers and leaders could understand the "inviolable biophysical principles that form the non-negotiable limitations of social-environmental sustainability," would you be re-energized about sustainability and be hopeful?

Or, if you found a well written, inspiring, thoughtful, and thought-provoking book on decision making, leadership, and social and environmental sustainability, would you just recommend the book to those who "get" it? Recommend it to those you "know" would benefit from the approach and language and examples presented in the book? Recommend it to those who would come back to you and say—'Thanks – That was great – I wish decision makers and leaders would read that book'?

It is not profound to recognize that "every decision has its antithesis," but it is profound to contemplate how decisions are made, why decisions are made that violate very simple ecological principles, how change makers could affect decision makers' decisions, and the structure of the decision making processes that really is, ultimately, dysfunctional. From decisions based on fighting "wars" against poverty and everything else (drugs etc.), to decisions that are based on an idea of "win-win," to decisions that are based on the mere structure of a given organization, hindsight can always reveal why a series of decisions lead to a given undesirable outcome. However, we must recognize that if those same decisions were made again, the outcome might be different, and therefore we need to be careful how we interpret "lessons-learned."

I read this book while in the "burned" out stage of working towards sustainability at a large Midwest U.S. university—partly burned out because I used some of the leverage of tenure and a few supportive administrators to "push" the envelope a bit, mostly to engage students with opportunities to make decisions and let them use their energy and minds. Apparently, this is not always appreciated by administrative leaders. For example, I worked for weeks with students so a truck full of coal could be parked on campus for a day, with educational signs made by students—but only after being approved by administrators. The signs explained how much coal the campus used for various activities because the decision to keep burning coal brought from Wyoming to Missouri was made because our leaders tell us we need "affordable energy so we can have affordable tuition and electric lights in the library to study." The demonstration was viewed as a success, not because it actually resulted in any changes on campus, but because the university was highlighted in the national press for putting on the demonstration. Is that really a "win?" The decision makers and president of the university said "thanks" after seeing all the press releases. Two years later an administrator has made the decision to discourage students from using campus facilities to participate in a nationwide beyond coal campaign. The students involved have been very affected by this "decision"—and not only because the administrator told the students that if they move forward with the campaign, he would do away with a student government committee on sustainability (which he cannot do!). This book is helping me understand this example and the positive lessons.

Ultimately, the challenge of this book to me is that it provides ideas, language, approaches, and examples to move a place of employment, a government, or a community toward what Chris Maser calls "psychological maturity for the social-environmental sake of all generations." I may not live up to the challenge of getting people with decision-making power in my community or place of employment to understand the message and lessons of this book (but I hope you can do better than I). But, with the help of this book, I will not do such things as send an email to the "stop your beyond coal campaign" administrator asking him to be more "psychologically mature," and I will not tell the students I wanted to use that language with the administrator. That is because I now have a greater understanding of the decision-making processes and how to act toward and with leaders. And, I am more ready to effect change in decision makers because of the lessons and wisdom and approach that Chris offers in this book. And, I understand positive messages and approaches.

I am going to get over my burn out and use what I have learned from this book to stay involved. That is not to say that Chris has once again inspired me as he first did with his book *The Redesigned Forest* when I started a degree in forestry in 1989; it is to say that this book provides me with intellectual tools (e.g., insights, language, approaches) that I can use as a person who does not have power, but who has access to people with leadership power who do not get it (and never will).

I will use this book in my courses—especially "conservation biology" and an honors freshman class entitled "sustainability in action and practice." I will use this book as the chair of a curriculum committee on sustainability to guide the process of bringing sustainability into the classroom. And I will use it to better work with faculty, staff, students and administrators. The book is replete with examples of decisions on the local, regional, and global scale. The examples bring together social and economic "concerns" with both simple and complicated biological concepts. Examples in this book can be used in any introductory sustainability course or in any boardroom (e.g., Ray Anderson, former CEO of Interface, would have loved this book), or any city council. And the idea of bringing children into the mix of decision making and priority making is brilliant!

Below I provide a simplistic example of personal decision making that I think abdicates personal social and environmental responsibility that I reworked through my mind after reading this book, and which I now know how to "understand."

I was asked to speak to new faculty about my university's public affairs mission and sustainability. The mission has three pillars: Ethical Leadership, Cultural Competence, and Community Engagement. I thought I gave good examples of how sustainability IS a public affair. After my presentation a faculty member in the construction management program came up to me to tell me that he was skeptical and thought the whole "sustainability thing" was a liberal new environmentalism that would soon die. I thanked him for listening to my talk and asked where he lived. He had some property surrounded by federal forest service land (with an easement for access); he was in the early stages of building a house that was totally off the grid (solar panels etc.), gray water system, drinking water from a well and federally protected stream, area for growing his own food and keeping some animals for food (using the waste for fertilizer etc.), and free wood from the federally protected forest for some of the building and heating. I was obsessed with the social and environmental aspects of such a project and how he was able to make the decisions. What are the true costs of the few that can do such projects—"because they want to be free from government manipulation of building codes and costs of energy and food"—to all societies? Read this book and you will be able to articulate to this highly educated professor where decision making is socially and environmentally sound ("functional"), where it is flawed ("dysfunctional," and, I would suggest, dangerous), and HOW to discuss it. At the time, I was left frustrated. However, I am going to read portions of this book a couple of more times and then hope I run into this construction management professor in our nice faculty lounge that provides many examples of poor decision making in consumption and waste (but making progress one fight at a time—which is not how progress needs to progress) and sit and have lunch with him. [Note: I believe this faculty member should build the house.] An icing on the too many layered cake story: the construction management program is adding "sustainability" to all the course names in the program because that is "where the industry is headed." Good decision? Good leadership?

I hope I can become a "psychologically mature decision-maker." I hope I can avoid making decisions at work that cause set-backs and back peddling on issues of social and environmental sustainability by university leaders. I hope that Chris Maser's illumination of a peaceful path for decision making and leadership toward social-environmental sustainability for all generations comes to fruition in my generation.

D. Alexander Wait
Missouri State University

Editor's Note

Virtually anyone can learn the mechanics of leadership and decision making, but its true heart and soul are born in a genuine love for people of all generations and the wisdom to protect their dignity and the social-environmental sustainability that nurtures them. Authentic leadership is embodied in four, self-reinforcing feedback loops of human behavior.

The first is an inner, metaphysical relationship, which determines how a person treats himself or herself. It is the individual's sense of spirituality, self-worth, personal growth, authenticity, inner strength, and so on. In short, it is degree of psychological maturity attained through personal growth that determines the level of consciousness a person exhibits in accepting accountability for his or her own behavior and its consequences.

The second relationship is how we treat one another. Not enough can be said for civility, respect, and hospitality toward other people. If we use these basic human behaviors to frame our leadership and the wisdom of compassion in our decisions, we can focus more on the mutual values that bind us and less on the tension between our beliefs and attitudes. In the final analysis, all we have to give one another of value—*ever*—is our love, trust, respect, and the benefit of our experience.

The third relationship is between people and their environment. Here, a basic principle is that community programs must be founded on local requirements and cultural values in balance with those of the broader outside world, which includes understanding environmental issues, long-term biophysical trends, and their social-environmental ramifications. After all, social-environmental sustainability is a reciprocal relationship between people and the land based on the constraints of nature's inviolable biophysical principles. Simply put, as we nurture the land, we nurture ourselves. As we abuse the land, we abuse ourselves. This is a critical nexus because every society and nation is rooted at the local level.

The fourth relationship is between people in the present and those of the future. Here, the question is: Do we who are living today owe anything to the future? If our answer is "No," then we are surely on course because we are consuming resources and polluting Earth as if there were no tomorrow. If, on the other hand, the answer is "Yes, we have an obligation to the future," then we must determine what we owe and how much because our present, *non-sustainable* course is rapidly destroying the environmental options for all generations—present *and* future. Meeting the obligation we say we have will require a renewed sense of morality, one committed to being other centered in caring for the welfare of those to come just as we wish vastly more of those before us had considered our welfare in mapping the outcome of their decisions and actions.

Finally, the essence of leadership is the imperative that we take personal responsibility for our words, deeds, decisions, and their consequences because the first step toward social-environmental sustainability begins with the respect and the quality of care we give ourselves. We must then extend that respect and care to our families, friends, and neighbors because the sustainability of a community, which is the foundation of every nation, is but a reflection of the health of its citizenry as measured by how people treat one another and thus protect the biophysical sustainability of their surrounding environment as part of everyone's birthright in all generations.

Chris Maser, Series Editor

Preface

The clear-sighted do not rule the world, but they sustain and console it.

Agnes Repplier, American essayist[1]

Every crisis in the world today—whether social or environmental—is a historical archive of human choices, decisions, and their subsequent actions, including those of yesterday. Despite all the evidence before us, we keep making the same types of unwise social-environmental decisions, ignoring volumes on volumes of historical evidence depicting their dire consequences, yet each time expecting a new and different outcome. We humans face a growing cataclysm of suffering from myriad causes, not the least of which are ideological strife and its wanton destruction of irreplaceable resources and the growing threat of global warming. Yet, those rare individuals who make decisions that would, in fact, further the social-environmental well-being of Earth are too often thwarted by the socially powerful minority who are afraid of losing the economic advantage garnered from the status quo.

Nevertheless, whosoever makes a social decision is simultaneously making an environmental decision and vice versa. This is an inescapable relationship because human society is an inseparable part of the environment, just as the environment is an indivisible part of human society. Therefore, every leader, regardless of his or her hierarchical level in government—local, national, or world—is a social-environmental decision maker, whether the leader understands it or not or accepts it or not. Thus, terming someone a "social-environmental decision maker," a "social decision maker," an "environmental decision maker," solely a "decision maker," or a "leader" matters not because society and the environment that cradles it are enveloped in an inescapable, self-reinforcing feedback loop of reciprocity.

Decision making—which defines all leadership—is an art, and like art, the quality of one's decisions depends on one's knowledge and perspective based on that knowledge, such as that of Abraham Lincoln, Mahatma Gandhi, Theodore Roosevelt, Adolph Hitler, Winston Churchill, Joseph Stalin, Jesus, Buddha, Golda Meir, or Nelson Mandela. With this in mind, I am the first to admit that any book on decision making—especially social-environmental decision making—is incomplete and always will be since definitive knowledge, with all its interactive ramifications, will forever elude us. That said, social-environmental decision making, which is one of the most multifaceted of human endeavors, is about to get even more so. In the days gone by, decision making was like playing a classic game of unidimensional chess, which was demanding enough, but those days have ended. As Vice President Hubert H. Humphrey acknowledged, "Leadership [decision making] in today's world requires far more than a large stock of gunboats and a hard fist at the conference table."[2]

The goals and objectives of the past were much simpler than those facing society today and in the future. Decision making to come is analogous to playing a multidimensional game of chess, wherein each of the three primary levels encompasses myriad sublevels, all of which interact. The three levels are inseparably interactive spheres—the atmosphere (air), the litho-hydrosphere (the rock that constitutes the restless continents and the water that surrounds them), and the biosphere (all life sandwiched in the middle, which includes human society). Today's game is at once a measure of complexity, uncertainty, interdisciplinary acuity, social-environmental sustainability, and social justice for all generations. As such, it demands an ever-greater systemic point of view seen with progressive clarity as one ascends to higher levels of consciousness, which comprises the integration of the intellect and the intuition, the material and the spiritual.

Some would argue this book is too idealistic and impractical in the face of an exploding human population with human needs that are real and immediate and question what I would propose to do about them. First, I would point out that the ideal is all that is worth striving toward and thus writing about. That said, the purpose of this book is to illuminate a peaceful path toward social-environmental sustainability for all generations. Second, we are bankrupting the global ecosystem, not only with our runaway human population that lives longer than in past eras but also through the destructive wastefulness of social conflicts of all scales, which nevertheless are justified by some segment of society as exercising the sound decision making of responsible leadership. Third, if we are to have social-environmental sustainability, we must accept a new paradigm, one that includes limiting our human population—through reproductive self-control, not through war. Fourth, we must control our material appetite within the nonnegotiable constraints of the biophysical principles that govern the universe and our place in it. This is the paradigm I am addressing, albeit many will resist any change that means altering their lifestyle and the point of view through which they justify it.

I have arrived at this dire necessity of a paradigm shift over a period of many years. After my childhood friend Billy Savage died at age 12 and I survived 2 years of boarding schools in Europe, I wandered in solitude the forest trails from valley bottom to mountain meadow and beyond in the Coast and Cascade Mountains of Oregon and Washington. And, always I found a deep stillness underlying the peace of nature.

Life was simple in those bygone days. The forest air was clean, and the streams ran pure with sweet, crystalline water. The wind sang in the treetops and carried the liquid melody of birds into the depths of my soul. For me, life was a constant treasure trove of novelty and wonder, as night gave way to day and the seasons flowed one into another wherein everything followed nature's rules of engagement in life. And everywhere, nature's exquisite beauty of form and function surrounded me, there to please my eye, inform my intellect, and heighten the consciousness of my being.

As the years of my youth passed into the shadow realm of history, I began to notice an insidious change creeping into the world of human society. War, as a way of life, claimed predominance in the human psyche, as evidenced by its progressive articulation in everyday speech: The U.S. Army Corps of Engineers is at war with the Mississippi River. Foresters and farmers are at war with weeds, insects, and anything else that dares to subvert the growth of a coveted plant. There is a righteous war on poverty and hunger, a war on cancer, and a war on obesity. There are holy wars and political wars of all kinds, as well as a war against nature in general. There is even a war on terror. In the end, we, as a highly competitive society, seem more and more willing—even driven—to making life into a battlefield to get what we want, when we want it and how much we want, whatever it is. And, the epitome of war, which always squanders social-environmental resources for all generations—resources vital to human survival and well-being—is the top-down, command-and-control mode of decision making in the guise of leadership, the chain of command, as it were, which has proven throughout history to be largely symptomatic in its objectives and irreversibly destructive in its outcomes.

Every time we think "war," we are a little closer to becoming what we profess to be against—violence in one form or another—because a war of any kind is violent by nature and incapable of being anything else. And, violence begins as simple fear that is fed by cultural custom and thus grows until it is acted out.

To become a truly peaceful society for the rest of the world to emulate, we, in the United States, must transcend violence; we must, as Mahatma Gandhi said, "become the change we want to see."[3] We can rise above violence only by focusing on peace with total concentration, dedication, and persistence. Peace will reign when there no longer is any thought of war at any scale. But, to eliminate war, its violence, and its flagrant waste of dwindling social-environmental resources, we must focus our thoughts on peace to the exclusion of all else for we cannot serve two masters—peace *and* war. We must choose one or the other. Which shall it be? But, remember, whichever we choose today, we inexorably bequeath the consequences to all generations for all time.

Where in the world today are there unequivocal voices among national and international leaders that speak—through consciously made, long-term decisions of social-environmental viability—for the children who must inherit the consequences of our other-centered wisdom or our self-centered folly? Where are there unequivocal voices that speak for protecting the productive capacity of the global ecosystem—our bequest to all children—present and future? Without such voices of consciousness, courage, and unconditional commitment to the future through the present in the high ranks of leadership in every nation, we, the adults of the world, are increasingly condemning the already desperate children of the world—as well as our children, grandchildren, and every generation yet unborn—to pay a progressively awful price for our petty, psychological immaturities as we bicker among ourselves about who will do what, who will garner what, rather than accept the sometimes difficult choices of our adult responsibilities as trustees of our home planet's social-environmental integrity.

Social insanity can be defined as doing the same thing over and over—despite the lessons of world history—while each time expecting new and dramatically different results. This insanity is called "symptomatic thinking" and is a summation of the way in which much of the world's burgeoning human society has navigated throughout history, culminating in the 20th century, a century in the deadly grapple of competition among people espousing various religious and political ideologies. The competitive root was to fulfill the immediate, materialistic wants and demands not only of the rich, industrialized nations at the expense of the poor, nonindustrialized countries but also at the expense of what the global environment can sustainably produce.

And what about the 21st century? Will it also be a century in deadly grapple among the polarized positions of corporate industrialists, the extremely rich 1% of society with the rest of society, religious/political ideologues, and the "preservationists," all of whom represent positions ensconced in the symptomatic thinking of a narrow segment of our global society—positions that in one way or another summarily discount not only the majority of humanity but also nature's interactive support systems? Is this to be an acceptable legacy of our mentorship to the young and those who must follow?

On the one hand, we change the world simply because we exist and use energy, which is both a fact and beyond our choice. On the other hand, how and why we change the world as we do *is* our choice, despite the fact that some people deem the global situation to be hopeless and thus dwell in a pit of despair. By way of example, there was virtually no looting in Japan, despite their desperation, following the magnitude nine earthquake and tsunami of March 2011—their choice.[4] Could that be said following a similar disaster in the United States, Europe, Asia, Middle East, or Africa? To all the generations of the future, beginning with today's children, we bequeath the consequences of our decisions and actions but allow them no voice in either.

To discuss the decision making of leadership in a manner that fosters social-environmental sustainability, it is necessary to offer a sound, rational, viable alternative to the chronic human disease of competitive, materialistic, symptomatic thinking, which not only has plunged the world into the current suite of synergistic, environmental crises but also continues to worsen them. To overcome industrialized society's flagrant disregard for nature's biophysical principles and the natural wealth they proffer humanity, it is essential to elucidate: (1) the inviolable biophysical principles that define the limitations of our human choices and the consequences they generate, (2) the inviolable rules of social-environmental decision making and the consequences they generate, (3) the current social-environmental crisis as it is unfolding, and (4) a solid foundation on which to base social-environmental decision making as a viable legacy to all generations. In addition, we must shift our symptomatic way of thinking and acting to a systemic consciousness of the world around us because the level of consciousness that caused the current social-environmental crises in the first place is not the level of consciousness that can fix it.

In terms of a foundation for environmentally sound decision making, I remember a time some years ago when I was employed as a research scientist, and I thought I had some answers. I now realize that I am not even sure of all the questions. But, there are a few things of which I am relatively certain, one of which is the equivocating, self-protective nature of most—*but not all*—leaders at all levels of the global society. The flip side of the leadership coin is the dictatorial, top-down, command-and-control mode of subjugating a given society. That said, there are isolated circumstances in which such authoritative behavior is indeed necessary—and may even be wise and benevolent, albeit exceedingly rare. Finally, *I am certain* that the long-term, biophysical integrity and ecological productivity of the global environment must take precedence over the short-term filling of monetary coffers and the maintenance of personal pillars of power before social-environmental-economic sustainability can prevail for all generations.

This book has three parts. Section I is nature's unequivocal message to humanity, especially its decision makers at all levels of government worldwide: Obey the inviolate biophysical principles that govern the universe and your place in it as the bedrock of leadership and its decision making for human survival—or reap your self-imposed consequences, a legacy you bequeath all generations everlasting. Chapter 1 is a discussion of the inviolable biophysical principles that form the nonnegotiable limitations of social-environmental sustainability within which humans can shape the environment to their benefit. When these limits are ignored, environmental degradation ensues, along with a diminished productive capacity of those ecosystems on which we rely for a good quality of life—a situation that is not only progressively irreversible but also passed forward to all generations. Chapter 2 expands the premise of Chapter 1 into the inviolable rules of decision making and thus leadership; these rules are vital to understand and follow if social-environmental sustainability is to prevail. Here, one must remember, as astronomer and author Carl Sagan pointed out, that, "In order to make an apple pie from scratch, you must first create the universe."[5]

Section II examines why a command-and-control form of decision making tends to become dysfunctional and fails and how to break the cycle of such behavior. Chapter 3 begins the discussion by examining the ways in which we carry our respective families—both the functional and dysfunctional aspects—with us throughout life. Further, it points out that the governments, agencies, and various organizations are as functional or dysfunctional as we are as individuals. Therefore, we—individuals acting as a single, collective body—are society's driving force and thus enable both its social and its environmental agencies to function as originally envisioned or limit their ability to so function by creating a military-style, command-and-control mode of leadership to the benefit of the corporate/political elite. Chapter 4 is a case study by Jessica K. La Porte about her experience with the environmental duplicity of the Southern Nevada Water Authority, which employed her to create a viable program of environmentally sustainable decision making, but from which she was ultimately fired for insisting the publicly touted plan of environmental sustainability be honored. Chapter 5 illustrates the ways in which dysfunctional behavior and homeostasis are corrected for the social-environmental benefit of all generations. These include understanding how job descriptions are used to control information and how positions of authority are used to control direction for political purposes. The correction culminates with the crafting of a vision, goals, and objectives, as well as understanding and implementing the seven steps in monitoring the vision's effectiveness to protect the environment's productive capacity for all generations.

Section III is a discussion of what it takes to be a psychologically mature decision maker who actually serves as a leader with the good of all generations in mind. Here, it is pointed out that (other than natural disasters) all social-environmental problems begin with our thinking. Put differently, we create the vast majority of the social-environmental problems that we end up facing. *We do it to ourselves.* Understanding both this and nature's inviolable biophysical limitations, the discussion is focused on how people must change their thinking through individual, personal growth if we, as a nation and world, are to have decision makers who actually lead from a position of psychological maturity for the social-environmental sake of all generations. Chapter 6 is a detailed enumeration of the essential qualities of a psychologically mature decision maker. Chapter 7 deals with the responsibilities and pressures faced by a social-environmental decision maker. These include such things as circumstances confronting decision makers as leaders, being and disclosing oneself, avoiding the trap of perfectionsim, inspiring performance, nurturing creativity, establishing realistic objectives, and how to deal with "burnout." Chapter 8 recognizes the fact that in a true democracy children deserve a voice in determining the kind of future we adults leave for them. After all, they must live with the social-environmental consequences of our decisions and actions. In that, they have no choice—unless we consciously, purposefully give it to them—which once again means changing our adult thinking.

NOTES

1. Agnes Repplier. http://www.dictionary-quotes.com/the-clear-sighted-do-not-rule-the-world-but-they-sustain-and-console-it-agnes-repplier/ (accessed January 12, 2011).
2. Hubert H. Humphrey. http://www.brainyquote.com/quotes/authors/h/hubert_h_humphrey.html (accessed April 3, 2011).
3. Louis Fischer. *The Essential Gandhi: An Anthology of His Writings on His Life, Work, and Ideas*. Random House, New York, 1962.
4. Christopher Beam. Japan earthquake 2011: Why so little looting in Japan [following the magnitude nine earthquake]? It's not just about honesty. http://www.huffingtonpost.com/2011/03/17/japan-earthquake-2011-why_n_837126.html (accessed March 22, 2011).
5. Carl Sagan. First Science.com. http://www.firstscience.com/home/poems-and-quotes/quotes/carl-sagan-quote_2284.html (accessed February 4, 2011).

Acknowledgments

I am deeply grateful to my friend Dr. Jack Ward Thomas for reviewing an early draft of the manuscript and helping me to improve it materially based on his many years of decision making in positions of leadership. Jack was the director of the U.S. Forest Service Research Laboratory in La Grande, Oregon, where I was headquartered from 1974 through most of 1980 while the U.S. Department of Land Management employed me as a research scientist. Jack subsequently became the 13th chief of the U.S. Forest Service during the Clinton administration.

I am also grateful to Jessica K. La Porte for writing an excellent case study based on her experience as an employee of the Southern Nevada Water Authority.

Once again, I gladly thank my wonderful wife, Zane, for her patience and loving support while I wrote this book.

Section I

Nature's Urgent Message to the World's Decision Makers

> We could learn a lot from crayons. Some are sharp. Some are pretty. Some are dull. Some have weird names, and all are different colors. But they all have to live in the same box.
>
> **—Robert Fulghum***

* Robert Fulghum. http//www.quotegarden.com/crayons.html (accessed March 18, 2011).

1 The Principles of Biophysical Sustainability

> We shall not cease from exploration. And the end of all our exploring will be to arrive where we started and know the place for the first time.
>
> **—T. S. Eliot**[1]

> Nothing in life is to be feared. It is only to be understood.
>
> **—Marie Curie**[2]

Change often occurs on the brink of disaster between need and fear. On the one hand, we know we need to do things differently. On the other hand, we are terrified of facing the unknown, unfamiliar, and uncertain. To change our direction for the future, however, we must suspend our conventional notions about change and our ability to learn because there are no problems to resolve other than those we perceive as manifestations of how we think and act. The problems we face are a matter of who we are consciously. And, many people prefer to err again and again rather than let go of some cherished belief, pet notion, deified assumption, or staunchly defended position. Others err because they are pessimistic in their outlook and are thus blind to viable options.[3]

Social-environmental sustainability demands that decision makers go beyond the immediate human valuation of a given resource to examine and disclose the fundamental issue of how its use will affect the long-term, biophysical sustainability of the ecosystem of which it is a component. One must also recognize and disclose the long-term, social-environmental issues that need to be dealt with concerning the method by which a resource is extracted. This is necessary because the overall integrity of an ecosystem, its productive capacity, and the sustainability of its resources will determine the array of options passed forward to future generations.

To accommodate the sustainability of the environment's ecological integrity, each person who would lead must understand the biophysical principles given in this chapter as a condition of wise decision making. These principles form the underpinnings through which nature operates and the social limitation we must understand and accept if we are to participate with nature in a sustainable fashion. Moreover, social-environmental justice dictates that leaders be held accountable for the outcome of their decisions (while in their position of authority instead of after the fact) because the effects of their decisions become the consequences for all generations—those present and those yet to come. This is particularly poignant in the face of an exploding human population; rapidly degrading ecosystems; and thus dwindling, per-capita resources.

THE WATERBED PRINCIPLE

As long as the human population was but a small fraction of its current size, Earth's resources were considered by most "civilized" people to be unlimited and free for the taking. But even then, "The history of almost every civilization," observed British historian Arnold Toynbee, "furnishes examples of geographical expansion coinciding with deterioration in [environmental] quality."[4] Today, there is much talk about "renewable" resources but no longer so much about unlimited resources.

Ultimately, however, all biophysical resources are finite—with the exception of solar energy, at least for the purposes of this book. We not only can run out of a resource by literally exhausting its earthly supply, such as oil or the extinction of a species and its attendant service to humankind (say the provision of coffee or chocolate), but also can alter an existing resource to render it useless to us, such as poisoning our drinking water through pollution of various kinds. And, *we are increasingly doing both.*

Moreover, as the burgeoning human population demands more and more material commodities from a rapidly dwindling supply of an increasing number of vital necessities (such as potable water), the ratio of resources apportioned to each human declines. Moreover, this decline is exacerbated by the progressively longer lives of today's humans and their prolonged demands for these same necessities. Further, those resources currently deemed "renewable" are only renewable as long as the system that produces them retains its biophysical integrity and is used in a sustainable manner—both ecologically and socially, as clearly stated by T. N. Narasimhan of the Berkeley Institute of the Environment:

> Historically, science has pursued a premise that Nature can be understood fully, its future predicted precisely, and its behavior controlled at will. However, emerging knowledge indicates that the nature of Earth and biological systems transcends the limits of science, questioning the premise of knowing, prediction, and control. This knowledge has led to the recognition that, for civilized human survival, technological society has to adapt to the constraints of these systems. Simultaneously, spurred by explosive developments in the understanding of materials (non-biological and biological), applied scientific research pursues a contrary goal of controlling the material world, with the promise of spectacular economic growth and human well-being. If adaptation to Nature is so important, why does applied research pursue a contrary course? … Also, in a world dominated by democratic ideals of freedom and liberty, the discipline required for adapting to Nature may often be overridden by competition among various segments of society to exercise their respective rights.[5]

Is it really so imperative to change my behavior, you might ask, if it infringes on my personal *rights*? That choice is yours, of course, but remember that you bequeath the consequences of your choice to *all* generations. Therefore, it would be well to consider the counsel of professor Johan Rockström and his interdisciplinary team of 29 scientists:

> Although Earth has undergone many periods of significant environmental change, the planet's environment has been unusually stable for the past 10,000 years. This period of stability—known to geologists as the Holocene—has seen human civilizations

> arise, develop and thrive. Such stability may now be under threat. Since the Industrial Revolution, a new era has arisen, the Anthropocene, in which human actions have become the main driver of global environmental change. This could see human activities push the Earth system outside the stable environmental state of the Holocene, with consequences that are detrimental or even catastrophic for large parts of the world.[6] [Holocene comes from the Greek *holos* ("whole") and *cene* ("new"). Anthropocene[7] comes from the Greek *anthropo* ("human") and *cene* ("new").]

Dawning of the Anthropocene Epoch represents "a new phase in the history of both humankind and of the Earth, when natural forces and human forces became intertwined, so that the fate of one determines the fate of the other. Geologically, this is a remarkable episode in the history of this planet."[8] Consider, for example, as you hike in a wilderness area today or wander through a national park, no matter how far removed you seem to be from the center of society, you are still breathing pollution. It is everywhere and will worsen as long as decisions to placate big industry by continually trumping a global pursuit of dramatically cleaning the world's air. Here, it must be stated in fairness that our material appetites feed the corporate drive for more, whereas the corporate drive for more—ever more—stimulates our material appetite for more—always more—through advertising in a self-reinforcing feedback loop.

We dare not kid ourselves about the importance of air quality. Our earthly survival, and that of our children and their children unto all generations, ultimately depends on clean air. Air is the interactive thread connecting soil, water, biodiversity, human population density, sunlight, and climate. (Biodiversity refers to the variety of living species and their biophysical functions and processes.) Think of this interconnectedness as the *waterbed principle*, which simply demonstrates that you cannot touch any part of a filled waterbed without affecting the whole of it.

Yet, we, as a society, with our myriad data bits and seemingly vast and ever-increasing knowledge, listen to the world's traditional economists and the corporate/political elite and *assume* they are correct when they take such ecological variables as air, soil, water, sunlight, biodiversity, genetic diversity, climate, and more and convert them, in theory at least, into economic constants whose values are unchanging—or discount them altogether as "externalities." Ecological variables are therefore omitted from consideration in most economic and planning models and even from our thinking—to say nothing of the political decisions rendered by today's global leaders. Biodiversity and genetic diversity, on the other hand, are euphemistically discounted as "externalities" when their consideration interferes with monetary profits.[9] On top of it all is the nagging problem of our rapidly growing human population. We talk about it and worry about it, but in the end we give only lip service to the one solution that can control it—total, real gender equality for women.

That notwithstanding, relationships among things are in constant flux as complex systems arise from subatomic and atomic particles in the giant process of evolution on Earth. At each higher level of complexity and organization is an increase in the size of the system and a corresponding decrease in the energies holding it together. Put differently, the forces that keep evolving systems intact, from a molecule to a human society, weaken as the size of the systems increases, yet the larger the system, the more energy it requires to function. Such functional dynamics are characterized by their diversity as well as by the constraints of the overarching laws and subordinate principles that govern them.

These principles can be said to *govern* the world and our place in it because they form the behavioral constraints without which nothing could function in an orderly manner. In this sense, the law of cosmic unification—the supreme law—is analogous to the Constitution of the United States, a central covenant that informs the subservient courts of each state about the acceptability of its governing laws. In this way, nature's rules of engagement inform society of the latitude whereby it can interpret the biophysical principles and survive in a sustainable manner. Beyond that, the global ecosystem, which cares not a whit about human beings, will always function in a manner sustainable unto itself, but not necessarily in a way that is favorable to human survival, let alone a life of dignity.

THE LAW OF COSMIC UNIFICATION

The *law of cosmic unification* is functionally derived from the synergistic effect of three universal laws: the first law of thermodynamics, the second law of thermodynamics, and the law of maximum entropy.

The *first law of thermodynamics* states that the total amount of energy in the universe is constant, although it can be converted from one form to another. Therefore, the amount of energy remains the same, even if you could go forward or backward in time. For this reason, the contemporary notion of either "energy production" or "energy consumption" is a non sequitur. The *second law of thermodynamics* states that the amount of energy in forms available to do useful work can only diminish over time. The loss of available energy to perform certain tasks thus represents a diminishing capacity to maintain order at a certain level of manifestation (say a tree) and so increases disorder or entropy. This "disorder" ultimately represents the continuum of change and novelty—the manifestation of a different, simpler configuration of order, such as the remaining ashes from the tree when it is burned. In turn, the *law of maximum entropy* says that a system will select the path out of available paths that maximizes entropy at the fastest rate given the existing constraints.[10]

The essence of maximum entropy simply means that, when any kind of constraint is removed, the flow of energy from a complex form to a simpler form speeds up to the maximum allowed by the relaxed constraint.[11] As it turns out, the law of maximum entropy freed early hominids from one of the basic constraints of nature when they adapted the intense entropy of burning wood to their everyday use. (A hominid, *hom-i-nid*, is any of the modern or extinct primates that belong to the taxonomic family Hominidae, *Hom-in-idae*, whereof we are members.) Control of fire gave hominids the ability to live in habitats that heretofore had been too cold or where the seasonal temperature variations had been too great. It also allowed them to cook food, making parts of many plants and animals palatable and digestible when they were baked, roasted, or boiled. The charred remains of flint from prehistoric firesides on the shore of an ancient lake near the river Jordan in Israel indicates that our ancient ancestors had learned how to create fire 790,000 years ago.[12] Moreover, the increased supply of protein embodied in cooked meat is thought to have facilitated evolution of increasing hominid brain capacity, ultimately leading to our mental abilities.[13]

We, in contemporary society, are all familiar with the basis of this law even if we do not understand it. For example, we all know that our body loses heat in cold weather, but our sense of heat loss increases exponentially when windchill is factored into the equation because our clothing has ceased to be as effective a constraint to the loss of heat as it was before the wind became an issue. Moreover, the stronger and colder the wind, the faster our body loses its heat—the maximum entropy (= loss) of our body's energy whereby we stay warm. If the loss of body heat to windchill is not constrained, hypothermia and death ensue, along with the beginnings of bodily decomposition—reorganization from the complex structure and function toward a simpler structure and function.

In other words, systems are by nature dissipative structures that release energy by various means, but inevitably by the quickest means possible. Suppose, for example, you have a wood-burning stove in your home with which you heat the 1,000 square feet of living space. To keep your house at a certain temperature, you must control the amount of energy extracted from the wood you burn. There are nine ways to accomplish this.

The first consideration is the kind of wood you choose, be it Douglas-fir, western redcedar, western hemlock, bigleaf maple, Pacific madrone, Oregon ash, Oregon white oak, red alder, or a combination. Your choice is important because each kind of wood has a different density and thus burns with a corresponding intensity. On one hand, the three coniferous woods (Douglas-fir, western redcedar, and western hemlock) are relatively soft, require less oxygen to burn than hardwoods, burn quickly, but produce only moderate heat. On the other hand, such hardwoods as bigleaf maple, Pacific madrone, Oregon ash, Oregon white oak, and red alder produce substantial heat—of which oak, madrone, and maple probably produce the most, followed by ash and alder. But these hardwoods, which have more calories stored in a cubic foot than do softwoods, also require more oxygen to burn than the softwoods, and they burn more slowly, thus lasting longer.

The second concern is the quality of wood that you burn. Sound, well-seasoned wood burns far more efficiently than either wet, unseasoned wood or wood that is partially rotten. In this case, the quality of the wood also determines the *effectiveness* whereby it heats your house. Good-quality wood is far more effective in the production of heat than is wood of poor quality.

The third determination is the size and shape of the wood. Small pieces produce a lot of heat but are quick to disappear. Large pieces take more time to begin burning, but last longer and may or may not burn as hot when they really get going, depending on the kind of wood. Split wood has more surface area per volume and burns more rapidly than do round pieces of wood of the same size, such as large branches, because the latter have more volume than surface area.

The fourth decision is how wide to open the damper and thereby control the amount of air (oxygen) fanning the flames, either increasing the intensity of burning (opening the damper) or decreasing the rate of burn (closing the damper). In each case, the length of time the damper is in a given position is part of the equation. The wider the damper is opened, the less the constraint, the hotter and faster the wood will burn, and the more rapidly heat will escape—the law of maximum entropy. This law also addresses the speed with which wood is disorganized as wood and reorganized as ashes—the second law of thermodynamics.

The fifth choice is how warm you want your house to be in terms of how cold it is outside. The colder it is outside, the more wood you must burn to maintain a certain level of heat—how much depends on the kind, shape, and condition of the wood you are burning. Conversely, the warmer it is outside, the less wood you must burn to maintain the same level of warmth.

The sixth consideration is how well your house is insulated against the intrusion of cold air and thus the escape of your indoor heat—both of which determine the amount of wood you must burn to maintain a given temperature. Another facet of how much wood you must burn depends in part on whether clouds are holding the heat close to Earth, thus acting as a constraint to the heat leaking out of your house, or whether clear skies allow heat to bleed from your home and escape into outer space.

The seventh option is how often you open the outside door to go in and out of the house and so let cold air flow in to replace the warm air rushing out. You could ameliorate this exchange by having an enclosed porch between the door opening into the house and the door opening directly to the outside. A well-insulated porch would act as a dead-air space and would be a functional constraint to the loss of heat from the house as you access the outside.

The eighth alternative is when to heat your house and for how long. You can, for instance, reduce the amount of heat required at night when snuggled in bed. If you go to bed early and get up early, it is about the same as going to bed late and getting up late. But if you go to bed early and get up late, you do not need to heat the house for as long as you would if you spent more time out of bed as opposed to in bed. Moreover, if it is well below freezing outside, you might have to keep the house warmer than otherwise to protect the water pipes from freezing.

And the ninth course of action, one that is both influenced by the other eight and influences them in turn, is how warmly you choose to dress while indoors. Whatever you wear constitutes a constraint to heat loss of a greater or lesser degree. Clearly, the warmer you dress, the less wood you must burn to stay warm and vice versa. It is the same with how many blankets are on your bed during the winter.

These nine seemingly independent courses of action coalesce into a synergistic suite of relationships; whereby a change in one automatically influences the other eight facets of the speed by which energy from the burning wood escapes your house. This said, the first and second laws of thermodynamics and the law of maximum entropy meld to form the overall unifying law of the universe—the law of cosmic unification—wherein all subordinate principles, both biophysical and social (decision making), are encompassed.

With respect to the functional melding of these three laws, Rod Swenson of the Center for the Ecological Study of Perception and Action, Department of Psychology, University of Connecticut, says these three laws of thermodynamics "are special laws that sit above the other laws of physics as laws about laws or laws on which the other laws depend."[14] Stated a little differently, these three laws of physics coalesce to form the *supreme* "law of cosmic unification," to which *all* biophysical and social principles governing nature and human behavior—decision making—are subordinate yet simultaneously *inviolable*. Inviolable means that we manipulate the effects of a principle through our decisions and subsequent actions, but we do not—and cannot—alter the principle itself.

Here, Professor T. N. Narasimhan asks a pertinent question:

> How then may humans sustain on a finite earth in which behavioral patterns of life-supporting natural resource systems can be understood, but cannot be predicted or controlled? Reason suggests that society must adapt its functioning to the constraints imposed by the nature of these resource systems. This general need to structure patterns of living within constraints that lie beyond human understanding or control may be referred to as "adapting to Nature."[15]

NATURE'S INVIOLABLE BIOPHYSICAL PRINCIPLES

Although I have done my best to present the principles in a seemingly logical order, it is impossible to be definitive because each principle is an ever-interactive strand in the multidimensional web of energy interchange that constitutes the universe and our world within it. Moreover, a different order can be found each time the principles are read, and each arrangement seems logical. Because each principle affects all principles (like a waterbed), every arrangement is equally correct in its overall function.

Principle 1—Everything Is a Relationship

The universe is a single, all-inclusive relationship constituted of an ever-expanding web of biophysical feedback loops, each of which is novel and self-reinforcing. Each feedback loop is a conduit whereby energy is moved from one place, one dimension, and one scale to another. And all we humans do—ever—is practice relationships with the flow of energy within this web because the existence of everything in the universe is an expression of its relationship to everything else within the web through the continual exchange of energy. Moreover, all relationships are forever dynamic, from the wear on your toothbrush from daily use to the rotting lettuce you forgot in the refrigerator. Herein lies one of the foremost paradoxes of life: The ongoing process of change is a universal constant over which, much to our dismay, we have no control.

Think, for example, what the difference is between a motion picture and a snapshot. Although a motion picture is composed of individual frames (instantaneous shots of the present moment = snapshots), each frame is entrained in the continuum of time and thus cannot be held constant, as Roman Emperor Marcus Aurelius observed: "Time is a river of passing events, and strong is its current. No sooner is a thing brought to sight than it is swept by and another takes its place, and this too will be swept away."[16]

Yet we, in our fear of uncertainty, are continually trying to hold the circumstances of our life in the arena of constancy through the decisions we make, as depicted in a snapshot—hence, the frequently used term *preservation* in regard to this or that ecosystem, this or that building. Yet, jams and jellies are correctly termed *preserves* because they are heated during their preparation to kill all living organisms and thereby, it is hoped, prevent noticeable change (deterioration) in their consistency.

Insects in amber are an example of true preservation in nature. Amberization, the process by which fresh tree resin is transformed into amber, is so gentle that it forms the most complete type of fossilization known for small, delicate, soft-bodied organisms, such as insects. In fact, a small piece of amber found along the south coast of England in 2006 contained a 140-million-year-old spider web constructed in the same orb configuration as that of today's black-and-yellow garden spiders. This is 30 million years older than a previous spider web found encased in Spanish amber. The web demonstrates that spiders have been ensnaring their prey since the time of the dinosaurs. And, because amber is three dimensional in form, it preserves color patterns and minute details of the organism's exoskeleton, allowing the study of microevolution, biogeography, mimicry, behavior, reconstruction of the environmental characteristics, the chronology of extinctions, paleosymbiosis,[17] and molecular phylogeny.[18] But, the same dynamic cannot be employed outside an airtight container, such as a drop of amber or canning jar. Nevertheless, whether natural or artificial, all systems are ultimately open because they are all part of one cosmic relationship with energy, which means a totally self-contained, closed-loop system is a physical impossibility.

Principle 2—All Relationships Are All Inclusive and Productive of an Outcome

I have often heard people say that a particular piece of land is "unproductive" and needs to be "brought under management," and management decisions are made accordingly. Here, it must be rendered clear that every relationship is productive of a cause that has an *effect*, and the effect, which is the cause of another effect, *is* the product. Therefore, the notion of an unproductive parcel of ground or an unproductive political meeting is an illustration of the narrowness of human valuation because such judgment is viewed strictly within the extrinsic realm of personal values, usually economics—not the intrinsic realm of nature's dynamics. This narrow valuation not only illuminates how little we understand nature's inviolable dynamics but also defies the certainty within which we hallow the validity of our knowledge and thus the ability of our economic assessments to address these profound issues.[19]

We humans are not, after all, so powerful a natural force that we can completely destroy an ecosystem because it still obeys the biophysical principles that determine how it functions at a given point in time. Nevertheless, we can, and have, so severely altered an ecosystem that it is incapable of providing—for many decades and even for all time—those goods and services we require to sustain human life and society. It is well to note, for example, that the total surface area covered in paved roads in the United States precludes the soil's ability to capture and store water. In addition, the ocean's ability to sequester carbon dioxide (one of the main greenhouse gases) is being severely impaired due to the dramatically disrupted population dynamics of marine fishes by systematically overfishing too many of the top predators.[20] All of the relationships we affect produce an outcome or product. Whether the product is beneficial to humans or even amenable to our long-term existence is another issue.

Principle 3—The Only True Investment in Our Global Ecosystem Is Energy from Sunlight

The *only true investment* in the global ecosystem is energy from solar radiation (materialized sunlight). Everything else is merely the recycling of already-existing energy. In a business sense, for example, one makes money (economic capital) and then takes a percentage of those earnings and *recycles* them by putting them back into the infrastructure of the enterprise for maintenance of buildings and equipment to facilitate making a profit by protecting the integrity of the initial outlay of capital over time. In a business, one recycles economic capital *after* the profits have been earned.

Biological capital, on the other hand, must be "recycled" *before* the profits are earned to stay even. This means forgoing some potential monetary gain by leaving enough of the ecosystem intact for it to function in a sustainable manner. In a forest, for instance, one leaves some proportion of the merchantable dead and live trees, the latter to gather solar radiation and produce living tissue that will subsequently die—and together with the already-dead trees—rot, recycle into the soil, and thereby replenish the fabric of the living system. In rangelands, one leaves the forage plants in a viable condition so they can capture solar radiation, grow, seed, die, decompose, and protect the soil from erosion as well as add organic material to the soil's long-term, ecological integrity and productive capacity.

In contrast, people speak incorrectly about fertilization as an *investment* in a forest, grassland, or agricultural crop when in fact it is merely recycling chemical compounds that already exist on Earth. In reality, people are simply taking energy (in the form of preexisting chemical compounds) from one place and putting them in another for a specific purpose. The so-called investments in the stock market are a similar shuffling of preexisting energy.

When people *invest* money in the stock market, they are really recycling energy from nature's products and services acquired through human labor. The value of the labor is transferred symbolically to a dollar amount, thereby representing a predetermined amount of labor. Let us say, for instance, you work for $10 an hour; then, a $100 bill would equal 10 hours of labor. Where is the *investment*? There isn't any, but there is a recycling of the energy put forth by one's labor and a symbolic recycling of the energy in the denomination of currency we spend.

Here, you might argue that people *invested* their labor in earning the money, and I would counter that whatever energy they put forth was merely a recycling of the energy they took in through the food they ate. Nevertheless, the energy embodied in the food may actually have simultaneously been a true investment and a recycling of already-existing energy.

It has long been understood that green plants use chlorophyll molecules to absorb sunlight and use its energy to synthesize carbohydrates from carbon dioxide and water. This process is known as photosynthesis, where *photo* means "light," *synthesis* means the "fusion of energy," and the combination is the basis for sustaining the life processes of all plants. The energy is derived from the sun (an original input) and combined with carbon dioxide and water, as well as nutrients from the soil in which the plants grow (all existing chemical compounds), to create a renewable source of usable energy. This process is analogous to an array of organic solar panels—the green plant.

Think of it this way, the plant (an array of solar panels) uses the green chlorophyll molecule (a *photoreceptor*, meaning receiver of light) to collect light from the sun within chloroplasts (small, enclosed structures in the plant that are analogous to individual solar panels). Then, through the process of photosynthesis, the sun's light is used to convert carbon dioxide and water to carbohydrates as usable energy for the plant (which is partially sustained by chemical nutrients in the soil, where, incidentally, most of the water is also derived), a process that is comparable to converting the sun's light in solar panels (derived from recycled energy in the form of materials) on the roof of a building into electricity for our use. These carbohydrates, in turn, are partly stored energy from the sun—a new input of energy into the global ecosystem—and partly the storage of existing energy from the amalgam of carbon dioxide and water, a process sustained in part by chemical nutrients from the soil.[21]

When, therefore, we eat green plants, the carbohydrates are converted through our bodily functions into different kinds of energy. By that is meant the energy embodied in green plants—which is partly a true investment of energy—is altered through digestion into the various types of energy our bodies require for their physiological functions. The *excess* energy (that not required for physiological functions) is expelled as bodily waste or expended in the form of physical motion, such as work. On the other hand, it is different when eating meat because the animal has already used the sun's contribution to the energy matrix in its own bodily functions and its own physical acts of living, so all we get from eating flesh is recycled energy.

In this sense, coal is stored energy from sunlight because it is formed from green plants, believed to have been mainly fern-like species that antedated the more modern plants and trees. Coal begins as layers of plant material that are protected from decomposition, usually in mud or acidic water, such as the shallow sea bottoms the Carboniferous period provided between 359 and 299 million years ago, give or take 2 to 3 million years. This burial trapped atmospheric carbon, along with various amounts of sulfur, hydrogen, oxygen, and nitrogen, which, over time, metamorphosed into sedimentary or metamorphic rock by changing the chemical and physical properties of the plant remains to form different kinds of coal.[22]

Oil, on the other hand, is only partly stored energy from sunlight because it is formed over millions of years from both tiny plants and animals that died, sank to the bottom of a shallow body of water, and were covered by sand, clay, and silt (which is fine-grained sediment of mud or particles of clay). As the dead plants and animals decomposed under their blanket of sand, clay, and silt, more dead plants and animals were being deposited, only to be buried as well—a process repeated over and over. Enormous amounts of pressure built up over millions of years, causing the bottom layer to become very hot, which transformed the clay, sand, and silt into rock and the decayed plants and animals between the rock into layers of a dark, heavy liquid we know as "oil."[23]

Principle 4—All Systems Are Defined by Their Function

The behavior of a system—any system—depends on how its individual parts interact as functional components of the whole, not on what an isolated part is doing. The whole, in turn, can only be understood through the relationships, the interaction of its parts. The only way anything can exist is encompassed in its interdependent relationship to everything else, which means an isolated fragment or an independent variable can exist *only on paper* as a figment of the human imagination—*of which it is, nevertheless, an interactive part.*

Put differently, the false assumption is that an independent variable of one's choosing can exist in a system of one's choice and that it will indeed act as an independent variable of one's choosing. In reality, all systems are interdependent and thus rely on their pieces to act in concert as a functioning whole. This being the case, no individual piece can stand on its own *and* simultaneously be part of an interactive system. Thus, *there neither is nor can there be an independent variable* in any system, be it biophysical, mechanical, or even intellectual because every system is interactive by its very definition as a system.

What is more, every relationship is constantly adjusting itself to fit precisely into other relationships that, in turn, are consequently adjusting themselves to fit precisely into all relationships, a dynamic that *precludes* the existence of an independent variable, so *no given thing can be held as a constant value beyond the number one* (the universal common denominator) because to do so would necessitate the detachment of the thing in question from the system as an independent variable. Therefore, all relationships are constituted by additions of *one* in all its myriad forms, from quarks, atoms, molecules, and proteins, which comprise the building blocks of life, to the living organisms, which collectively form the species and communities. The only way the number one can exist, as the sole representative of any form on Earth, is to be the last, living individual of a species—something intimated on the tribal level by James Fennimore Cooper's 1826 book, *The Last of the Mohicans*[24]—because extinction is forever.

Therefore, to understand a system as a functional whole, we need to understand how it fits into the larger system of which it is a part and so gives us a view of systems supporting systems supporting systems supporting systems ad infinitum.

Principle 5—All Relationships Result in a Transfer of Energy

Although technically a "conduit" is a hollow tube of some sort, the term is used here to connote any system employed specifically for the transfer of energy from one place to another. Every living thing, from a virus to a bacterium, fungus, plant, insect, fish, amphibian, reptile, bird, mammal, and every cell in our body, is a conduit for the collection, absorption, transformation, storage, transfer, and expulsion of energy. In fact, the function of the entire biophysical system is tied up in the collection, absorption, transformation, storage, transfer, and expulsion of energy—one gigantic, energy-balancing act, or perhaps more correctly *energy-juggling act.*

For example, prior to the invasion of the "New World" by Europeans, human impact on the environment was not simply a process of increasing change in response to the growth of the indigenous populations. Instead, the landscape was given time to rest and recover as people moved about; cultures collapsed; populations declined because of periodic starvation, disease, and war; and habitations were abandoned.

Human activities may be constructive, benign, or destructive, all of which are subjective concepts based on human values, but change—the dance of energy—is continual, albeit at various rates and in various directions. All changes are, in addition, cumulative. Even mild, slow change can show dramatic effects over the long term. Although there was, of course, some localized European impact prior to 1750, thereafter (but especially after 1850) populations of European Americans expanded tremendously; they severely exploited the resources, greatly accelerating the modification of the environment in the process.

In particular, the settlers who introduced domestic livestock in the San Juan Mountains of Colorado left an eolian dusty signature. (*Eolian dust* is composed of sand, silt, or clay-size, fragmented material transported and deposited primarily by wind.)

Mineral aerosols in wind-blown dust from the soil's surface are an important influence on climate, as well as on terrestrial and marine biogeochemical cycles. The amount of dust can be affected by human activities, which alter surface sediments. The changes in the flow of regional- and global-scale wind-borne dust following the rapid expansion and settlement of human populations are not well understood, but a study of sediment cores from alpine lakes in the San Juan Mountains from the past 5,000 years has yielded some regional-scale answers.

The chemical composition of the dust is not comparable to that of the surrounding bedrock, an indication that it came from hundreds of miles away. Such transport is not surprising because the winds of winter are known to blow dust from the deserts of California and Nevada to Colorado. Dating carbon and lead revealed that the average annual amount of dust deposited in the lakes after the 1800s was 500% more than the dust deposited prior to that time. This sudden increase appeared around the time of a boom in ranching, when cattle and sheep spread across the landscape, overgrazing the native, erosion-controlling plants. There was a subsequent drop in the level of dust deposited, however, with the passage in 1934 of the Taylor Grazing Act.

Nevertheless, the dust has yet to settle. The past century has seen more than a five-fold increase in the deposition of agricultural chemicals from fertilizers: potassium, magnesium, calcium, nitrogen, and phosphorus. The infusion of these chemicals into the alpine ecosystem will have serious implications with respect to the alkalinity of surface waters, aquatic productivity, and terrestrial nutrient cycling.[25]

Principle 6—All Relationships Are Self-Reinforcing Feedback Loops

Everything in the universe is connected to everything else in a cosmic web of interactive feedback loops, all entrained in self-reinforcing relationships that continually create novel, never-ending stories of cause and effect, stories that began with the Eternal Mystery of the original story, the original cause. Everything, from a microbe to a galaxy, is defined by its ever-shifting relationship to every other component of the cosmos. Thus, "freedom" (perceived as the lack of constraints) is merely a continuum of fluid relativity. In contraposition, every relationship is the embodiment of interactive constraints to the flow of energy—the very dynamic that perpetuates the relativity of freedom and thus all relationships, which by definition precludes absolute freedom.

Hence, every change (no matter how minute or how grand) constitutes a systemic modification that produces novel outcomes. A feedback loop, in this sense, comprises a reciprocal relationship among countless bursts of energy moving through specific strands in the cosmic web that cause forever-new, compounding changes at either end of the strand, as well as every connecting strand.[26] And here, we humans often face a dichotomy with respect to our interests.

On the one hand, while all feedback loops are self-reinforcing, their effects in nature are neutral because nature is impartial with respect to consequences. We, on the other hand, have definite desires as far as outcomes are involved and thus assign a preconceived value to what we think of as the end result of nature's biophysical feedback loops. A simple example might be the response of North American elk in the Pacific northwestern United States to the alteration of their habitat. In this case, the competing values were (and still are) elk as an economically important game animal (commodity) versus timber as an economically important commodity.

In the 1940s and 1950s in western Oregon, the timber industry often used the adage: Good timber management is good wildlife management. At the time, that claim seemed plausible because elk populations were growing in response to forests being clear-cut. By the mid-to-late 1960s and throughout the 1970s, however, elk populations began to exhibit significant declines. Although predation was run out as the obvious reason, it did not hold up under scrutiny since the large predators, such as wolves and grizzly bears, had long been extirpated, and the puma (mountain lion) population was being decimated due to bounties placed on the big cats.

As it turned out, the cause of the decline in elk numbers was subtler and far more complicated than originally thought. The drop in elk numbers was in direct response to habitat alteration by the timber industry. This is not surprising since elk, like all wildlife, have specific habitat requirements, which consist of food, water, shelter, space, privacy, and the overall spatial connectivity of these components in a specific configuration. When any one of these elements is in short supply, it acts as a limiting factor or constraint with respect to the viability of a species' local population.[27]

By way of illustration, here is a simplified example: In the early days, before extensive logging began, the land was well clothed in trees, making food the factor that limited the number of elk in an area. As logging cleared large areas of forest, grasses and forbs grew abundantly, and elk, being primarily grazers, became increasingly numerous. This relationship continued for some years until—for an instant in time—the perfect balance between the requirements of food and shelter was reached. The proximity to water did not play as important a role in this balance because of the relative abundance of forest streams and because elk can travel vast distances to water. Thus, hunters, loggers, and big-game biologists initially perceived clear-cut logging as the proverbial win-win situation (a positive, self-reinforcing feedback loop by their valuation).

But as it turned out, the main interplay among the potential factors limiting the number of elk was between food and shelter. At first, food was the limiting factor because elk were constrained in finding their preferred forage by the vast acres of contiguous forest. In contraposition, continued logging started to shift the habitat configuration in a way that proved detrimental to the elk because, while the habitat for feeding continued to increase with clear-cutting, that for shelter declined disproportionately. Accordingly, the shelter once provided by the forest became the factor that increasingly reversed the elk's growth in numbers. Here, it must be understood that shelter for elk consists of two categories—one for hiding in the face of potential danger (simply called *hiding cover*) and one for ameliorating the ambient temperature in the heat of summer and the cold of winter (called *thermal cover*).

Thermal cover often consists of a combination of forest thickets or stands of old-growth trees coupled with topographical features that block the flow of air. As such, thermal cover allows the elk to maintain their bodies' core temperatures in dense, cool shade in summer and get into areas of calm, out of the bitterly cold winds in winter, thereby markedly reducing the windchill factor and thus conserving their body heat.[28] At length, the hunters and big-game biologists began to see the systematic, widespread clear-cutting of the forest as a losing situation for huntable populations of elk (now a negative, self-reinforcing feedback loop by their valuation), although they did not equate the loss of thermal cover as the cause.

At a wider scale, populations of tiny bark beetles are exploding due to the combination of human folly and global warming. First, vast areas of North American forests were simplified through a century of fire suppression and more than a century of logging old-growth timber—often by clear-cutting. Simplifying the inherently complex forest structure created an all-you-can-eat smorgasbord of roughly even-aged, closely spaced trees that do not produce enough resin to create a gooey, chemical, protective barrier against the beetles that kill a tree in a matter of days by dining on its living tissue, thereby girdling the trunk and starving the tree. These indigenous beetles include the spruce bark beetle (which has swept British Columbia and most of the western United States), the mountain pine beetle (in the more southerly lodgepole pine, ponderosa pine, and white-bark pine—from the Pacific Northwest to the Rocky Mountains), and the piñon engraver beetle (which has taken most of the piñon pine out of thousands of acres of piñon-juniper forests in the American Southwest). Second, add to this self-reinforcing feedback-loop scenario more than a century and a half of air pollution from the burning of fossil fuels and the consequential warming of the global climate, and the result is trees stressed by heat and drought that cannot muster the magnitude of defense necessary to combat the growing onslaught of bark beetles.[29]

On yet a wider scale, consider the overexploitation of the large, predatory marine fishes, such as sharks and tuna, which allows the populations of smaller, plankton-feeding fishes to proliferate. At some point, the numbers of plankton-feeders become large enough to dramatically reduce the amount of phytoplankton and thus the ocean's ability to absorb atmospheric carbon dioxide, which in turn affects global warming.[30] The overall result of this self-reinforcing feedback loop is warming oceans that affect major wind patterns, which affect the direction of ocean currents, which is shifting dead zones in the oceans and causing them to grow.[31]

In turn, the stronger, more persistent winds that are expected to accompany a warming climate will shift the dynamics of the biosphere through such phenomena as increased areas of drought and their prolonged duration. Could this self-reinforcing feedback loop bring a repeat of the 1930s, when arguably the most severe drought of the past century gripped almost two-thirds of the United States, as well as parts of Mexico and Canada, and when numerous dust storms occurred in the southern Great Plains?[32]

On a still-wider scale, self-reinforcing, biophysical feedback loops are ultimately controlled by Earth's climate and so are greatly influenced by the levels of atmospheric carbon dioxide over time. Evidence from ice cores and marine sediments indicates that changes in carbon dioxide over timescales beyond the glacial cycles are finely balanced and act to stabilize global temperatures.[33] What is more, the long-term balance between the emissions of carbon dioxide into the atmosphere through such events as volcanic eruptions and the removal of carbon dioxide from the atmosphere through such processes as its burial in deep-sea sediments holds true despite glacial-interglacial variations on relatively short timescales. Today, on the other hand, that part of the feedback loop whereby carbon dioxide (CO_2) is removed from the atmosphere by the chemical breakdown of silicate rock in mountains (termed *weathering*), as well as carbonate minerals (those containing carbon trioxide, CO_3) that are buried in deep-sea sediments, is being severely disrupted—even overwhelmed—by human activities that are raising the level of carbon dioxide emissions through excessive industrial pollution, including the burning of fossil fuels.[34]

Principle 7—All Relationships Have One or More Trade-offs

All relationships have trade-offs that may not be readily apparent or immediately understood. To illustrate, for most of the past 900 years, the buildings in London, many of which had cream-colored limestone façades, were clean. But, then things began to change as a result of the introduction of coal-burning stoves. That notwithstanding, the rate of change was so slow, the cumulative effects were not readily apparent until a threshold of visibility had been crossed, and the protracted exposure to the sooty pollution of city air began to turn the buildings dark gray and black. And so it is that smutty buildings dominated the cities of Europe and the United States for most of the 19th and 20th centuries.[35] In fact, archival photographs show that the limestone Cathedral of Learning on the University of Pittsburgh campus in Pennsylvania, built during a period of heavy pollution in the 1930s, became soiled while still under construction.[36]

Reductions in Pittsburgh's air pollution began in the late 1940s and 1950s.[37] Rain has slowly washed the soiled areas of the 42-story Cathedral of Learning since then, leaving a white, eroded surface. The patterns of whitened areas in archival photographs show the greatest rates of cleansing occurred on the high corners of the building, predominantly where the impact of both rain and wind is most intense. It is also clear that the discoloration of buildings is a dynamic process, whereby the deposition of pollution is a relatively consistent but is simultaneously washed away to varying degrees and patterns over the building's surface. Moreover, sooty pollutants soiled buildings, such as the Cathedral of Learning, much more rapidly in the past than they are being cleaned by wind and rain in the present.[38]

In this century, though, the buildings will gradually become more colorful as the city air is cleaned through the promulgation of pollution-control laws and wind-swept rain that is washing away the encrusted soot. The outcome of such cleaning may well be multicolored buildings as the natural reddish of some limestone is accentuated or a yellowing process occurs as a result of pollutants that are more organic in constitution. What is more, the switch from coal to other fuels has cast the Tower of London in hues that are slightly yellow and reddish-brown. As the atmosphere is cleaned and thus dominated more by organic pollutants, a process of yellowing on stone buildings due to the oxidation of organic compounds in the fumes of diesel and gasoline may become of concern.[39] The oxidation of this increased organic content from the exhaust of motor vehicles may have overall aesthetic consequences for the care of historic buildings—namely, recognizing a shift away from the simple gypsum crusts of the past to those richer in organic materials and thus warmer tones, particularly browns and yellows.[40]

And, this says nothing about plant life growing on cleansed buildings, a phenomenon made possible because vehicular exhaust emits less of the sulfate than is present in the pollution from coal, pollution that suppresses the growth of algae, lichens, and mosses. Consequently, buildings may come to exhibit greens, yellows, and reddish-brown in different places and various patterns because, while lichens and algae prefer humid environs, such as cracks, they can grow on flat surfaces as well.[41]

The foregoing deals only with the dynamics of nature in response to soiling such limestone buildings as the Cathedral of Learning by different types of pollution and the long-term cleansing effects of wind and rain. Added to the trade-offs among these variables is the diversity of preferences espoused in 2003 by employees of the university.

Whereas some university officials were in favor of scrubbing the building with baking soda to remove the black, 70-year-old industrial grime, Cliff Davidson, the environmental engineer from Carnegie Mellon University who studied the building, prefers to let nature do the work. Although the whiter spots have been scrubbed by wind-driven rain over decades, according to Davidson, the darker spots in nooks and crannies might well remain for centuries, if they could be cleaned at all. In contrast, Doris Dyen, director of cultural conservation for the Rivers of Steel National Heritage Area, expressed appreciation of how buildings in Pittsburgh were being spruced up. "At the same time," she said, "you can lose a little bit of a sense of what Pittsburgh was like for 100 years when all the buildings were showing the effects of the 24-hour-a-day operation of the steel mills in the area." G. Alec Stewart, dean of the University Honors College, took yet a different tack, "It would make a stunning addition to the night skyline of Pittsburgh if we were able to illuminate it [the Cathedral of Learning] as significant monuments are in other major cities," comparing it to the Washington Monument.[42]

So, what are some of the significant decision trade-offs with respect to the Cathedral of Learning?

1. Clean the building artificially in the short term *or* let nature do it over time.
2. Clean the building to blend into the cityscape and thus forgo the sense of familiarity *or* maintain the soot-derived appearance and thus avoid rapid change.
3. Purposefully trade the sooty, vegetation-free exterior for an exhaust-enriched, vegetation-covered exterior of the building and thereby give up a sense of Pittsburgh's 100-year history *or* keep the status quo, meaning within the realm of nature's dynamics.
4. Illuminate the Cathedral of Learning from the outside to create a monument-like effect, such as the Washington Monument in the District of Columbia, *or* keep the status quo.

In the end, each of these trade-offs is couched in terms of a decision to change or not, based largely on some culture value that blends naturally into an emotional criterion of choice and thus a decision to act—or not.

Other relationships have much more discernible trade-offs. Take the springtime ozone hole over Antarctica as illustrative. It is finally shrinking after years of growing. As the hole grew in size due to the human-induced, ozone-destroying chemicals in the stratosphere, the risk of skin cancer increased because more ultraviolet radiation reached Earth. Although today the good news is that the ozone hole is now shrinking and, through a complicated cascade of effects, could fully close within this century, what about tomorrow? Because the hole in the stratospheric ozone layer does not absorb much ultraviolet radiation, it keeps the temperature of Antarctica much cooler than normal. A completely recovered ozone layer, on the other hand, could significantly boost atmospheric warming over and around the icy continent and ostensibly augment its melting.[43] In this case, what is good for humans may not be good for Antarctica, and vice versa.

Principle 8—Change Is a Process of Eternal Becoming

Change, as a universal constant, is a continual process of inexorable novelty. It is a condition along a continuum that may reach a momentary pinnacle of harmony within our senses. Then, the very process that created the harmony takes it away and replaces it with something else—always with something else. Change requires constancy as its foil to exist as a dynamic process of eternal becoming. Without constancy, change could neither exist nor be recognized.

Immediate Change

We all cause change of some kind every day. I remember a rather dramatic one I made inadvertently along a small stream flowing across a beach on its way to the sea. The stream, having eroded its way into the sand, created a small undercut that could not be seen from the top. Something captured my attention in the middle of the stream, and I stepped on the overhang to get a better look, causing the bank to cave in and me to get a really close-up view of the water. As a consequence of my misstep, I had both altered the configuration of the bank and caused innumerable grains of sand to be washed back into the sea from whence they had come several years earlier riding the crest of a storm wave.

Understanding Historical Change

Whereas mine was a small, personally created change in an infinitesimal part of the world, others are of gigantic proportions in their effects. People of civilizations that collapsed centuries ago are a good example of such gargantuan effects because they were probably oblivious to the impact that could be wrought by long-term shifts in climate. Although not likely to end the debate regarding what caused the demise of the Roman and Byzantine empires, new data suggest that a shift in climate may have been partly responsible. The plausibility of this notion has been given a scientific boost of credibility through studying the stalactites of Soreq Cave in Israel.[44]

Stalactites are the most familiar, bumpy, relatively icicle-shaped structures found hanging from the ceilings of limestone caves. They are formed when water accumulates minerals as it percolates through soil before seeping into a cave. If the water's journey takes it through limestone, it typically leaches calcium carbonate and carbon dioxide in its descent. The instant the water seeps from the ceiling of a cave, some of the dissolved carbon dioxide in the fluid escapes into the cave's air. This gentle, soda-pop-like fizzing process causes the droplet to become more acidic and so results in some of the calcium carbonate crystallizing on the cave's ceiling, thereby initiating a stalactite. As this process is performed over and over, the separation of calcium carbonate from within the thin film of fluid flowing down its surface allows the stalactite to grow. The procedure is so slow it typically takes a century to add four-tenths of an inch to a stalactite's growth.[45] Moreover, stalactites, like tree rings, can tell stories of paleoclimatic events, such as the severe drought that took place on the Colorado Plateau in the mid-1100s.[46]

By using an ion microprobe, it has become possible to read the chemical deposition rings of the Soreq Cave stalactites with such precision that even seasonal increments of growth can be teased out of a given annual ring. The results indicate that a prolonged drought, beginning in the Levant region as far back as 200 years BCE and continuing to AD 1100, coincides with the fall of both empires.[47] (Levant is the former name of that region of the eastern Mediterranean that encompasses modern-day Lebanon, Israel, and parts of Syria and Turkey.) Although determining why civilizations collapse is always more complicated than one might imagine, an inhospitable shift in climate might well be part of the equation that either *forces people to adapt by changing their behavior or eliminates them*—the challenge squarely before today's decision makers in all countries in all levels of government.

Change as a Historical Continuum

People began to have an influence on Earth's climate thousands of years before the Industrial Revolution and the burning of fossil fuels to power machinery commenced spewing carbon dioxide into the air. Clearing land through the use of fire, first for hunting and then for farming, removed trees that otherwise would have sequestered carbon dioxide rather than allowing it to escape into the atmosphere. As well, the growing human population and the farming practice of disturbing the soil also allowed sequestered carbon to escape. During these early millennia of human interaction with their environment, their activities are estimated to have released about 386 billion tons of carbon into the atmosphere by 1850. Between 1850 and 2000, however, the Industrial Revolution and the burgeoning number of people added 485 tons of carbon, primarily from burning fossil fuels, which more than equaled in a century and a half what had previously taken humankind eight millennia to do.[48]

The foregoing set the stage for the historical record of the Asian monsoon's activity, which is archived in a 1,800-year-old stalagmite found in Wanxiang Cave in the Gansu Province of north-central China. Mineral-rich waters dripping from the cave's ceiling onto its floor year after year formed the stalagmite (a mirror image of a stalactite) that grew continuously for 1,800 years, from AD 190 to 2003. Like trees and the stalactites in the Soreq Cave of Israel, stalagmites have annual growth rings that can provide clues about local environmental conditions for a particular year. Chapters in the Wanxiang Cave stalagmite, written over the centuries, tell of variations in climate that were similar to those of the Little Ice Age, Medieval Warm Period, and the Dark-Age Cold Period recorded in Europe. Here, the critical point is that the warmer years were associated with stronger East Asian monsoons. (A monsoon is a large-scale system of wind that blows seasonally in opposite directions and thus determines the climate of a large region, such as the summer period of heavy rainfall over Southeast Asia.)

By measuring the amount of oxygen-18 (a rare form of "heavy" oxygen) in the growth rings of the stalagmites, the years of weak summer monsoons with less rain can be pinpointed due to the large amounts of oxygen-18 in the rings. The information secreted within the life of the stalagmite tells the story of strong and weak monsoons, which in turn chronicle the rise and fall of several Chinese dynasties. This is an important deliberation because monsoon winds have for centuries carried heavily rain-laden clouds northward from the Indian Ocean every summer, thereby providing nearly 80% of the annual precipitation between May and September in some parts of China—precipitation critical to the irrigation of crops.

In periods when the monsoons were strong, dynasties, such as the Tang (618–907) and the Northern Song (960–1127), enjoyed increased yields of rice. In fact, the yield of rice during the first several decades of the Northern Song dynasty allowed the population to increase from 60 million to as many as 120 million. But, periods of weak monsoons ultimately spelled the demise of dynasties.

The Tang dynasty, for example, was established in AD 618 and is still determined to be a pinnacle of Chinese civilization, a kind of golden age from its inception until the ninth century, when the dynasty began to lose its grip. The Tang was dealt a deathblow in AD 873 when a growing drought turned horrific, and widespread famine took a heavy toll on both people and livestock. Henceforth, until its demise in AD 907, the Tang dynasty was plagued by civil unrest.

Weak monsoon seasons, when rains from the Indian Ocean no longer reached much of central and northern China, coincided with droughts and the declines of the Tang, Yuan (1271–1368), and Ming (1368–1644) dynasties, the last two characterized by continual popular unrest. Weak monsoons with dramatically diminished rainfall may also have helped trigger one of the most tumultuous eras in Chinese history, called the Five Dynasties and Ten Kingdoms period, during which time five dynasties rose and fell within a few decades, and China fractured into several independent nation-states.

Data from the stalagmite indicate that the strength of past Asian monsoons was driven by the variability of natural influences—such as changes in solar cycles and global temperatures—until 1960, when anthropogenic activity appears to have superseded natural phenomena as the major driver of the monsoon seasons from the late 20th century onward. In short, the Asian monsoon cycle has been disrupted by human-caused climate change[49]—based on decades of symptomatic decision making. Here, an observation by the British biologist Charles Darwin is apropos: "It is not the strongest of the species that survive, nor the most intelligent, but *the one most responsive to change*" [emphasis added].[50]

The variability of the Asian monsoons is not the only climatic event to cause human strife, but it is indicative of other past and potential future conflicts. It is thought, for example, that a dire shortage of food caused the genocide at a Pueblo village called Sacred Ridge, near the town of Durango, Colorado, between AD 700 and AD 900.[51] Moreover, increasing global warming and the potential for shortages of food may portend growing civil conflicts to come, as the exploding human population is increasingly confronted with and stressed by a progressively unpredictable environment.[52]

Principle 9—All Relationships Are Irreversible

Because change is a constant process orchestrated along the interactive web of universal relationships, it produces infinite novelty that precludes anything in the cosmos from ever being reversible. Take my misstep on the aforementioned stream's edge. One moment I was standing on the level beach, and the next I was conversing with the water. At the same time, the sand I had knocked into the stream was being summarily carried off to the sea. What of this dynamic was reversible? Nothing. Why not? Simply put, I could not go back in time and make a different decision of where to place my foot. And, because we cannot go back in time, nothing can be restored to its former condition. All we can ever do is repair something that is broken so it can continue to function, albeit differently than in its original form.

A simple example of repairing something is mending a hole in a sock, a lesson my mother taught me over 50 years ago. To mend a sock, she had three items: a wooden darning egg (although an old-fashioned light bulb also works), a darning needle, and darning thread. With patience and dexterity, she wove the thread back and forth across the hole. Then, she turned the sock around far enough to weave the thread through the existing strands until the hole was repaired in a neat crosshatch. At this point, the mended portion of the sock was often stronger than the original fabric had been, which meant it took me longer to wear it out a second time. The sock was repaired but not restored to its original condition. Its physical structure, however, was mended in a way that allowed the sock to continue functioning as a sock. If you want a detailed discussion of this principle, read *Earth in Our Care*.[53]

Principle 10—All Systems Are Based on Composition, Structure, and Function

We perceive objects by means of their obvious structures or functions. Structure is the configuration of elements, parts—composition of constituents—be it simple or complex. The structure can be thought of as the organization, arrangement, or makeup of a thing. Function, on the other hand, is what a particular structure either can do or allows to be done to it or with it.

To maintain biophysical functions means that we humans must maintain the characteristics of the ecosystem in such a way that its processes are sustainable. The characteristics we must be concerned with are: (1) composition, (2) structure, (3) function, and (4) nature's disturbance regimes, which periodically alter an ecosystem's composition, structure, and function.

We can, for example, change the composition of an ecosystem, such as the kinds and arrangement of plants in a forest, grassland, or agricultural crop. This alteration means that composition is malleable to human desire and thus negotiable within the context of cause and effect. In this case, composition is the determiner of the structure and function in that composition is the cause, rather than the effect, of the structure and function.

Composition determines the structure, and structure determines the function. Thus, by negotiating the composition, we simultaneously negotiate both the structure and the function. On the other hand, once the composition is in place the structure and function are set—unless, of course, the composition is altered, at which time both the structure and function are altered accordingly.

In this sense, the composition or kinds of plants and their age classes within a plant community create a certain structure that is characteristic of the plant community at any given age. It is the structure of the plant community that in turn creates and maintains certain functions. In turn, it is the composition, structure, and function of a plant community that determine what kinds of animals can live there, how many, and for how long.

Thus, if an owner of forestland changes the composition of the forest, he or she changes the structure, hence the function, and thus affects the animals. The animals living in the forest are not just a reflection of its composition at any given point in time but are ultimately constrained by it.

If townspeople want a particular animal or group of animals within their urban growth boundary, let us say a rich diversity of summering birds and colorful butterflies to attract tourist dollars from bird-watchers and visitors in general, members of the community would have to work backward by determining what kind of function to create. To do so, they would have to know what kind of structure to create, which means knowing what type of composition is necessary to produce the required habitat(s) for the animals the community wants. Thus, once the composition is ensconced, the structure and its attendant functions operate as an interactive unit in terms of the habitat required for the animals.

For example, the connectivity—accessibility—of habitat components is particularly important for the resident population, regardless of species, because each habitat has a biological carrying capacity, meaning a finite number of individuals that can live in a particular area without altering it to their detriment. If, however, you, the reader, either grew up in a city or now live in one, you may be familiar with habitat corridors in urban settings.

Birds in urban landscapes generally occupy parks, which are analogous to forest fragments, whereas tree-lined streets form linear corridors that connect the fragments within the urban matrix. To understand the species-habitat dynamics of an urban setting, a study conducted in Madrid, Spain, examined the effects of street location within the urbanscape, vegetative structure along the streets, and human disturbance (pedestrian and automotive) on the richness of bird species within the street corridors. In addition, the birds' temporal persistence, density of feeding and nesting guilds, and the probability of a street's being occupied by a single species were also taken into account.

The number of species increased from the least-suitable habitats (streets without vegetation) to the most suitable habitats (urban parks), with tree-lined streets an intermediate landscape element. Tree-lined streets that connected urban parks influenced the number of species within the streets' vegetation positively, as well as species persistence, population density, and the probability that the individual species would continue to occupy the streets. That said, human disturbance did exert a negative influence on the same variables.

Wooded streets could potentially function as corridors that would allow certain species to fare well by supporting alternative habitats for feeding and nesting, particularly those birds that feed on the ground and nest in trees or tree cavities. Local improvements in quality and complexity of the vegetation associated with certain streets, as well as a reduction in the disturbance caused by people, could exert a positive influence on the regional connectivity of streets as a system of urban corridors for birds. Because of the differential use of corridors by species with various habitat requirements, streets as habitat corridors could be further improved by taking the requirements of different species into account.[54]

People and nature are continually altering the structure and function of this ecosystem or that ecosystem by manipulating the composition of its plants, an act that subsequently changes the composition of the animals dependent on the structure and function of the resultant habitat. By altering the composition of plants within an ecosystem, people and nature alter its structure and, in turn, affect how it functions, which in turn determines not only its potential ecosystem services but also what uses humans can derive from those ecosystem services.

Principle 11—All Systems Have Cumulative Effects, Lag Periods, and Thresholds

Nature, as previously stated, has intrinsic value only and so allows each component of an ecosystem to develop its prescribed structure, carry out its biophysical function, and interact with other components through their evolved, interdependent processes and self-reinforcing feedback loops. No component is more or less important than another—except in human valuation based on personal desire for a particular outcome. Each component may differ from the other in form, but all are complementary in function.

Our intellectual challenge in decision making is to recognize that *no given factor can be singled out as the sole cause of anything.* All things operate synergistically as cumulative effects that exhibit a lag period before fully manifesting an outcome of some type. Cumulative effects, which encompass many little, inherent novelties, cannot be understood statistically because ecological relationships are far more complex and far less predictable than our statistical models lead us to believe—a circumstance Francis Bacon may have been alluding to when he said, "The subtlety of Nature is greater many times over than the subtlety of the senses and understanding."[55] In essence, Bacon's observation recognizes that we live in the "invisible present" and thus cannot always recognize cumulative effects.[56]

The invisible present is our inability to stand at a given point in time and see the small, seemingly innocuous effects of our actions as they accumulate over weeks, months, and years. Obviously, we can all sense change—day becoming night, night turning into day, a hot summer changing into a cold winter, and so on. But, some people who live for a long time in one place can see longer-term events and remember the winter of the exceptionally deep snow or a summer of deadly heat.

Despite such a gift, it is a rare individual who can sense, with any degree of precision, the changes that occur over the decades of their lives. At this scale of time, we tend to think of the world as being in some sort of ongoing, relatively steady state, with the exception of technology and periodic natural disasters. Moreover, we typically underestimate the degree to which slow, seemingly innocuous change has occurred—such as global warming. We are unable to sense slow changes directly, and we are even more limited in our abilities to interpret the relationships of cause and effect in these changes. To illustrate:

> The government's colorful map of planting zones is being updated for a warmer 21st century.
>
> The official guide for 80 million gardeners and a staple on seed packets reflects a new reality: The coldest day of the year isn't as cold as it used to be. So some plants that once seemed too vulnerable to cold can now survive farther north.
>
> It's the first time since 1990 that the U.S. Department of Agriculture has updated the map and much has changed. Nearly entire states, such as Ohio, Nebraska and Texas, are in warmer zones.
>
> ...
>
> For example, Des Moines, Iowa, used to be in zone 5a, meaning the lowest temperature on average was between minus 15 and minus 20 degrees Fahrenheit. Now it's 5b, which has a coldest temperature of 10 to 15 degrees below zero.[57]

Nevertheless, these subtle processes, acting quietly and unobtrusively over decades, reside cloaked in the invisible present, such as gradual declines or improvements in habitat quality. Thus, from a leadership point of view, the invisible present can be fraught with the tyranny of many, little, seemingly unimportant and unrelated social-environmental decisions that invite disaster—especially when abetted by informed denial, despite the warnings of history and people with clear foresight.

For example, virtually all the world's roughly 3,000 professional climate scientists (and other professionals who work closely with them) foresee a more inhospitable world for the generations of young children already born, such as higher prices for food; growing numbers of plants and animals becoming extinct, along with their ecological functions and services; as well as other ecological disruptions linked directly to the warming climate. Yet, despite all the evidence observed and irrefutably measured worldwide, such as melting glaciers and rising ocean levels, the fossil fuel industry and ideological groups are trying to delay regulation of greenhouse emissions by funding campaigns that describe confusion and intimidation in an effort to discount the data.[58]

At length, however, cumulative effects, gathering themselves below our level of conscious awareness, suddenly become visible. By then, it is too late to retract our decisions and actions, even if the outcome they cause is decidedly negative with respect to our intentions. So it is that cumulative effects from our decisions and resultant activities can multiply unnoticed until something in the environment shifts dramatically enough—even for people without social-environmental foresight—to see the outcome through casual observation. That shift is defined by a threshold of tolerance in the system, beyond which the system as we knew it suddenly, visibly, irreversibly becomes something else. Within our world, this same dynamic takes place in a vast array of scales in all natural *and* artificial systems, from the infinitesimal to the gigantic.

At a personal level, everyone experiences cumulative effects, lag periods, and thresholds when they become ill, even if it is just a common cold. For instance, if you go to a social function, you may become infected with the cold virus, something you would not know. In fact, you would be unaware of the virus multiplying in your body, a phenomenon that may continue unnoticed for some days (the cumulative effects within the lag period, or in parlance of medicine, the *incubation period*). At length, you begin to sense something is wrong; you just do not feel "up to snuff" (the threshold); shortly thereafter, you have the full-blown symptoms of the classic cold. In this case, the entire process encompasses a few days—from infection to expression.

A shorter-term example of cumulative effects, lag period, and threshold is the cutting down of my neighbor's dying walnut tree. Initially, a man from the tree service sawed off the small branches with intact twigs. The effect was barely discernible at first, even as they began to pile up on the ground. Each severed branch represented a cumulative effect that would have been all but unnoticeable had they not been accumulating under the tree.

After an hour or so (lag period) of removing the small limbs on one side of the tree, the cumulative effects gradually became visible as they crossed a structurally irreversible threshold. Had the same volume of twigs been removed from throughout the tree and simultaneously gathered and removed from the ground, the cumulative effects would not have been as apparent. Nevertheless, the tree was gradually transformed into a stark skeleton of larger branches and the main trunk. Then, the large branches were cut off a section at a time, with the same visual effect as when the small ones had been removed, until only the trunk remained. The piecemeal removal of the tree created a slowly changing vista of my neighbor's house until an unobstructed view of it appeared for the first time as another stark threshold was crossed.

If we now increase the spatial magnitude to encompass the formation of a river's delta, the timescale involved for the cumulative effects to cross the threshold of visibility may well require centuries to millennia. When a river reaches the sea, it slows and drops some of its load of sediment. As the amount of sediment accrues on the seabed, it diverts the river's flow, causing it to deposit additional sediment loads in other areas (cumulative effects). Thus, over many years (lag period), the accumulated sediment begins to show above the water (threshold) and increasingly affects the river's flow as it forms a classic delta. The speed with which the delta grows has numerous variables, such as the amount of precipitation within the river's drainage basin in any given year, as well as the amount of its annual sediment load. Many of today's extant river deltas began developing around 8,500 years ago, as the global level of the seas stabilized following the end of the last ice age.[59] And so the process of change and novelty continues unabated in all its myriad and astounding scales.

Here is an ideal situation to point out the dynamics of informed denial. In response to global warming and rising sea levels, which are submerging deltas and oceanic islands and slowly inundating coastal cities such as New Orleans, the engineering solution—based on the U.S. Army Corps of Engineers' symptomatic decisions—has been to disregard the ongoing change and build higher levees, a patently futile response in the long-term effects of global warming and its oceanic consequences.

Principle 12—All Systems Are Cyclical, But None Are Perfect Circles

While all processes in nature are cyclical, no cycle is a perfect circle, despite such depictions in the scientific literature and textbooks. They are, instead, a coming together in time and space at a specific point, where one "end" of a cycle approximates—*but only approximates*—its "beginning" in a particular time and place. Between its beginning and its ending, a cycle can have any configuration of cosmic happenstance. Clearly, there will be change, but the degree and form of the change will be determined by the "original" conditions. Biophysical cycles can thus be likened to a coiled spring insofar as every coil approximates the curvature of its neighbor but always on a different spatial level (temporal level in nature), thus never touching.

The size and relative flexibility of a metal spring determines how closely one coil approaches another—such as the small, flexible, loosely coiled spring in a ballpoint pen juxtaposed to the large, stiff, coiled spring on the front axle of an 18-wheel truck. The smaller and more flexible the spring of the pen, the closer its coils when juxtaposed to those of the 18-wheel truck. In this sense, the cycles of annual plants in a backyard garden or a mountain meadow are relatively rapid and thus close together in time. Conversely, the millennial cycles of Great Basin bristlecone pines growing on rocky slopes in the mountains of Nevada, where they are largely protected from fire, or a Norway spruce growing on a rocky promontory in the Alps of Switzerland have more protracted cycles over centuries to millennia.

Regardless of its size or flexibility, a spring's coils are forever reaching outward. With respect to nature's biophysical cycles, they are forever moving toward the next level of novelty in the creative process and so are perpetually embracing the uncertainty of future conditions—never to repeat the exact outcome of an event as it once happened. This phenomenon occurs even in times of relative climatic stability. Be that as it may, today's progressive global warming will only intensify tomorrow's uncertainties—such as the increasingly quick loss of groundwater beneath the Central Valley of California, Argentine Patagonia, the Middle East and Russia, northeastern China, northern India, and the Canning Basin of Australia because it is being pumped out for agriculture faster than it can be replenished[60]; reduced production of global grain crops[61]; rising sea levels[62]; and collapsing coastlines in the Arctic.[63]

In human terms, life is composed of rhythms or routines that follow the cycles of the universe, from the minute to the infinite. We humans most commonly experience the nature of cycles in our pilgrimage through the days, months, and years of our lives wherein certain events are repetitive—day and night, the waxing and waning of the moon, the march of the seasons, and the coming and going of birthdays, all marking the circular passage we perceive as time within the curvature of space—although many may perceive life as a linear progression from birth to death. In addition to the visible manifestation of these repetitive cycles, nature's biophysical processes are cyclical in various scales of time and space, a phenomenon that means all relationships are simultaneously cyclical in their outworking and forever novel in their outcomes.

Some cycles recur frequently enough to be well known in a person's lifetime, like the winter solstice. Others are completed only in the collective lifetimes of several generations, like the life cycle of a 3,000-year-old giant sequoia in California's Sequoia National Park—hence the notion of the invisible present. Still others are so vast that their motion can only be assumed. Yet, even they are not completely aloof because we are kept in touch with them—knowingly or not—through our interrelatedness and interdependence. Regarding cycles, farmer and author Wendell Berry said, "It is only in the processes of the natural world, and in analogous and related processes of human culture, that the new may grow usefully old, and the old be made new."[64]

Principle 13—Systemic Change Is Based on Self-Organized Criticality

When dealing with scale (a small, mountain lake as opposed to the drainage basin of a large river, such as the Mississippi in the United States or the Ganges in India), scientists have traditionally analyzed large, interactive systems in the same way that they have studied small, orderly systems, mainly because their methods of study had proven so successful to their sense of understanding. The prevailing wisdom has been that the behavior of a large, complicated system could be predicted by studying its elements separately and by analyzing its microscopic mechanisms individually—such reductionist-mechanical thinking is predominant in Western society and tends to view the world and all it contains through a lens of intellectual isolation ensconced in symptomatology. During the last few decades, however, it has become increasingly clear that many complicated systems, like forests, oceans, and even cities do not yield to such traditional analysis.

Instead, large, complicated, interactive systems seem to evolve naturally to a critical state in which even a minor event starts a chain reaction that can affect any number of internal elements and can lead to a dramatic alteration in the system. Although such systems produce more minor events than catastrophic ones, chain reactions of all sizes are an integral part of system dynamics. According to the theory called "self-organized criticality," the mechanism that leads to minor events (analogous to the drop of a pin) is the same mechanism that leads to major events (analogous to an earthquake).[65] With respect to earthquakes, for example, the weight of seasonal rains or snow on the soil, or flood waters, such as might collect in a lake—and by analogy a reservoir—in areas of geological faults can be heavy enough to slightly depress a fault zone and thus cause initial movement along the fault.[66]

Not understanding this, analysts have typically and erroneously blamed some rare set of circumstances (some exception to the rule) or some powerful combination of mechanisms when catastrophe strikes. Now in my 70s, I understand the fallacy I was taught in high school and in college, namely, that *the exception proves the rule.*

Nevertheless, every ecosystem moves inevitably toward a critical state that alters it in some dramatic way. This dynamic makes ecosystems dissipative structures in that energy is built up through time only to be released in a disturbance of some kind, such as a fire, flood, or landslide in some scale, ranging from a freshet in a stream to the eruption of a volcano, after which energy begins building again toward the next release of pent-up energy somewhere in time.

These "disturbances," as ecologists think of these events, can be long term and chronic, such as large movements of soil that take place over hundreds of years (termed an *earth flow*), or acute, such as the crescendo of a volcanic eruption that sends a pyroclastic flow sweeping down its side at amazing speed. (A *pyroclastic flow* is a turbulent mixture of hot gas and fragments of rock, such as pumice, that is violently ejected from a fissure and moves with great speed down the side of a volcano. *Pyroclastic* is Greek for "fire-broken.")

Here, you might interject that neither a movement of soil nor a volcano is a living system in the classical sense. Although true, all disturbance regimes are part and parcel of the living systems they affect. Thus, interactive systems, from the habitat of a gnat to a tropical rainforest, perpetually organize themselves to a critical state in which a minor event can start a chain reaction that leads to a catastrophic event—as far as living things (such as humans) are concerned, after which the system begins organizing itself toward the next critical state. Therefore, systems never reach a state of equilibrium but rather evolve continuously from one semi-stable state to another. This dynamic is precisely why social-environmental sustainability is a moving target for decision makers—*not a fixed end point or a steady state*, as intimated in so-called "steady-state economy" and "closed-loop technology," both biophysical impossibilities.

Principle 14—Dynamic Disequilibrium Rules All Systems

If change is a universal constant in which nothing is static, what, then, is a natural state? In answering this question, it becomes apparent that the "balance of nature" in the classical sense (disturb nature and nature will return to its former state after the disturbance is removed) is a fond imagining. In fact, the so-called balance of nature is a romanticized figment of the human imagination, something we conjured to fit our hoped-for, snapshot image of the world in which we live. In reality, nature exists in a continual state of ever-shifting *dis-equilibrium*: Ecosystems are entrained in the irreversible process of change and novelty, thereby altering their composition; structure; function; and the resultant, interactive feedback loops in which the available resources are used—irrespective of human influence. Perhaps the most outstanding evidence that an ecosystem is subject to constant change and disruption, rather than remaining in a static balance, comes from studies of naturally occurring external factors that dislocate ecosystems, and climate appears to be foremost among these factors.

After a fire, earthquake, volcanic eruption, flood, hurricane, or landslide, for example, a biological system may eventually be able to approximate what it was through resilience—the ability of the system to retain the integrity of its basic relationships. In this way, an old forest that is burned, blown over in a hurricane, or smashed in a tsunami can be replaced by another—albeit different—old forest on the same acreage. Thus, despite a repetitive disturbance regime, a forest ecosystem can remain a forest ecosystem, though much changed in composition, structure, and function.[67] So it is that ancient forests around the world have been evolving from one critical, biophysical state to the next, from one natural catastrophe to the next through a process of what is termed *autogenic* (produced from within, self-generating) succession. Autogenic succession can be characterized by "successional stages," a concept that refers to the characteristic developmental stages that a forest goes through from bare ground to an ancient or "old-growth" forest.

Autogenic succession works as follows: Grasses and other herbaceous plants are the first to grow in a burned area and so constitute the first successional stage following a fire. In their occupation of a given site and in their growing, they gradually alter the characteristics of the soil, such as pH, until it is no longer optimum for their survival and growth. Their offspring may germinate but not survive, creating areas occupied only by parent plants. As the offspring succumb to the changes in the soil and the parent plants age, die, and are not replaced, openings appear in the vegetative cover that allow shrubs to become established.

In this way, the herbaceous stage changes to an early shrub stage. In turn, the shrub stage follows the same process as the herbaceous stage and gives way to tree seedlings that, in combination, become the shrub-seedling stage. As the shrubs die, the shrub-seedling stage gives way to the sapling stage, which grows into a young forest stage, a mature forest stage, and finally an old-growth forest stage—that is, until succession starts over. The six generalized autogenic, successional stages that a western coniferous forest goes through can be characterized as follows: herbaceous → shrub-seedling → sapling → young forest → mature forest → old-growth forest → *fire* or other disturbance, which starts the cycle over and eventually over again. However, rather than "discrete" stages of development that precede in an orderly succession, one after another, the "stages" of forest development form a complex continuum in which each stage builds on the dynamics and biophysical nuances of the preceding one. Therefore, no two forested areas develop alike in an absolutely repetitive fashion.[68]

But, regardless of how closely an ecosystem might approximate its former state following a disturbance, the existence of every ecosystem is a tenuous balancing act because every system is in a continual state of reorganization that occurs over various scales of time, from the cycle of an old forest to a geological phenomenon, such as Mauna Loa, the active volcanic mountain in Hawaii. Whereas people can manipulate a forest to some extent, Mauna Loa is entrained in an eternal flux of physical novelty over which no human has a smidgen of control.

Finally, we, the human component of the world, must understand and accept that the foregoing biophysical principles are an interactive thread in the tapestry of the natural/cultural world that must be accounted for—and honored—if society is to become a sustainable partner with its various environments. As such, they are an essential and unavoidable part of the "commons," which includes such things as clean air, pure water, fertile soil, viable forests, and bountiful oceans as the birthright of every human being and life in general—the "global commons." Protecting the long-term biophysical viability of the global commons is the foremost responsibility of every designated decision maker—*regardless* of the nomenclature by which they are known.

SUMMARY

Chapter 1 was a discussion of the inviolable biophysical principles that form the non-negotiable constraints whereby humans can practice conscious, purposeful social-environmental sustainability to their benefit. When, however, those limits are ignored, environmental degradation ensues, along with a diminished productive capacity of those ecosystems on which we rely for a good quality of life—a situation that is not only progressively irreversible but also passed forward to all generations.

Chapter 2 expands the premise of Chapter 1 into the inviolable rules of decision making within the role of leadership, which are necessary if social-environmental sustainability is to prevail.

NOTES

1. T.S. Eliot. BrainyQuote. http://www.brainyquote.com/quotes/quotes/t/tseliot109032.html (accessed March 18, 2011).
2. Marie Curie. ThinkExist. http://thinkexist.com/quotation/nothing_in_life_is_to_be_feared-it_is_only_to_be/14155.html (accessed March 18, 2011).
3. Russ Beaton and Chris Maser. *Economics and Ecology: United for a Sustainable World.* CRC Press, Boca Raton, FL, 2011.
4. Arnold J. Toynbee. *A Study of History*, Volumes 1–6 (abridgement by D. C. Somervell). Oxford University Press, New York, 1987.
5. N. Narasimhan. Limitations of science and adapting to Nature. *Environmental Research Letters*, 2 (July–September 2007):034003 doi:10.1088/1748-9326/2/3/034003 (accessed October 1, 2009).
6. Johan Rockström, Will Steffen, Kevin Noone, and others. A safe operating space for humanity. *Nature*, 461 (2009):472–475.
7. Jan Zalasiewicz, Will Steffen, and Paul Crutzen. The new world of the Anthropocene. *Environmental Science and Technology*, 44 (2010):2228–2231.
8. Dawn of the Anthropocene epoch? Earth has entered new age of geological time, experts say. http://www.sciencedaily.com/releases/2010/03/100326101117.htm (accessed February 4, 2011).
9. Beaton and Maser, *Economics and Ecology*.
10. **(1)** Rod Swenson. Emergent evolution and the global attractor: the evolutionary epistemology of entropy production maximization. *Proceedings of the 33rd Annual Meeting of the International Society for the Systems Sciences,* P. Leddington (ed.), 33(3), 46–53, 1989; and **(2)** Rod Swenson. Order, evolution, and natural law: fundamental relations in complex system theory. In *Cybernetics and Applied Systems*, C. Negoita (ed.), 125–148. New York: Dekker, 1991.
11. Rod Swenson and Michael T. Turvey. Thermodynamic reasons for perception-action cycles. *Ecological Psychology*, 3 (1991):317–348.
12. **(1)** Wolfgang Haber. Energy, food, and land—the ecological traps of humankind. *Environmental Science and Pollution Research,* 14 (2007):359–365; and **(2)** David Robson. Proto-humans mastered fire 790,000 years ago. *ABC News*, October 28, 2008 (accessed February 27, 2009).
13. Ann Gibbons. Food for thought. *Science* 316 (2007):1558–1560.
14. Rod Swenson. Spontaneous order, autocatakinetic closure, and the development of space-time. *Annals of the New York Academy of Sciences*, 901 (2000):311–319.
15. N. Narasimhan. Limitations of science.
16. Marcus Aurelius. BrainyQuote. http://www.brainyquote.com/quotes/authors/m/marcus_aurelius.html (accessed December 30, 2008).
17. G.O. Poinar, A.E. Treat, and R.V. Southeott. Mite parasitism of moths: examples of paleosymbiosis in Dominican amber. *Experientia*, 47 (1991):210–212.
18. The general discussion of amberization is based on **(1)** George O Poinar, Jr. Insects in amber. *Annual Review of Entomology*, 46 (1993):145–159; **(2)** Anonymous. Scientist: Earth's oldest spider web discovered. London. *Corvallis Gazette-Times*, Corvallis, OR. December 16, 2008; and **(3)** Enrique Peñalver, David A. Grimaldi, and Xavier Delclòs. Early Cretaceous spider web with its prey. *Science*, 312 (2006):1761.
19. Beaton and Maser, *Economics and Ecology*.

20. Chris Maser. *Earth in Our Care: Ecology, Economy, and Sustainability*. Rutgers University Press, Piscataway, NJ, 2009.
21. **(1)** Yuan-Chug Cheng and Graham R. Fleming. Dynamics of light harvesting in photosynthesis. *Annual Review of Physical Chemistry* 60 (2009):241–262; **(2)** Paul May. Chlorophyll. http://www.chm.bris.ac.uk/3motm/chlorophyll/chlorophyll_h.htm (accessed January 5, 2009).
22. Coal. http://en.wikipedia.org/wiki/Coal (accessed March 23, 2011).
23. Nhu Lee. What is oil? http://www.greenscreen.org/articles_sr/Planet%20Earth%20and%20Solar%20System/Natural%20Resources/Oil-%20Sr.pdf (accessed March 23, 2011).
24. James Fennimore Cooper. *The Last of the Mohicans* (Bantam Classic edition). Bantam Dell, New York, 1981.
25. The foregoing discussion of dust in the alpine lakes is based on J.C. Neff, A.P. Ballantyne, G.L. Farmer, and others, Increasing eolian dust deposition in the western United States linked to human activity, *Nature Geoscience* 1 (2008):189–195.
26. Maser, *Earth in Our Care*.
27. Chris Maser. *Social-Environmental Planning: The Design Interface between Everyforest and Everycity*. CRC Press, Boca Raton, FL, 2009.
28. The discussion of elk is based in part on Jack Ward Thomas, Hugh Black, Jr., Richard J. Scherzinger, and Richard J. Pederson. Deer and elk. In *Wildlife Habitats in Managed Forests: The Blue Mountains of Oregon and Washington*. Jack Ward Thomas (technical ed.), 104–127. Agricultural Handbook No. 553. Portland, OR: U.S. Department of Agriculture, Forest Service, Pacific Northwest Range and Experiment Station, 1979.
29. **(1)** Reese Halter. *The Insatiable Bark Beetle*. Rocky Mountain Books, Heritage House, Surrey, BC, Canada, 2011; **(2)** Andrew Nikiforuk. *Empire of the Beetle: How Human Folly and a Tiny Bug Are Killing North America's Great Forests*. Greystone Books, Vancouver, BC, Canada. *Homer Tribune*, 2011. http://homertribune.com/2011/11/reading-andrew-nikiforuk's-empire-of-the-beetle/ (accessed November 9, 2011); and **(3)** Ed Berg. Reading Andrew Nikiforuk's *Empire of the Beetle*. *Homer Tribune*, 2011. http://homertribune.com/2011/11/reading-andrew-nikiforuk's-empire-of-the-beetle/ (accessed November 9, 2011).
30. **(1)** Ransom A. Myers and Boris Worm, Extinction, survival or recovery of large predatory fishes. *Philosophical Transactions of the Royal Society of London: Biological Sciences* 360 (2005):13–20; **(2)** Peter Ward and Ransom A. Myers, Shifts in open-ocean fish communities coinciding with the commencement of commercial fishing, *Ecology* 86 (2005):835–847; and **(3)** Kenneth T. Frank, Brian Petrie, Jae S. Choi, and William C. Leggett, Trophic Cascades in a formerly cod-dominated ecosystem. *Science* 308 (2005):1621–1623.
31. **(1)** D. Pauly and V. Christensen, Primary production required to sustain global fisheries, *Nature,* 374 (1995):255–257; **(2)** John A. Barth, Bruce A. Menge, Jane Lubchenco, and others, Delayed upwelling alters nearshore coastal ocean ecosystems in the northern California current, *Proceedings of the National Academy of Sciences of the United States of America,* 104 (2007):3719–3724; **(3)** F. Chan, J. A. Barth, J. Lubchenco, and others, Emergence of anoxia in the California current large marine ecosystem, *Science,* 319 (2008):920; and **(4)** Ryan R. Rykaczewski and David M. Checkley, Jr., Influence of ocean winds on the pelagic ecosystem in upwelling regions, *Proceedings of the National Academy of Sciences of the United States of America,* 105 (2008):1965–1970.
32. Siegfried D. Schubert, Max J. Suarez, Philip J. Pegion, and others, On the cause of the 1930s Dust Bowl, *Science,* 303 (2004):1855–1859.
33. David Archer. Carbon cycle: checking the thermostat. *Nature Geoscience*, 1 (2008):289–290.

34. Richard E. Zeebe and Ken Caldeira. Close mass balance of long-term carbon fluxes from ice-core CO_2 and ocean chemistry records. *Nature Geoscience*, 1 (2008):312–315.
35. The discussion of color changes in buildings is based on **(1)** Carlotta M. Grossi, Peter Brimblecombe, Rosa M. Esbert, and Francisco Javier Alonso. Color changes in architectural limestone from pollution and cleaning. *Color Research and Application*, 32 (2007):320–331; and **(2)** Catherine Brahic. Cleaner air to turn iconic buildings green: with atmospheric changes, limestone buildings will turn yellow, reddish-brown and green. *New Scientist*, December 4, 2008. http://www.newscientist.com/article/dn16198-cleaner-air-to-turn-iconic-buildings-green.html (accessed December 8, 2008).
36. C.I. Davidson, W. Tang, S. Finger, V. Etyemezian, M.F. Striegel, and S.I. Sherwood. Soiling patterns on a tall limestone building: changes over sixty years. *Environmental Science and Technology*, 34 (2000):560–565.
37. Cliff I. Davidson. Air pollution in Pittsburgh: a historical perspective. *Journal of the Air Pollution Control Association*, 29 (1979):1035–1041.
38. **(1)** V. Etyemezian, C.I. Davidson, M. Zufall, W. Dai, S. Finger, and M. Striegel. Impingement of rain drops on a tall building. *Atmospheric Environment*, 34 (2000):2399–2412; and **(2)** Vicken Etymezian, Cliff I. Davidson, Susan Finger, and others. Vertical gradients of pollutant concentrations and deposition fluxes at the Cathedral of Learning. *Journal of the American Institute for Conservation*, 37 (1998):187–210.
39. **(1)** Grossi et al., Color Changes; and **(2)** Alessandra Bonazza, Peter Brimblecombe, Carlota M. Grossi, and Cristina Sabbioni. Carbon in black crusts from the Tower of London. *Environmental Science and Technology*, 41 (2007):4199–4204.
40. Grossi et al., Color Changes.
41. **(1)** Grossi et al., Color Changes; and **(2)** Bonazza et al., Carbon in Black Crusts.
42. Bill Zlatos. Foes, lack of funds may scrub Cathedral of Learning cleaning. *Pittsburgh Tribune-Review*, July 6, 2003.
43. **(1)** Sid Perkins. As ozone hole heals, Antarctic could heat up. *Science News*, July 5 (2008):10; **(2)** S.-W. Son, L.M. Polvani, D.W. Waugh, and others. The impact of stratospheric ozone recovery on the Southern Hemisphere westerly jet. *Science*, 320 (2008):1486–1489; and **(3)** Perlwitz J., S. Pawson, R.L. Fogt, and others. Impact of stratospheric ozone hole recovery on Antarctic climate. *Geophysical Research Letters*, 35 (2008):L08714, doi:10.1029/2008GL033317 (accessed December 17, 2008).
44. Lee Dye. Did climate change kill the Roman Empire? December 10, 2008. http://abcnews.go.com/Technology/JustOneThing/story?id=6428550&page=1 (accessed December 10, 2008).
45. **(1)** Sid Perkins. Buried treasures. *Science News*, 169 (2006):266–268; **(2)** Martin B. Short, James C. Baygents, and Raymond E. Goldstein. Stalactite growth as a free-boundary problem. *Physics of Fluids*, 17 (2005):083101, 12 pp. (accessed December 17, 2008); and **(3)** M. B. Short, J. C. Baygents, J. W. Beck, and others. Stalactite growth as a free-boundary problem: a geometric law and its platonic ideal. *Physical Review Letters*, 94 (2005):018510, 4 pp. (accessed December 17, 2008).
46. D. Meko, C.A. Woodhouse, C.A. Baisan, and others. Medieval drought in the upper Colorado River Basin. *Geophysical Research Letters*, 34 (2007):L10705, doi:10.1029/2007GL029988 (accessed December 17, 2008).
47. **(1)** Ian J. Orland, Miryam Bar-Matthews, Noriko T. Kita, and others. Climate deterioration in the eastern Mediterranean as revealed by ion microprobe analysis of a speleothem that grew from 2.2 to 0.9 Ka in Soreq Cave, Israel. *Quaternary Research*, 71(2009):27–35; **(2)** A. Kaufman, G.J. Wasserburg, D. Porcelli, and others. U-Th isotope systematics from the Soreq Cave, Israel and climatic correlations. *Earth and Planetary Science Letters*, 156 (1998):141–155; and **(3)** Avner Ayalon, Miryam Bar-Matthews, and Eytan

Sass. Rainfall-recharge relationships within a karstic terrain in the eastern Mediterranean semi-arid region, Israel: δ ^{18}O and δD characteristics. *Journal of Hydrology*, 207 (1998):18–31.

48. **(1)** Jed O. Kaplan, Kristen M. Krumhardt, Erle C. Ellis, and others. Holocene carbon emissions as a result of anthropogenic land cover change. *The Holocene*, 21 (2011):775–791; **(2)** Jonathan A. Foley, Chad Monfreda, Navin Ramankutty, and David Zaks. Our share of the planetary pie. *Proceedings of the National Academy of Sciences of the United States of America*, 104 (2007):12585–12586.
49. **(1)** Pingzhong Zhang, Hai Cheng, R. Lawrence Edwards, and others. A test of climate, sun, and culture relationships from an 1810-year Chinese cave record. *Science*, 322 (2008):940–942; **(2)** Kallie Szczepanski. The monsoons fail, and China's dynasties fall. http://asianhistory.about.com/od/asianenvironmentalhistory/a/ChinaMonsoon.htm (accessed January 5, 2009); **(3)** Ker Than. Chinese kingdoms rose, fell with monsoons? *National Geographic News*, November 2, 2008. http://news.nationalgeographic.com/news/2008/11/081106-monsoons-china.html (accessed January 10, 2009); and **(4)** Yongjin Wang, Hai Cheng, R. Lawrence Edwards, and others. Millennial- and orbital-scale changes in the East Asian monsoon over the past 224,000 years. *Nature*, 451 (2008):1090–1093.
50. Charles Darwin. *On the Origin of Species*. Modern Library, a Division of Random House, New York, 1998.
51. James M. Potter and Jason P. Chuipka. Perimortem mutilation of human remains in an early village in the American Southwest: a case for ethnic violence. *Journal of Anthropological Archaeology*, 29 (2010):507–523.
52. **(1)** Solomon M. Hsiang, Kyle C. Meng, and Mark A. Cane. Civil conflicts are associated with the global climate. *Nature*, 476 (2011):438–441; and **(2)** J. P. Reganold, D. Jackson-Smith, S. S. Batie, and others. Transforming U.S. agriculture. *Science*, 332 (2011):670–671.
53. Maser, *Earth in Our Care*.
54. The three previous paragraphs are based on Esteban Fernández-Juricic, Avifaunal use of wooded streets in an urban landscape. *Conservation Biology*, 14 (2000):513–521.
55. Francis Bacon. http://Science.prodos.org (accessed January 2, 2009).
56. John J. Magnuson. Long-term ecological research and the invisible present. *BioScience*, 40 (1990):495–501.
57. Seth Borenstein. New federal map for what to plant reflects warming. http://abcnews.go.com/Technology/wireStory/federal-map-plant-reflects-warming-15438721#.TyBJcBw0i4A (accessed January 25, 2012).
58. Bill Blakemore. Shakespeare, global warming, sunset, and you. November 5, 2011. http://abcnews.go.com/blogs/technology/2011/11/shakespeare-global-warming-sunset-and-you/ (accessed November 7, 2011).
59. **(1)** Sid Perkins. O river deltas, where art thou? *Science News*, 172 (2007):118; and **(2)** Pippa L. Whitehouse, Mark B. Allen, and Glenn A. Milne. Glacial isostatic adjustment as a control on coastal processes: an example from the Siberian Arctic. *Geology*, 35 (2007):747–750.
60. **(1)** V. M. Tiwari, J. Wahr, and S. Swenson. Dwindling groundwater resources in northern India, from satellite gravity observations. *Geophysical Research Letters*. 36 (2009):L18401, 5 pp., doi:10.1029/2009GL039401; and **(2)** Devin Powell. Satellites show groundwater dropping globally. *Science News*, 181 (2011):5–6.
61. Reganold et al., Transforming U.S. agriculture.
62. Andrew Kemp, Benjamin P. Horton, Jeffrey P. Donnelly, and others. Climate related sea-level variations over the past two millennia. *Proceedings of the National Academy of Sciences of the United States of America*, 108 (2011):11017–11022.

63. M. Jones, C. D. Arp, M. T. Jorgenson, and others. Increase in the rate and uniformity of coastline erosion in arctic Alaska. *Geophysical Research Letters*, 36(2009):L03503, 5 pp., doi:10.1029/2008GL036205.
64. Wendell Berry. The road and the wheel, *Earth Ethics,* 1 (1990):8–9.
65. Per Bak and Kan Chen. Self-organizing criticality. *Scientific American,* January (1991):46–53.
66. **(1)** Devin Powell and Alexandra Witze. Weather affects geologic activity. *Science News*, 180 (2011):8; **(2)** Daniel Brothers, Debi Kilb, Karen Luttrell, and others. Loading of the San Andreas fault by flood-induced rupture of faults beneath the Salton Sea. *Nature Geoscience*, 4 (2011):486–492; and **(3)** Thomas J. Sheeran. Expert: wastewater well in Ohio triggered quakes. http://abcnews.go.com/US/wireStory/expert-wastewater-ohio-triggered-quakes-15275844#.TwJKDhw0gpc (accessed January 3, 2012).
67. **(1)** Peter H. Morrison and Frederick J. Swanson. Fire history and pattern in a Cascade Range landscape. USDA Forest Service General Technical Report. PNW-GTR-254. Pacific Northwest Research Station, Portland, OR, 1990; and **(2)** Charles Grier Johnson, Jr. Vegetation response after wildfires in national forests of northeastern Oregon. USDA Forest Service, Pacific Northwest Region. R6-NR-ECOL-TP-06-98, 1998.
68. The discussion of autogenic succession is based on Chris Maser. *Our Forest Legacy: Today's Decisions, Tomorrow's Consequences*. Maisonneuve Press, Washington, DC, 2005.

2 Decision Making and Nature's Rules of Engagement

> No person, institution, or nation has the right to participate in activities that contribute to large-scale, irreversible changes of the Earth's biogeochemical cycles or undermine the integrity, stability, and beauty of the Earth's ecologies—the consequences of which would fall on succeeding generations as an irrevocable form of remote tyranny.
>
> **—David Orr**[1]

Before I discuss the rules of decision making, it is necessary to revisit the waterbed principle: to touch one part is to touch all parts simultaneously. Yet, with our myriad data bits and seemingly vast knowledge, most, but not all, economists assume—often through "informed denial"—that air, soil, water, sunlight, biodiversity, and climate are economically constant entities that, when possible, are omitted from consideration in our economic models and to a large extent in planning models[2] unless required by law. The fallacy in this notion is that no ecological variable—including air, soil, water, sunlight, biodiversity, or climate—can be converted into an illusionary economic externality without dire biophysical consequences. Again, unless required by law, biodiversity is discounted when its consideration interferes with monetary profits.[3] Then, there is the nagging problem of human population, which we talk about and worry about, but in the end give only lip service to the one thing that can actually control it—total, real, gender equality. Finally, we—the human component of the world—must understand and accept that each of these entities is an interactive variable in the tapestry of the natural/cultural world that must be accounted for if society, as we know it, is to garner the wisdom to become a sustainable partner with its various environments. Be forewarned, to achieve social-environmental sustainability means that we must consciously, purposefully control our human population—or nature will, in ways most frightening and painful (witness the ongoing droughts and famines in Africa).

EVERY DECISION HAS ITS ANTITHESIS

In today's world, human consciousness can be likened to the key in the Chinese proverb: To every man is given the key to the gates of heaven, and the same key opens the gates of hell. In this case, those with enlightened awareness who act to repair and protect our environmental quality for all generations turn the key toward *heaven*, whereas those who steadfastly remain committed to "informed denial" to protect their economic status quo turn the key increasingly to *hell on Earth* for all generations to come. To enliven our understanding of the proverbial key, let us briefly examine each entity individually and then each entity in the collective of their relationship.

Air: The Breath of Life—And of Death

It is late afternoon on a clear, warm, sunny September day. A tiny spider climbs a tall stalk of grass in a subalpine meadow and raises its body into the air, almost standing on its head. From spinnerets on the tip of its abdomen, it ejects a mass of silken threads into the breeze. Suddenly, without visible warning, the spider is jerked off its stalk and born aloft to join other spiders riding the warm afternoon air flowing up the mountainside, all casting their fortunes to the wind. Like their ancestors in centuries and millennia past, they float on currents of air from the far corners of Earth and become the first inhabitants of newly formed South Sea Islands.

Spiders are not the only things circumnavigating the world on currents of air. In 1883, Krakatoa, a small Indonesian island between Java and Sumatra, was virtually obliterated by explosive eruptions that sent volcanic ash high enough above Earth to ride the world's airways for more than a year, affecting the climate by filtering the sun's light.[4] Air also carries the reproductive spores of fungi and the pollen of various trees and grasses. It carries dust and microscopic organisms.[5] And, it carries life-giving oxygen and water and death-dealing pollution—the legacy of human society.[6]

Although air currents carry life-giving oxygen, water, and life-sustaining dust from Africa to the Amazonian rainforest,[7] they also transport the "key to death"—a human legacy made visible.[8] In addition to the carbon dioxide affecting the world's oceans, toxins from such areas as the notoriously polluted air of Mexico City hitchhike on the wind across the Gulf of Mexico toward the United States, where the forest edges in fragmented landscapes function as significant traps for airborne nutrients and pollutants from both near and afar.[9] Nearby areas include local agricultural fields, such as monocultures of corn raised to produce food for humans and animals and for the production of biofuels, whereas faraway places include such urban settings as Beijing, China. What is more, forest edges effectively concentrate these chemical fluctuations below the canopy, where they can have cascading effects on soil-nutrient cycling, microbial activity, seedling dominance, and other ecological processes.[10] Even this result can be judged as mild compared with what is happening in some of our national parks.

A report of a $6 million study by the U.S. National Park Service, the *Western Airborne Containments Assessment Project*, documents the fact that pesticides, heavy metals, and other airborne contaminants—70 in all—are literally raining down on 20 national parks and monuments from Denali in Alaska and Glacier in Montana to Big Bend in Texas and Yosemite in California. Over time, the pollutants enter the atmosphere, are flushed out with rain and snow, only to reenter it again and be flushed out again—and again and again—but each time at a higher elevation. The toxins range from mercury produced by power plants to such industrial chemicals as PCBs (polychlorinated biphenyls), dieldrin, and DDT (both dieldrin[11] and DDT are banned insecticides in the United States,[12] but not worldwide).

In addition, contaminants in fish from the eight parks studied all exceed the safe threshold for human consumption. The parks most affected are Sequoia and Kings Canyon (California), Mount Rainier and Olympic (Washington), Glacier (Montana), Rocky Mountain (Colorado), Gates of the Arctic and Denali (Alaska), as well as Alaska's Noatak National Preserve. Mercury also exceeds the safe limits for fish-eating wildlife at all eight parks, and DDT is in dangerous amounts for fish-eating wildlife in Glacier, Sequoia, and Kings Canyon national parks.[13]

On the other hand, the roughly 440,000 trees within the urban growth boundary of my hometown (Corvallis, OR) annually remove about 70 tons of air pollutants; 3,000 tons of carbon dioxide, the greenhouse gas; and 100 million gallons of storm-water runoff as compared to pavement. They also save about 4 trillion joules of energy in cooling buildings during summer's heat.

In terms of dollars and cents, the models estimate the largest annual economic saving of $3 million in processing storm waters (compared to pavement), $600,000 in annual increased real estate values, and $400,000 worth of reductions in air pollution. These are annual benefits, from trees that would cost $450 million to replace.[14]

Water: A Captive of Gravity

Water is the most important commodity from the world's forests. Most of the usable water in North America north of Mexico comes from snow. When snow melts in the mountains, where there is mostly rock and scant soil, water is shed quickly and with little percolation. Where there is soil, water is purified and stored by flowing slowly through soil healthy with nature's processes.[15] Water picks up containments when flowing through soil stripped of nature's processes and polluted with chemicals, both naturally occurring and artificially produced.[16] In addition, water, bearing tons of toxic effluents, flows directly into streams, rivers, estuaries, and the open ocean. Water is a captive of gravity, so all the pollutants it accumulates on its downhill journey eventually end up in oceans worldwide,[17] where they accumulate in ever-increasing concentration—first because the ocean has no outlets and cannot be flushed and second due to the surface evaporation.

The preceding refers to water that, flowing for a time through the soil, completes its journey to the ocean above ground. There is, however, a belowground analog in subterranean seeps, trickles, and rivulets, which coalesce into invisible streams and rivers that flow from the mountains to the ocean entirely below ground. On reaching the oceans, they enter the marine environment through porous soils along the beach, just below the salty surface, or erupt as freshwater springs on the ocean floor of the continental shelf near many of the world's shores, where the freshwater influences the dynamics of the marine ecosystem.[18] In turn, rising sea levels due to global warming will pollute these aquifers with seawater, to the detriment of humans inhabiting coastal areas.[19]

These hidden waterways are worldwide conduits that, like their aboveground counterparts, today increasingly deliver myriad human-made, toxic, and carcinogenic chemical compounds from agricultural fields and tree farms, from urban settings, industrial complexes, and fractured-rock drilling for natural gas to the oceans of the world. But, around 2,000 years ago, the Roman geographer Strabo wrote about the residents of Latakia, Syria, who rowed their boats 2½ miles out into the salty Mediterranean, where they dove a few yards to the ocean floor and collected fresh, safe drinking water in goatskin containers for the residents of their city.[20]

Soil: The Great Placenta

Soil is the nexus between the living (plant and animal) and nonliving components of the landscape, an exchange membrane much like the placenta through which a mother nourishes her child. Derived from rock, soil is built up by plants that live and die on and in it. It is also enriched by animals that feed on plants, void their bodily wastes, and eventually die, decay, and return to the soil as organic matter. Soil is by far the most alive and biologically diverse part of any terrestrial ecosystem.[21] In addition, soil organisms are the regulators of most of its processes, which translate into plant and then animal productivity.[22]

Many cultures have emphasized in their religion and philosophy that humans must be trustees of the soil. In his covenant with God, Abraham was instructed: "Defile not therefore the land which ye shall inhabit, wherein I dwell."[23] An indigenous American proverb states: "Treat the earth well: it was not given to you by your parents, it was loaned to you by your children. We do not inherit the Earth from our Ancestors, we borrow it from our Children."[24] And, Franklin D. Roosevelt spoke an undeniable truth when he said, "A nation that destroys its soils destroys itself."[25]

In spite of such long-held beliefs, most people do not grasp their importance, due, perhaps, to their intangibility. So, most humans pay little attention to the soil, which not only is underfoot but also is more and more under concrete and asphalt and thus increasingly "invisible" to the masses of the industrialized countries and, therefore, like air, is taken for granted.

It is important to realize, however, that all human society is tied to the soil for reasons beyond measurable materialistic riches.[26] Yet, in the name of short-term profits, we rob the soil of the very organic material necessary to sustain its fertility. We then use artificial chemicals that not only poison the soil and alter the way its many hidden processes function but also pollute the water flowing through it into the streams, rivers, estuaries, and the oceans of the world.

Soil is also the stage on which the entire human drama is enacted. If we continue to destroy the stage on which we depend for life, we will play a progressively ebbing role in a terminal tragedy of human society.

Biodiversity: The Variety of Life

Biodiversity refers to the variety of living species and their biological processes. It acts as an ecological insurance policy for the flexibility of future choice and viable decisions in terms of human options for life because every ecosystem adapts in various ways to changes in its environment. In turn, the degree of a system's adaptability depends on the richness of its biodiversity, which creates backup processes that provide alternative, functional channels in case of a failure and so retains the system's ability to respond to continual change.[27]

Biological backups strengthen the ability of an ecosystem to retain its integrity. For example, an ecosystem can have more than one species that performs a similar function, which means the loss of one or two species is not likely to result in such severe functional disruption that it causes ecosystem collapse because other species can make up for the functional loss. At some point, however, the loss of one more species will tip the balance and cause a change that will alter the quality and productivity of the system. This point of irreversibility is an unknown biological threshold. In other words, we do not know which species' extinction will trigger such an event.[28] That is why it pays to save every species we can. As Aldo Leopold said, "The first rule of intelligent tinkering is to save all the pieces."[29]

Species richness is important because each species has a shape, structure, and role that facilitate certain functions. These functions interact with those of other species to create a viable system through self-reinforcing feedback loops.[30] All biodiversity is ultimately governed by the genetic code, which replicates species' character traits and thus maintains backups into each ecosystem.

A stable ecosystem may recover from disturbances to which it is adapted but conversely may be vulnerable to disturbances to which it is not adapted. The diversity of plants and animals, therefore, buffers an ecosystem from disturbances. When species are lost from an ecosystem, it involves not only their combined structural and functional diversity but also their genetic diversity, eventually resulting in the simplification of a complex ecosystem to the point it is rendered of lesser value to humans.

Some forests, for example, are living entities that complete a cycle of interdependent processes over many years or several centuries—spanning many human generations. And yet, with grossly incomplete and shortsighted knowledge at our disposal, we often seem to have unquestioning faith in that information. So certain are we of our data that we blithely predict the sustained-yield capability of such things as tree farms far into an unforeseeable future.

Based on these erroneous predictions, we justify clear-cutting the old-growth forest and converting it to a biologically simplified plantation, completely ignoring biodiversity, *especially that which sustains the infrastructure of the forest soil*.[31] Such is a gamble, pure and simple, to which we will not know the answer for decades—perhaps centuries. Destroy the soil, and the forest ceases to be. Destroy the forest, and the soil becomes further impoverished and increasingly subject to erosion, which progressively degrades its capacity to store and purify one of humanity's primary sources of water.

Human Population: A Matter of Equality

We have been warned for decades that the human species is overpopulating Earth, yet our population continues to explode, and the usable portion of Earth per individual shrinks, as does the allotted proportion of its resources, all of which—except solar radiation—are finite. We have tried many symptomatic remedies for this situation: education, birth control, feeding the hungry, shipping technology to poor nations, and so on. But, we have not addressed the primary cause of overpopulation, either at home or in so-called third-world countries: the inequality between men and women. To wit: "Egyptian soldiers were caught on video beating and disrobing a veiled female protester. At least three soldiers are seen on tape exposing the young woman's midriff and bra under the black robe she wears as part of her religious [Islamic] veil, kicking her, stomping on her stomach and hitting her head with batons."[32]

Men, who justify their behavior in the name of patriarchal religion, have long dominated women. "In Egypt's conservative, male-dominated society, women are not supposed to express themselves so openly in public. The violence, according to critics, serves as punishment not just for defying authority, but for violating the rules for women, and is meant to deter other women from joining the protests."[33]

And in Israel:

> Thousands of ultra-Orthodox Jews are protesting in Jerusalem to defend their contentious effort to ban the mixing of the sexes on buses, sidewalks and other public spaces.
>
> These practices are rejected by mainstream Judaism and efforts by a minority group to impose its lifestyle on others has provoked anger among secular Israelis and among the wider ultra-Orthodox community.
>
> In one city near Jerusalem, radical ultra-Orthodox recently accosted school girls and spit at women and children they believe are dressed immodestly. That prompted a large protest rally.
>
> Israel's Supreme Court has ruled against segregation.[34]

Through such domination, women in some nonindustrialized, poor countries physically produce most of the food yet are allowed to own but an infinitesimal part of the land. And, women have had only one way to be uniquely valued—by having male babies.

Women who have a good education have fewer children and have them later in life. Why? Education takes time and often delays marriage. Knowledge makes birth control understood, and improved economic status, which goes with education, provides the wherewithal to afford the means of birth control. Moreover, educated women also make more money and may find it less attractive to stay out of the workforce.

I submit, therefore, that if we are serious about controlling the human population, women must have unequivocal access to opportunities for finding personal and social value beyond childbearing—and not treated like the foregoing news stories from Egypt and Israel portray. On the surface, this means such things as equal opportunities for education, jobs, and pay, which few have today, even in the industrialized countries. But at its root, it means changing the insufferable attitude of superiority that the world's men arrogantly superimpose on women—a difficult, but vital task if human society is to live in sustainable harmony and dignity with Earth.

Sunlight: The Source of Global Energy

The sun, considered to be an average star, is a fusion reactor that has been burning over 4 billion years and has been worshiped by various cultures around the world for millennia. And yet, it contributions to life on Earth are often taken for granted in our increasingly urbanized world. *Sunlight (solar radiation) powers all Earth's processes.* In fact, solar radiation provides in 1 minute enough energy to supply the world's energy requirement for 1 year. At this rate, it could supply in 1 day all the energy used by the current human population for 27 years. Moreover, in just 3 days, the amount of solar radiation reaching Earth is equivalent to the energy stored in all sources of fossil fuel.[35]

We, in turn, harvest the sun's energy through the fruits, vegetables, and grains we eat. But, what happens when air pollution—smog—filters sunlight before it reaches the plants? What happens if the ozone shield were to disappear, allowing deadly ultraviolet rays to bombard Earth? What would happen if both these events occur simultaneously, which could potentially happen?

Air pollution now directly affects the quality of the sun's light and energy reaching the surface of the Earth and can have an impact on virtually all life. Like everything else, the quantity and quality of the sun's light reaching Earth today in a given location is an ecological variable—thanks to air pollution and climate change—that must be taken into account as we plan for social-environmental sustainability.

Climate: The Global Arbiter

Human societies of all types continue to develop or utilize a broadening array of technologies while assuming, or not caring, that these technologies will harm the interactive atmosphere, litho-hydrosphere, and biosphere, which together sustain all life—including us humans. The day has come, however, when the life-support systems themselves are being disrupted by increased use of technologies with the cumulative power to disarrange ecosystems to such an extent that the biophysical services we require are in peril—unlike anything faced by humanity until the 20th century. Such is the newest and greatest threat faced by the world we inhabit.

The irrefutable increase in global warming is caused by an array of gases in the atmosphere that trap heat radiated outward from Earth's surface—the "greenhouse effect." Carbon dioxide has to date received the most attention, but other gases are involved, including water vapor, ozone, nitrous oxide, chlorofluorocarbons (CFCs), and methane.

The consensus among scientists is that global warming, at some unknown intensity, is occurring at an unprecedented rate, as evidenced by the massive melting of the world's continental glaciers and oceanic ice. Further, sea levels are rising as the ice melts and ocean temperatures warm, the latter causing the water to expand and contribute to rising sea levels. Further discussion is redundant for those who accept the compounding evidence of global warming and a waste of time for those who cling steadfastly to informed denial.

Turning the Key

> Only a people serving an apprenticeship to nature can be trusted with machines. Only such people will so contrive and control those machines that their products are an enhancement of biological needs, and not a denial of them.
>
> **—Herbert Read, British philosopher[36]**

We humans directly affect five of the seven variables necessary for social-environmental survival: air, soil, water, biodiversity, and population density, and we indirectly influence the other two—sunlight and climate. If, for example, we choose to clean the world's air, we will, at the same time, begin cleaning the soil and water because airborne pollutants will no longer add poisons to them. If we then choose to treat the soil in such a way that we can grow what we desire without the use of artificial chemicals and if we stop using the soil as a dumping ground for toxic wastes, the soil can once again begin to filter and purify the water. If we stop dumping waste effluents into the water (streams, rivers, estuaries, and oceans), the world's oceans can again evolve toward a more productive state in supplying the goods and services required to support human life.

With clean air, soil, and water, we can also have an atmosphere free of most greenhouse gases and smog, resulting in a more benign—and perhaps predictable—climate in which to live. Then, with the ozone shield repaired, we will again have safer sunlight to power Earth. On a healthy home planet, we can have a high-quality environment for an ecologically sustainable population.

A population in balance with its habitat would reduce demands on Earth's resources. With reduced competition for resources can come cooperation and coordination that would allow landscapes to provide the maximum possible biodiversity. Protecting such biodiversity translates into the gift of choice, which in turn translates into hope and dignity for all generations.

For discussion's sake, add to this scenario the end of wars and their weapons. Such a world, a wonderful place in which to live and raise families, is possible. But, it ultimately hinges on one thing—our unequivocal decision to clean the air, which is the foundation of the "global commons." And, every aspect of the commons constitutes the birthright of every human being—indeed of all life on planet Earth.

If we, for example, achieve a world of lasting peace, a balanced and sustainable human population, and food in abundance, as well as gender and racial equality and democratic governance, but we do not clean the air—then all is for naught. If we do everything outlined—except clean the air—we will still pollute the entire Earth from the blue arc of the heavens to the bottom of the deepest sea in every corner of the globe.

Clean air is the absolute "bottom line" for human survival. Without clean air, there eventually will be no difference in the way we commit suicide, either directly by nuclear war or indirectly through air pollution. Regardless of our choice, the consequence is the same. We will be the collective authors of our own demise—the "Greek tragedy" of modern society.

But, there are alternatives, and the choice is ours. To our children, their children, and their children's children unto all generations we bequeath the consequences of our leadership—the perpetual basis of which is the quality and farsightedness of our social-environmental decisions.

Richard Buckminster Fuller, a man of many skills, including being an engineer and author, saw the necessity of change when he wrote, "With orthodox science now—there are many band-aids stuck on in order to make things fit. I think what is going to happen … is that the whole lot is going to collapse, and the major philosophy is going to be based on … the interconnectedness of everything, and the cyclic nature of everything, and how things are reproducing at many levels without scale, time or size."[37]

THE INVIOLABLE RULES OF DECISION MAKING

Just as there are inviolable biophysical principles that govern nature and the universe, so there are inviolable rules of decision making. I use the term *inviolable* advisedly because, should one attempt to cheat in applying nature's rules of engagement in life, all generations pay the price exacted by the cheater. *There are no exceptions.* To put it plainly, there are no problems in the world or the universe outside our own thinking and desire for quick gains at the expense of others. All the problems plaguing the globe today are the products of our self-centered thinking, just as they have been throughout history, which prompted Winston Churchill to say:

> When the situation was manageable it was neglected, and now that it is thoroughly out of hand we apply too late the remedies which then might have effected a cure. There is nothing new in the story. ... It falls into that long, dismal catalogue of the fruitlessness of experience and the confirmed unteachability of mankind.
>
> Want of foresight, unwillingness to act when action would be simple and effective, lack of clear thinking, confusion of counsel until the emergency comes, until self-preservation strikes its jarring gong—these are the features which constitute the endless repetition of history.[38]

This quotation is part of Winston Churchill's speech to the British Parliament in 1935, as he saw with clear foreboding the onrushing threat of Nazi Germany to international peace.

Rule 1—Everything Is a Relationship

From a decision-making perspective, there are four basic relationships that must be accounted for consciously because they are all interactive in space through time: (1) intra-personal, (2) inter-personal, (3) between people and the environment, and (4) between people in the present and those of the future.

Intra-Personal

An *intra-personal* relationship is an individual's inner sense of self-worth, personal growth, authenticity, and so on. In short, it is degree of psychological maturity that makes a person conscious of and accountable for his or her own behavior and its consequences. The more conscious we are, the more other-centered we are, the more self-controlled our behavior, and the greater our willingness to be personally accountable for the outcomes of our behavior with respect to the welfare of fellow citizens, present and future, and Earth as a whole.

People who succeed in changing their lives and growing beyond their present limitations of consciousness do so because they not only know that success is the interpretation of an event and not the event itself but also because they recognize their own inner power, which gives them a sense of inner authority.

Each decision is a fork in the road of life; each presents at least two choices—right-hand fork or left-hand fork. The direction of our lives is determined by cumulative results of many little decisions. Some are remembered, but most are not because they were made unconsciously. We tend to remember the "big decisions," while seldom realizing that a single big decision is merely the sum of the many little decisions made along the way. We give just a little here and again a little there, and eventually we have pointed ourselves in a new direction.

It is imperative that we take personal responsibility for our words and deeds because the first step toward sustainability begins with the respect and quality of care we give ourselves. Thereafter, we must extend that respect and care to our families, friends, and neighbors because the sustainability of a community is but a reflection of the health of its citizenry as measured by how people treat one another and thus protect the global commons as everyone's birthright.[39]

Inter-Personal

Not enough can be said for civility, respect, and hospitality toward other people. If we use these basic human behaviors to frame our decision making, we can focus more on the mutual values that bind us and less on the tensions between our beliefs and attitudes. Here, an old saying is apt: Milk and honey are more easily swallowed than are sour grapes.

If, for example, someone is rushing blindly to get somewhere and shoves you out of the way, you have at least two choices in how to respond. You might express resentment and anger, or you can be patient, kind, and understanding. Consider, for example, the comments of an old man knocked down by a rude young man in a hurry. On rising from the floor, the old man regarded the younger for a moment and then said: "Son, as you now are, so I once was. As I now am, so you shall one day be."

Our thoughts and actions are the seeds we sow each time we make a choice, and they will, sooner or later, come home to roost. If we do not like the outcome of our decision, we can always choose to choose again.

In fact, while we always have a choice, we must choose—in that we have no choice. We are not, therefore, victims of our circumstances but rather the consequential product of our choices and decisions. And the more we are able to choose love and peace over fear and violence, the more we gain in wisdom and the more we live in harmony and social-environmental sustainability. This is true because what we choose to think about determines how we choose to act, and our thoughts and actions set up self-reinforcing feedback loops—or self-fulfilling prophecies, as it were—that become our individual and collective realities.

It is just such self-reinforcing behavioral feedback loops based on competition for money through the exploitation of resources—often including one another—that are destroying our environment and thus our society. As long as competition is the overriding principle of our social-economic system, we can only destroy our environment because it has become the battlefield in which the war of competition is fought. Our *over-emphasis* on competition in nearly everything fosters the material insecurity that manifests as greed, which is just the fear of not having enough to feel secure materially.

Our obsession with competition, with having to be "right" at the expense of someone else's having to be "wrong," brings forth another human tendency—defending a point of view when faced with a perceived threat to our sense of material survival. There are, however, as many points of view as there are people, and each person is indeed right from his or her vantage point. Therefore, no resolution is possible when people are committed only to winning agreement with their respective positions.

The alternative is to recognize that right versus wrong is a judgment about human values and is not a winnable argument, albeit each decision has a particular outcome that is more or less acceptable to a particular individual or society, at least at the moment of decision. It is best, therefore, to define and focus on the principle involved in the discussion as the fundamental issue. An issue, often perceived as a crisis, becomes a question to be answered. In struggling toward the answer, both positive and negative options become apparent; once apparent, they become a choice; once a choice, they become a decision; once a decision, they become an action; once an action, they become an outcome.

We must, therefore, approach the decision making of leadership within the context of a sustainable landscape design by mutually supporting our common desire for a good-quality style of life. Too many efforts stagnate and wither because people fail to nurture the social nature of human interactions. Leadership that builds truly sustainable communities must include celebrations and create opportunities simply to have fun together. A community that works and plays together stays together and has an easier time forming partnerships to achieve a commonly held vision of a sustainable future. And, just as sustainability begins with individual leaders as wise decision makers, so do the health and success of partnerships, which foster a sustainable environment.

Strong partnerships require reciprocity, which is, after all, the essence of sustainable social networks and communities. Leaders must therefore pay particular attention in their decision making to serving the necessities and addressing the concerns of everyone in the group because people are more likely to give when they feel relatively assured of help when they need it. This kind of caring and trust, which amounts to being one another's keeper, lends tremendous creativity toward resolving problems and accepting the inevitable setbacks.

Leadership in the form of true partnerships, like marriage, is a tough business. We must, with humility, make the commitment to take the time and have the patience to share power and resources—something as simple as lettuce from our gardens for the neighborhood. And, we must remain true to our commitments—a step most people either forget or do not understand.

In addition, when we assign values to nature's actions through politics based on our perceptions of "good" and "bad," we interject the artificial variable of partiality, which often clouds our vision of how the biophysical principles function. When this happens, we rob ourselves of our ability to predict the future with any degree of accuracy by rejecting nature's impartial neutrality expressed through the inviolability of the biophysical principles.

Social-environmental leadership can only be created on a person-to-person basis. It must grow from the bottom up. It cannot be imposed from the top down. Social-environmental leadership is an ongoing, ever-evolving process of decision making, not a fixed end point that once achieved is sufficient unto itself. The purpose of such leadership is to help people and communities empower themselves and one another as they struggle toward sustainability.

For social-environmental leadership to exist, self-centeredness must blend into other-centered teamwork. Setting aside egos and accepting points of view as negotiable differences while striving for the common good over the longer term are necessary for teamwork. Unyielding self-centeredness represents a narrowness of thinking that prevents cooperation, coordination, collaboration, possibility thinking, and the resolution of issues. Teamwork demands the utmost personal discipline of a true democracy, which is the common denominator for lasting success in any social endeavor.

But, even if we exercise personal discipline in dealing with current social-environmental problems, most of us have become so far removed from the land that sustains us that we no longer appreciate it as the embodiment of continuous processes and self-reinforcing feedback loops. Attention is focused instead on a chosen product, the success or outcome of management efforts, and anything diverted to a different product is considered a challenge to the economic bottom line. It is time, therefore, to reevaluate the philosophical underpinnings that gird our notions of nature, decision making, partnership, community, and society and how they can be sustainably integrated into the common future of leadership that lies ahead—generation after generation.[40]

Between People and the Environment

Sustainability means that development programs must, to the extent possible, integrate the local people's requirements, desires, motivations, and identity in relation to the surrounding landscape. It also means that local people, those responsible for development initiatives and their effect on the immediate environment and the surrounding landscape design, must participate equally and fully in all debates and discussions, from the local level to the national. Here, a basic principle is that programs must be founded on local requirements and cultural values *in balance* with those of the broader outside world, which includes understanding environmental issues, long-term ecological trends, and their social-environmental ramifications—all the while honoring the limitations of nature's inviolable biophysical principles and their counterpart, inviolate rules of decision making. After all, social-environmental sustainability is a reciprocal relationship between people and the land. As we nurture the land, we nurture ourselves. As we abuse the land, we abuse ourselves and threaten the future of all generations, beginning with ours.

For example, a hillside meadow that bakes annually under the summer sun and turns brown is seen as having no value to a community because it does not produce a visible product the community wants. The meadow is therefore subdivided for a housing development. Within a year of completing the housing development, many of the community's wells begin drying up because the meadow had in fact been the water catchment that not only collected the water from rain and melting snow but also filtered, purified, and stored it deep in the soil prior to filling the wells with this most precious of liquids.

Second, short-term trends must be viewed in relation to long-term trends and long-term trends in relation to even longer-term trends. The more we trace the present into the past, the better we understand the present. The more we project the present into the future, the more humble we need to be in our notion that we understand the present. Knowledge of the past tells us what the present is built on and what the future may be projected on. But, this is true *only* if we accept past and present as a cumulative collection of our understanding of a few finite points along an infinite continuum—the trends that may, and likely do, point to the future.

Thinking and knowledge in Western society have become so linear either we do not recognize or we ignore the fact that everything is defined by its relationship to everything else. Nothing exists in isolation. Everything is interactive with something else, which negates the possibility of absolute freedom. In the example, water for the community's wells existed in relationship to the hillside meadow, the precipitation that fell on it, the plants that held the soil in place, and the soil that gathered, stored, and released the water, as well as the gravity that pulled it downward into the wells. But, these relationships were not understood until the housing development irreparably altered them.

Failing to account for a community's long-term supply of water in the face of short-term dollars to be made by a few people is dangerous because changes in the spatial patterns of land use, which grossly alter habitats through time, may well be crucial to understanding the dynamics of landscape design and will have implications for many biophysical processes. Changes in the patterns of landscapes also relate to the flows of materials and energy across landscapes, such as the processes of erosion, the movement of water and sediments, and the trapping of heat by urban areas as heat islands. Characterization of the relationships among changing patterns on the landscape and how those changes affect ecological processes is particularly important if we are to develop a more complete understanding of landscape dynamics, our effects on them, and their reciprocal effects on our human communities as the inevitable, self-reinforcing feedback loops that we must increasingly live with and that we pass to future generations.

Moreover, wise social-environmental decision making requires understanding that the setting of a community helps define it because people select a community for what it has to offer them within the context of its landscape. A logging community is therefore set within a context of forest, a ranching community within a context of lands for grazing, and a community of commercial fishers along a coastline, be it a lake or an ocean. The setting helps create many characteristics that are unique to the community. By the same token, the values, decisions, and development practices of a community alter the characteristics of its surrounding environment.

Much of what a community says about itself is reflected in the aesthetics of the physical structures with which it chooses to surround itself. This includes buildings, zoning, design of transportation systems, and the allowance of natural occurrences, such as forested areas, open spaces, streams, springs, wetlands, and rock outcroppings, within and around the artificially structured setting.

In turn, a community's worldview defines its collective values, which determine how it treats its surrounding landscape. As the landscape is altered, so are the community's biophysical and social options altered in like measure. A community and its landscape are thus engaged in a mutual, self-reinforcing feedback loop of reciprocity as the means by which their processes reinforce themselves and one another.

Each community has physical, cultural, and political qualities that make it unique and more or less flexible. The degree of flexibility of these attributes in a community is important because sustainable systems must be ever flexible, creative, and adaptable. The process of sustainable development must therefore remain flexible because what works in one community may not work in another or may work for different reasons.

Beyond this, the power of sustainable development comes from the local people as they move forward through a process of growing self-realization, self-definition, self-determination, and the evolution of their decision making. Such personal growth opens the community to its own evolution within the context of the people's sense of place, as opposed to coercive pressures applied from the outside.

Sustainable development encompasses any process that helps people meet their requirements, from self-worth to food on the table, simultaneously creating a more biophysically and culturally sustainable and just society for the current generation *and* those that follow. Due to its flexibility and openness, sustainable community development is perhaps more capable than other forms of development of creating such outcomes because it integrates the requirements of a local community with those of the immediate environment and surrounding landscape, as well as neighboring communities.[41]

Between People in the Present and Those of the Future

We, the citizens of these United States and the world, must now address a question of human consciousness: Do those living today owe anything to the future? If our answer is "No," then we are surely on course because we are consuming resources and polluting Earth as if there were no tomorrow. If, on the other hand, the answer is, "Yes, we have an obligation to the future," then we must determine what and how much we owe because our present, *non-sustainable* course is rapidly destroying the environmental options for all generations to come. Meeting the acknowledged obligation will require a renewed sense of consciousness—to be other-centered in caring for the welfare of those to come, as we wish vastly more of those before us had considered our welfare as an outcome of their decisions and actions.[42]

To change anything, we must reach beyond where we are, beyond where we feel safe. We must dare to move ahead, even if we do not fully understand where we are going, because we will never have perfect knowledge. We must ask innovative, other-centered, future-oriented questions to make necessary changes for the better.

True progress toward an ecologically sound environment and an equitable world society will be expensive in both money and effort. The longer we wait, however, the more unlikely success becomes, as the environmental condition continues to deteriorate, and the more expensive, difficult, and unlikely become the necessary social changes—until the situation becomes so dire it is "do or die."

No biological shortcuts, technological quick fixes, or political promises can mend what is increasingly broken. Dramatic, fundamental change—frightening, painful, and increasingly expensive—is necessary if we are really committed to the world's children, present and future. It is not a physical question of whether we can change, but rather one of whether we will choose to change *and* do so. Whatever our decisions, we, the adults of the world, bequeath to our successors the consequences of our actions taken through our democratic system of government for which *we are all responsible as decision makers, whether we choose to participate consciously or not.*[43]

Rule 2—All Relationships Are All Inclusive and Productive of an Outcome

Clearly, all relationships are productive simply because they produce an outcome of some sort. The challenge for humanity, and thus its decision makers, is what kind of outcome will result. Every functional part of government, from communities to nations, is influenced by how every other part of government functions—the waterbed principle.

Relearning—dredging the recesses of memory for things long forgotten—and the development of things new is an ongoing process of multidimensional learning. Such learning encompasses both theoretical and practical conceptualization, decision making, action, and the deeper aspects of reflection—including the intellect, intuition, and imagination. Multidimensional learning is important because overemphasis on action—one part of which is competition—simply reinforces fixation on short-term, quantifiable results. Our overemphasis on action can preclude reflection, a persistent practice of deeper learning that can produce wiser decisions and thus far better outcomes over time.

Many of today's problems resulted from yesterday's solutions, and many of today's solutions are destined to become tomorrow's problems because we too frequently insist on little, *symptomatic* approaches that yield *immediate* results without regard for what happens to the social-environmental system itself. Consider, an example, Calgary-based TransCanada Corporation's 1,700-mile, $7 billion Keystone XL pipeline, which would transport as much as 700,000 barrels of oil a day from Canada's tar sands, across the United States, to the Gulf of Mexico.

Extracting and refining oil from these sands not only generates two to four times the amount of greenhouse gases per barrel of final product as a similar amount of conventional oil would but also is fraught with potential environmental hazards from leaking or broken pipes as it flows from central and northern Alberta to the Gulf of Mexico, where spills can occur. Moreover, from its extraction to using it in an automobile, oil extracted from tar sands emits 10–45% more greenhouse gases than conventional crude oil.[44]

Nevertheless, speaking on behalf of the American Petroleum Institute, the executive vice president Marty Durbin said:

> There's been extensive review, extensive opportunities for state and local governments, environmentalists and landowners all along the pipeline to voice their concerns. … It is time to move forward so that *we can create jobs* [emphasis mine] and further strengthen our relationship with America's number one trading partner and number one source of imported oil, Canada. At a time of stubborn unemployment when Americans across the country are clamoring for work, even contemplating saying no to this project should be out of the question.[45]

The American Petroleum Institute is the only national trade association that represents the totality of our nation's oil and natural gas industry—producers, refiners, suppliers, pipeline operators, and marine transporters, as well as service and supply companies, which account for 9.2 million American jobs and 7.7% of the U.S. economy. The institute consists of more than 480 corporate members, from the largest major oil company to the smallest of independent companies.[46]

The following quotation is from a piece on the American Petroleum Institute's Web site, "No Unnecessary Delays on Keystone XL":

> *Why wait on jobs—20,000 U.S. jobs* [emphasis mine] during the 1,700-mile pipeline's construction phase alone? Why unnecessarily delay, by even a week, a $13 billion project the Canadian Energy Research Institute says will help fully utilize Canada's oil sands—projected to support *600,000 new U.S. jobs* [emphasis mine] by 2035 and more than $775 billion (Canadian dollars) in GDP [gross domestic product] from 2010 to 2035?
>
> …
>
> At the same time, why wait on bringing more oil from our No. 1 foreign supplier? Why unnecessarily delay a project that will bring up to 830,000 barrels of oil per day to U.S. refineries? Why not take steps now that will secure America's energy future, one in which 92 percent of our liquid fuel needs would be supplied by the U.S. and Canada by 2030?[47]

Now comes the political pressure for a *potential*, symptomatic, quick fix to create jobs, which has long been used to persuade the public into acquiescing to corporate/political desires.

On December 16, 2011, House Speaker, Republican John Boehner guaranteed that a controversial *pipeline will be part of a year-end jobs package*. According to Boehner, "If that bill comes over to us [without the pipeline], we will make changes to it, and I will guarantee you that the Keystone Pipeline will be in there when it goes back to the United States Senate." To which Republican Harold Rogers, chairman of House Appropriations, added, "I would imagine that if the Senate sends over a bill that does not include the pipeline that it would be defeated here. It has to include the pipeline."[48]

January 9, 2012, ABC News carried a similar story:

> The Interior Department is moving forward with a plan to ban new mining claims on 1 million acres near the Grand Canyon [which the George W. Bush administration opened to new mining], even as congressional Republicans try to block efforts to limit mining operations in an area known for high-grade uranium ore.
>
> …
>
> Conservation groups hailed the 20-year ban, first announced in October, as a crucial protection for an American icon. The mining industry and some Republican members of Congress called the ban detrimental to Arizona's economy and the nation's energy independence.
>
> …
>
> Republican members of Arizona's congressional delegation have lambasted temporary bans imposed by [Interior Secretary Ken] Salazar in 2009 and again last year. They say a permanent ban on the filing of new mining claims would *eliminate hundreds of jobs* [emphasis mine] and unravel decades of responsible resource development. Rep. Jeff Flake (R-AZ) and other GOP lawmakers are backing legislation to prevent Salazar from moving forward with the 20-year ban.[49]

And, on January 18, 2012, headlines on ABC News read, "Keystone Pipeline Delay Sparks Outrage on Capitol Hill."

> House Speaker John Boehner this morning pressed President Obama to approve the Keystone XL Pipeline. …
>
> The speaker said the project would "create 100,000 new jobs," and he warned that further delay on a decision about whether to go forward on the project could result in losing out to China.
>
> "The Keystone Pipeline has been through three years and it's passed every approval process as required by the law. Even the president's own State Department has indicated that this thing ought to move forward," Boehner said. "The Canadians are in conversations with the Chinese, and if we don't build this pipeline to bring that Canadian oil and pick up the North Dakota oil and deliver it to our refineries in the Gulf Coast, that oil is gonna get shipped out to the Pacific Ocean and will be sold to the Chinese."
>
> "This is not good for our country," he continued. "The president wants to put this off until it's convenient for him to make a decision. That means after the next election. The fact is the American people are asking the question right now: Where are the jobs? The president's got an opportunity to create 100,000 new jobs almost immediately. The president should say yes."[50]

Thus, the predominant, symptomatic, quick-fix scenario continues, despite the fact that new mining poses risks of irreversible degradation to the Colorado River, lands considered sacred by American Indian tribes, wildlife habitat, and pollution of the soil and groundwater from toxins within close proximity to one of the nation's most-visited parks—which is part of our national commons.

What society really needs are "big fixes" in the form of systemic ideas that promote and safeguard social-environmental sustainability while understanding that any "fix" is only as good as the integrity of its biophysical foundation and must be subject to alteration based on experience and new knowledge.

Where, asked the late publisher Robert Rodale, are the "big ideas," those that change the world? Likely, they exist but remain unrecognized because our quick-fix, symptomatic social trance does not allow the time, patience, or intellectual freedom and breadth of understanding necessary for *systemic solutions*.

A big idea, according to Rodale, must:

1. be generally useful in good ways—leadership that is multidimensional in its decision making and translates into social-environment sustainability, as opposed to unidimensional decision making, which is strictly focused on a position to defend, be it a corporate bottom line or a military objective;
2. appeal to generalists and give them a leadership advantage over specialists—true social-environmental decision making requires an understanding of the system as a whole *and* so necessitates an amalgamation of broad, systemic-oriented thinkers *and* narrow, more symptomatic-oriented thinkers, with systemic thinkers in charge;
3. exist in both an abstract and a practical sense—decision making, as seen in item 1, not only is practical in its outcome, but also is abstract in that its practical outcome requires people to work together toward the conceptualized

ideal of a shared vision with love, respect, humility, wonder, and intuition, as well as their intellect;

4. be of some interest at all levels of human concern—social-environmental sustainability requires the continual building of relationships, which requires sensitive, systemic-thinking leaders who make wise decisions because they touch all levels of society, both within itself and with nature;
5. be geographically and culturally viable over extensive areas—systemic-thinking leaders are a general necessity if the natural world is to remain viable and habitable for future generations;
6. encompass a multitude of academic disciplines—even systemic-thinking leaders require the counsel of learned advisors (many of whom are specialists) who help to represent the integration of all disciplines, such as soil science, mycology, philosophy, sociology, theology, education, politics, ecology, forestry, and economics, to make the wisest decisions for the long-term public good if social-environmental sustainability is to become a reality; and
7. have a life over an extended period of time—social-environmental sustainability is, by definition, a journey of ever-wiser decision making across generations that requires a succession of skilled, systemic-thinking leaders who know how to facilitate and guide a shared vision toward fruition, as experience accumulates and new knowledge emerges.[51]

Leadership that fits Rodale's requirements can help people to understand that life is not condensable, that any model is an operational simplification—a working hypothesis, which is always ready for and in need of improvement. When we accept that there are neither shortcuts nor concrete facts other than direct observation (such as glaciers melting worldwide), we will see how communication functions as a connective tool through which we can and must share experience, invention, cooperation, and coordination in all aspects of social-environmental decision making.

When people speak from and listen with their hearts, they unite and produce tremendous power to formulate new realities and bring them into being through collective decisions and actions. While today's exploiters of the environment with their narrow, special interests will have passed away by the end of this century, all of the biophysical necessities for social-environmental sustainability will remain. Therefore, leadership that honors the productive capacity of an ecosystem within the limitations of its biophysical principles is not only critically important but also a big idea. After all, social-environmental sustainability is only a choice—our choice, but one that must be carefully and humbly planned if it is to endure the ongoing, short-sighted, contradicting political vagaries of human decisions and actions.

To protect the best of what we have in the present—for the present *and* the future—we must continually adjust our thinking and behavior to accommodate to new circumstances, knowledge, and understanding. Society's saving grace is that we have choices. Accordingly, much of what needs to be done *can* be—if enough people want it to be done *and* decide to do it.

Rule 3—The Only True Investment Is Energy from Sunlight

It must be understood and accepted by all decision makers that the only true investment of energy in our world is light from the sun, which is 93,205,679 miles from Earth, yet delivers 120 trillion watts of energy to Earth's surface—thereby providing enough energy in 1 hour to satisfy the requirements of the global human society for an entire year.[52] Everything else is either a *re-allocation* or a *re-investment* of energy—including every coinage that changes hands in the global marketplace. This makes solar energy the only truly renewable resource. Moreover, capturing this energy and converting it to usable forms is dependent on viable communities of green plants worldwide. After all, fossil fuels are at least partly ancient solar radiation archived over millennia in the earth. It is, therefore, absolutely critical for decision makers to understand and accept that other earthly resources are limited because they already exist and obey the three thermodynamic laws, which make *them not only finite but also exhaustable.* At the moment, however, that does not seem to be universal knowledge.

Consider, for example, the overexploitation of the large predatory marine fishes, such as sharks and tuna, which allows the populations of smaller, plankton-feeding fishes to proliferate. At some point, their numbers become large enough to reduce the amount of phytoplankton dramatically and thus the ocean's ability to capture sunlight and absorb atmospheric carbon dioxide; in turn, these changes affect global warming.[53] That said, curbs on fishing these predators until their populations are larger than those required to maintain a sustainable yield and thus control the plankton feeders could, within biological limits, lead to sustainable profits from fisheries.[54] But then, warming oceans affect the major wind patterns, which affect the direction of ocean currents, which is shifting dead zones in the oceans and causes them to grow.[55] (Dead zones are areas in aquatic environs that are virtually depleted of dissolved oxygen, and from which aquatic life must either flee or suffocate.) Moreover, what happens in the oceans affects global climate change, which in turn affects the world's people—something for which global decision makers will ultimately be accountable.

Rule 4—All Relationships Involve a Transfer of Energy

All we humans ever do is transfer energy from one place to another, from one level of intensity to another. A wise decision maker uses energy carefully because the conservation of energy allows more work to be done with a minimum of disruption to the functional processes. Such care not only constrains material costs but also allows each individual to be fully valued for the energy he or she expends in the furtherance of a shared vision.

Many of today's decisions, however, are based on quick fixes rather than the wisdom of long-term biophysical sustainability. Consider the historic treatment of floodplains as a case in point.

Conduits transport things from one place to another, as well as concentrate the energy of that which they transport. By this I mean, given a constant volume of water, the force of it coming out of a hose increases (up to some point) with the decreasing diameter of the hose. Anyone who has ever turned a nozzle knows this as a technique used in "power washing" a sidewalk or the outside of a house prior to painting it.

The same dynamic applies to a broad river being suddenly restricted by a narrow canyon to flow through. As the volume is constrained, the velocity increases. Here, the positive lesson from the concentrating action of a conduit is used in narrowing a stream or river with a concrete structure to increase the water's velocity and therefore its ability to turn turbines and convert the force of water into electrical power for use by humanity.

There is, of course, the antithesis, the negative dynamic of constraining the volume of water and increasing its velocity. I remember seeing this dynamic at work in a mountain village in eastern Slovakia, where the people had lined the stream flowing through their village with concrete. Clearly, all they could do under the circumstances was to effect a symptomatic fix because the recent, clear-cut logging had denuded hillsides and was causing frequent flooding; heretofore, the stream had rarely done so because the forested slope had not only intercepted the precipitation, thus slowing its velocity, but also held the soil in place, thus allowing the water to infiltrate. The concrete confined the water in a narrower, smooth channel, as opposed to the one of cobbles and boulders in years past. The consequence of this action was far greater erosion of the stream's channel where the concrete ended than would otherwise have been the case. Moreover, the deleterious effects of such constraint were obvious—the stream's velocity was simply exported downstream, beyond the artificial conduit, where it did much unnecessary damage.

Conversely, a floodplain, as the name implies, is a relatively flat, open expanse that frequently floods. These are areas where storm-swollen streams and rivers spread out, decentralizing the velocity of their flow by encountering friction caused by the increased surface area of their temporary bottoms, which dissipate much of the floodwater's energy. I saw this phenomenon in 1967 while working in the Tarai of Nepal, where the Trisuli River had a floodplain more than a mile wide. I watched the floodplain fill within 2 days during the monsoon rains.

Although this principle of friction slowing velocity is well known to humanity and is the idea behind disk brakes on automobiles and hand brakes on bicycles and wheelchairs, as well as the design of the parachute, people refuse to accept it whenever they covet the land claimed seasonally by a river. But, when we refuse to accept a lesson from nature's biophysical principles, it forms an intellectual blockage in our potential understanding of those principles that ultimately govern social-environmental sustainability. The inertia of "informed denial" is the usual procedure when the immediate, economic cost of rectifying a mistake is thought to be great, which is nothing less than passing the debt to some other generation. Informed denial, as a remedy, is based on the same level of social-environmental consciousness and its decision making that initiated the problem in the first place (a *reaction* instead of a *response*), which can only compound the problem.

The refusal to accept the biophysical constraints of nature as they affect social-environmental decisions is a serious issue in Western industrialized countries and those adopting Western ways—particularly when it comes to the control of energy. The U.S. Army Corps of Engineers is a prime example of the dynamic.

The corps is one of the oldest, largest, and most unusual agencies in the federal government of the United States. As a bureaucracy of the executive branch that takes its marching orders from Congress, it is a military organization with an overwhelmingly civilian workforce. It is also an environmental regulator that is generally viewed as suspect by environmentalists because members of Congress often authorize projects to steer federal money to their districts, and the corps frequently justifies them with questionable cost–benefit analyses and in the process has reconfigured—artificially engineered—the American landscape. After all, the motto of the corps is *Essayons*, French for "Let us try." The motto indicates that, throughout its history, the corps has seen nature as an enemy to subjugate by equating engineering and control with *social-environmental progress.*

For instance, the Corps of Engineers not only designs and builds structures to control flooding and to improve navigation but also issues permits for the alteration of such bodies of water as streams, rivers, marshlands, and estuaries. Although the corps leaders today speak of "working in harmony with nature," the corps still proudly mobilizes its "Annual Campaign against the Mighty Mississippi." This ongoing battle caused Burton Kemp, a former corps geologist in Mississippi, to say it is not surprising when the corps takes a militaristic approach to the environment. "I'm afraid it's not a Corps of Scientists. It's not a Corps of Biologists," he says with a sigh. "It's a Corps of Engineers."[56]

One aspect of the program conducted by the Army Corps of Engineers is the practice of "channel improvement" in streams and rivers. These "improvements" lead to a straightening of a stream's channel and a change of its shape, which in turn destabilizes the channel. Destabilization of the channel causes downstream effects—erosion of the banks; alterations of the channel's bed; degradation of the aesthetics; and frequently undesirable changes in the composition of the plants and animals that can and do inhabit the stream. From these numerous "improvements," each planned on its own, isolated rationale (symptomatic thinking), comes the next, larger-order magnitude of massive flooding when the constrained waters release their pent-up energy through a breach in the levees designed to imprison their flow.[57]

To better understand what is meant by the sudden release of pent-up energy, let us go back to the floods in California during the winter of 1996–1997, when 16 crews worked to shore up some of the 1,100 miles of levees in the Sacramento-San Joaquin Delta. In other areas, crews used sandbags and plastic sheets to shore up critical sections in the 6,000-mile network of levees in northern California.[58]

A dozen major breaks in the levees occurred along the San Joaquin, Mokelumne, and Consumnes Rivers, and many other places were threatened because of the pressure of the water day after day. "It's a race with Mother Nature, but right now we're ahead," said Captain Mark Bisbee of the state Forestry Department.[59]

A levee failed on the San Joaquin River on January 10, 1997, sending work crews fleeing through dense fog as a 90-foot gap opened, and water began rushing into neighboring fields, where it swamped up to 5,000 acres. A break in a levee near Lathrop allowed the flooding of more than 25 square miles and damaged as many as 400 homes.[60]

Water from a ruptured levee on the Feather River, 100 miles northeast of San Francisco, flooded a farm. The farmer and his wife lost $300,000 worth of cattle.

"The sheriff's department just wouldn't let us in," said the farmer, "so 200 head died a slow death. It was gruesome."[61] Some of the farmer's cattle were ensnared in ditches or fences; one cow, snagged on a small gate, had to be burned free with torches. The stench of the rotting animals was everywhere. Thus, floods gave way to fields of death across northern California, where hundreds of drowned cows, horses, and other farm animals—their bloated carcasses tangled in barbed wire or mired in ditches—lay strewn across the soggy landscape.

Although major losses of farm machinery, barns, homes, and wells appeared to be the immediate headache for farmers (including the loss of livestock for some), others were concerned about the survival of their crops. While an estimated 150,000 acres of red winter wheat was likely damaged, wheat farmers were most worried about the potential loss of topsoil to erosion from running water.

With at least three reservoirs nearing their capacities, California water officials said the danger from flooding was far from over as runoff from rain and melting snow continued to build. Because of water released from dams, some rivers (including the Central Valley's Stanislaus and San Joaquin) would continue bulging with runoff inside their eroding levees until February, well after the storm event had passed. "It's unfortunate," said Jeff Cohen, of the California Water Resources Department, "that there is damage downstream, but it's the requirement [draining water from the reservoirs]." He added that the corps requires water to be drained from reservoirs before they reach capacity.[62]

You may be wondering where the people's responsibility lies in this story about flooding. It has to do with priority and people's choices. In bygone times, people lived in California's Central Valley and farmed the land in concert with the rivers, including their periodic floods, because they knew where the floodplain was and respected it. Despite the wisdom exhibited in yesteryear, it has long been American tradition to wrest every usable acre from nature, lest an acre be thought of as "unproductive" and therefore a "wasted economic opportunity."

If the rivers could be controlled and the flooding stopped, then more of the *unproductive acres* could be made to produce that which Americans thought desirable. So, dams and levees were constructed. If they failed to produce the desired control, more dams and levees were built. In the end, however, they are proving no match for nature, as the story illustrates. But, there is something left unsaid by the story: It is the choices people made that brought all this about.

Let us look at just four possible choices people *could have made* prior to the floods of 1996–1997: do not (1) live or farm in the floodplain; (2) live and farm in the floodplain *without* dams and levees and plan for, be prepared for, and accept the risk of periodic flooding; (3) live and farm in the floodplain *with* dams and levees in place, thinking the problem of flooding is solved, but move after a levee breaks once or twice; (4) live and farm defiantly in the floodplain *with* dams and levees in place, regardless of the dire consequences of periodic flooding, and depend on the federal government to bail you out with federally subsidized flood insurance.

All these choices represent different levels of self-imposed constraints on one's own behavior based on different perceived values for monetary gain and lifestyle. The choice that seems to have been generally accepted over time is the last one: *Live and farm defiantly in the floodplain regardless of consequences and depend on the federal government to bail you out with federally subsidized flood insurance*, which means the people are products of their choices and their decisions based on those choices—***they are patently not victims of circumstances***. This, then, becomes the primary social-environmental constraint or "fixed point" around which all human residential, rural, and commercial development is made to revolve, despite the fact that sooner or later the rivers will remember their floodplains and reclaim them—at least temporarily and at tremendous financial cost.

When their irresponsible risk-taking fails, people want the government (hence society at large) to rescue them, despite their having known the ill-advised risk of building in the floodplain with its inevitable consequences. But, why should the people at large, through personal taxes paid to the government, be expected to bail out those individuals who make unwise choices when they gamble with the inevitability of such high stakes? Do we, through our taxes, bail someone out of financial trouble when they lose heavily in a high-stakes game of craps in a casino in Las Vegas? Building in a floodplain and wagering in a game of craps are both gambling, so what is the difference? Where is personal responsibility? How is one to learn responsibility if one does not have to accept the full measure of the consequences of their choices?

It was not always this way, according to Scott Faber, director of floodplain programs for American Rivers, based in Washington, D.C. According to Faber, "Floods may be acts of God, but flood losses are acts of hubris."[63] Predictable, natural events, like floods, have turned into natural "disasters" that people try to control with dams and levees because housing and commercial development have "flooded" the floodplains.

"At the turn of the century," writes Faber, "there was virtually no development in floodplains. Over the last 60 years, government programs have assumed responsibility for flood 'control' by building and repairing levees, providing relief, and subsidizing flood insurance. These programs actually put people in harm's way by eliminating incentives for local communities to direct new development away from flood-prone areas."[64]

He goes on to say that levees and dams create a false sense of security, which encourages people to build in flood-prone areas and thus increases the potential for financial and personal catastrophe when a levee inevitably fails. Thus, thousands of flood-weary Midwesterners decided in 1993 to stop "playing chicken" with the Mississippi and Missouri rivers. They opted instead for a voluntary program that relocated more than 8,000 homes and businesses, even whole towns, onto the bluffs, so that thousands of people were literally high and dry when floodwaters returned in 1995.

As development continually encroaches on floodplains, says Faber, the rainfall that once was absorbed slowly and naturally by the land now courses rapidly into rivers, which get higher and faster as they flow toward centers of human population. An isolated decision to drain a wetland, till a farm, pave a parking lot, or put in a new street has little measurable effect on flooding in and of itself, but when combined with thousands of other seemingly innocent, unrelated decisions as a cumulative effect, the results can be devastating.

Rather than work together to solve regional problems (systems thinking), Faber says most communities and rural landowners simply pass the water downstream as fast as possible.[65] More dams and levees cannot eliminate human problems associated with flooding because dams and levees fostered the problem in the first place, and more of the same is hardly the cure. Periodic flooding, at times of mammoth proportions, is one of nature's nonnegotiable constraints, especially during wet periods in the weather cycle, which heretofore has been another of nature's nonnegotiable constraints—and is becoming more dramatic with global warming. Hence, a wise community will both recognize and bow to nature when nature is clearly beyond human control.

Government agencies in the United States have not been interested in the long-term future of the landscape because the computed, but often unrealistic, cost-benefit ratio is on the side of utility.[66] Nonetheless, for a system to be viable through some scale of time, its processes must remain functional.

How might such social-environmental functionality be maintained when seasonal flooding is the most damaging natural disaster worldwide, and the flood-vulnerable population is expected to grow in coming decades? This is a valid question because the risk of flooding is set to increase due to both climate change and shifting uses of the land, such as filling in wetlands and expansion of impervious surfaces, which prevent rain and melting snow from infiltrating the soil and thus lead to more rapid runoff into rivers.[67]

In the United States, for example, annual losses to flooding continue to rise, punctuated by major events in the Midwest ($30 billion in total costs in 1993; $15 billion in 2008) and California's Central Valley (1995 and 1997; $4 billion each event). Meanwhile, political pressure to continue locating new housing developments and associated shopping malls in river floodplains continues apace, despite the chronically underfunded maintenance of the levee system.[68]

Dikes, levees, and dams to prevent high flows from entering floodplains diminishes both the floodplains' natural storage capacity for season high water and the processes that sustain viable riparian forests and wetlands. The continuing assault on floodplains makes them among the planet's most threatened ecosystems, even though functioning floodplains—those connected to rivers—are among the most valuable ecosystems for supporting biodiversity and providing biophysical goods and services to society. In the face of increasing uncertainty with respect to global warming and the potential for increased flooding worldwide, it is imperative to rethink how we treat floodplains. If, for example, levees were strategically removed or placed in such a way that rivers could reclaim their floodplains, the floodplains would be allowed to perform their biophysical function of slowing the flood's velocity, storing water, and conveying floodwaters in such a way that their unsustainable uses would be largely curtailed while maximizing their biophysical benefits for both society and private landowners.[69]

Under natural conditions, connected floodplains, as ecosystems, sustain high levels of biological productivity and diversity, thereby providing myriad ecosystem services, second only to estuaries and mangrove swamps in terms of value per acre. For example, perennial plants, in either repaired habitats or biomass crops (the mass of material in a living system that is usually measured in terms of dry weight), increase carbon sequestration and improve water quality through the reduction of soil erosion, the entrapment of fine sediments, and the increased retention of chemical nutrients.[70]

An alternative strategy for using mangrove swamps is illustrative of the principle being suggested for dealing with floodplains. A mangrove swamp is a coastal area, which can be thought of as a tidal flat forming a direct link to the sea and thus the waxing and waning of the tides, which includes those along the Gulf of Mexico. They are composed of evergreen trees with massive, stilt-like roots that are tolerant of highly saline conditions created by the continual evaporation of the seawater during low tides and inundation during high tides.[71]

One study in Bintuni Bay, in Indonesia, found that leaving the mangrove forests intact would be more productive than cutting them. When the nontimber uses of the mangrove forests, such as fisheries, locally used products, and the control of soil erosion, were included in the calculation, the most economically profitable strategy was to retain the forests. Maintaining healthy mangrove forests yielded $4,800 per 2.5 acres annually over time, whereas cutting the forests would yield a one-time value of $3,600 per 2.5 acres. Maintaining the forests would ensure continued local uses of the area worth $10 million per year and provide 70% of the local income while protecting a fishery worth $25 million a year.[72]

We humans can no longer assume that the services nature inherently performs are always going to be there because the consequences of our frequently unconscious actions often affect nature negatively in unforeseen and unpredictable ways. Yet, we can be sure that the loss of individual species and their habitats through the simplification of ecosystems can, and will, impair the ability of nature to provide the services we need to survive with any semblance of human dignity and well-being. Losses are just that—irreversible and irreplaceable.

Reconnecting floodplains is not a call to empty floodplains of human activity but rather an approach that would accomplish three primary objectives: (1) reduce the risk of flooding, (2) increase the production of nature's free goods and services in floodplains, and (3) increase the social-environmental resiliency to the potential impacts of climate change. To this end, it would be wise to focus on the strategic reconnection of large areas of floodplains currently used for agriculture, as opposed to reconnecting such areas in densely populated floodplains, which would incur considerably more expense. Despite the inevitable socioeconomic and political challenges, reconnecting large floodplains is consistent with private ownership and a vibrant agricultural economy.

The primary element to the economic and political viability of this approach is that reconnected floodplains could by and large remain in private ownership, thereby generating revenue through productive agriculture. Agricultural practices consistent with periodic inundation include pasture for livestock, commercial timber, and the cultivation of flood-tolerant crops, such as switchgrass and willows, which can be grown as biomass for fuel. In areas where floods are predominant, annual crops can be grown during the dry season. Even where floods are coincident with the growing season, a broader variety of crops could be cultivated on land that is inundated infrequently—less than once per decade, for example.[73]

Rule 5—All Systems Are Based on Composition, Structure, and Function

It is imperative for a decision maker to understand that, like every system, the organization he or she creates is based on the composition of people and their various areas of expertise and levels of psychological maturity, which defines how the organizational system can and will function. Grasping this dynamic is critical because the degree to which organizational members are behaviorally functional or dysfunctional, systemic or myopic in their points of view, team players or grandstanders, will determine the nature of the team and the kind and quality of its products. Therefore, a leader who is conscious enough to make social-environmental sustainability the bedrock of his or her administration must begin with the necessary function of his or her team and work backward to find the interdisciplinary expertise (composition) required for creating the necessary structure, which in turn will allow the functional fulfillment of the vision: social-environmental sustainability.

Rule 6—All Relationships Have One or More Trade-offs

Every decision one makes involves trade-offs. Let us take a simple example from a city park in a Mississippi community. The challenge for the city council's leadership is to determine what kind of recreation will take place, considering that different factions of the community want different things: Some want to play baseball (which requires a diamond of bases, a backstop, and a wide-open space); some soccer (which requires a wide-open space and goals at both ends of the field); some volleyball (which requires a relatively small space filled with sand or asphalt, two poles and a net stretched between them); and some tennis (which, like volleyball, requires a relatively small space, but an asphalt foundation with a low net in the middle, as well as a high surrounding fence to prevent the ball from escaping). Since space is limited, how might more than one group be accommodated?

Volleyball and tennis would be an easy match to accommodate because the space required by each is relatively small and thus fits within the park's boundaries. Moreover, the composition of each area is similar enough to create roughly the same configuration, which allows for the different functional activities of each sport to be carried out at different times in the same space. Baseball and soccer, on the other hand, would each use the entire park not only because of the large area required by each sport but also because the dynamics of each sport is dramatically different and thus incompatible within the confines of the park. Therefore, the council has three immediate choices: volleyball and tennis, baseball, or soccer.

But, now suppose a decade passes, and the townspeople change their minds about which sports they want in the park. If it had been used for a soccer field, it could easily be converted to a baseball field. However, changing a baseball field to a soccer field would be more difficult because the backstop would have to be removed. Whereas redesigning a volleyball area into either a baseball field or a soccer field would be simple (provided it had sand pit construction instead of asphalt), converting a tennis court to either would require considerably more effort because the asphalt surface and the high metal fence would have to be removed before anything else could be done. And with time, the park might even be converted into a shopping mall.

Each of these decisions represents a self-reinforcing feedback loop in terms of the land's ecological viability as a living system. In that sense, each decision is the beginning of a never-ending story of cause and effect, which touches all local generations throughout time.

But, what would happen if the city park were made into an agricultural field in which corn is grown as biofuel? "Corn-based ethanol is the worst among the alternatives that are available at present, although this is the biofuel that is most advanced for commercial production in the United States."[74] Corn, it turns out, is one of the most energy-intensive crops when it comes to the amount of fertilizer and water it requires.

To wit, farmers are not only applying seven times the amount of synthetic nitrogen as they did in the late 1960s but also the natural gas used in its production accounts for 90% of the cost of the ammonia, which is the basis for the nitrogen fertilizer applied to corn. The pesticides and herbicides required to produce these vast monocultures are also gas-based petrochemicals. Then there is the substantial amount of diesel fuel needed to operate the farm machinery, which says nothing of the enormous quantity of water this exceedingly thirsty crop requires—1,700 gallons for each gallon of ethanol produced.[75] Therefore, an increased reliance on biofuel is merely trading an oil problem for a water problem that goes from the field into the cities and into the oceans of the world.[76]

To illustrate, as the water of the Mississippi River flows toward the Gulf of Mexico, collecting runoff from the Appalachian Mountains to the Rocky Mountains and everywhere in between, it passes through 10 states, through massive agricultural fields and by numerous towns and cities. Along its journey, it gathers fertilizers and pesticides from the Corn Belt and leached sewage from the urban areas. By the time the Mississippi enters the gulf, its current has been transformed into a conduit for chemical nutrients. The enriched current stimulates massive blooms of algae every summer that strip the water of oxygen, thereby creating a huge dead zone. (A *dead zone* is an area that is virtually depleted of dissolved oxygen, from which aquatic life either flees or suffocates.) This dead zone, which is the size of Massachusetts (7,900 square miles), has existed since the 1970s and supports almost no life beyond algae and bacteria.

Today, after years of inaction, the problem is severe. What is more, much of the water entering the Mississippi comes from massive fields of corn, which is grown in soil with tile drains. Consequently, more nitrogen seeps into the river from cornfields than from fields of crops without drainage tiles. Therefore, making ethanol from corn not only will cause more cornfields to be planted but also will exacerbate the dead zone in the Gulf of Mexico—perhaps beyond repair.[77]

In the Great Plains, where new ethanol plantations are established, an unexpected environmental cost is exacted because groundwater is the only source for irrigation. As water soaks through the soil, it collects carbon dioxide from decomposing organic matter in the soil through which it percolates. According to Gwen L. Macpherson, a hydrogeologist at the University of Kansas in Lawrence, groundwater holds, on average, from 10 to 100 times as much carbon dioxide as water in lakes and rivers.[78]

Thus, when groundwater is pumped to the surface, the carbon dioxide escapes into the air, where it adds to the growing supply of greenhouse gases. Nonetheless, people have been pumping about 178 cubic miles of water from below ground annually and thereby have been releasing approximately 331 million tons of carbon dioxide into the atmosphere every year. Although the volume of carbon dioxide released from the groundwater is a small percentage of that produced from the combustion of fossil fuels, it is about three times the amount spewed from the throats of volcanoes, which are a natural source of the greenhouse gas.[79]

As it turns out, the global oceans—the largest natural reservoir of carbon dioxide—absorb about one-third of the carbon dioxide we humans spew into the atmosphere every year. Although this process is extremely slow, taking hundreds to thousands of years, once dissolved in the water, a carbon atom can remain there for decades or centuries depending on the depth in the ocean in which it is located. However, anthropogenic carbon dioxide now penetrates the whole water column of the North Atlantic Ocean.

Moreover, there is a strong possibility that dissolved carbon dioxide in the ocean's surface waters will double over its pre-industrial levels by mid-century and will be accompanied by greater acidity as well as by a decrease in the carbonate ion. When carbon dioxide reacts with seawater, it produces carbonic acid, which can be thought of as the soda-water effect. This change in seawater chemistry will have profoundly negative effects on those calcium-secreting organisms in the world's oceans that depend on calcium carbonate for the production of their shells (mollusks, including planktonic mollusks, and marine algae) and skeletons (corals).

On a global scale, the alterations in surface-water chemistry from the anthropogenic deposition of nitrogen, sulfur, and dissolved inorganic carbon are relatively slight compared with the acidification caused by the oceanic uptake of anthropogenic carbon dioxide. The impacts are more substantial in coastal waters, however, than in the deep ocean. In coastal areas the ecosystem responses to acidification could have severe implications for people, especially those who rely on the seas of the world for food.

Over time, these changes will send ripples throughout the marine food web, from the microscopic plankton to the plankton-feeding whales and all life in between. As the ocean gets warmer and more acidic, the amount of dissolved oxygen will diminish accordingly, as will the building blocks for coral and other calcium-secreting organisms.

In fact, these species already have a reduced ability to produce their protective shells and supportive skeletons. In addition, the increase in carbonic acid is even now beginning to dissolve the shells and skeletons once they are produced and is also making them increasingly susceptible to wear and erosion. Decreased calcification will no doubt compromise survival of these organisms and could shift marine flora and fauna toward non-calcifying species. For example, the common periwinkle (a small marine snail) normally grows extra-thick shells when living among crabs, but if the water is too acidic, the snail's ability to produce a thicker-than-normal protective shell is disrupted.[80]

Therefore, while we have control of what we introduce into the environment, once introduced, it is forever out of our control. Of this, every decision maker and leader must be cognizant, perform a risk analysis with due diligence, and be willing to accept responsibility for the outcome of his or her decisions—be they symptomatic to "fix" an immediate, short-term problem, as the Los Angeles Metropolitan Water District wants, or systemic for the sake of social-environmental sustainability for all generations over time.

Consider that, already a decade ago, the Los Angeles Metropolitan Water District wanted to drain a Mojave Desert aquifer, which is the size of Rhode Island, under the 35,000-acre Cadiz ranch to quench urban thirst. But, the plan was rejected due to widespread environmental opposition. However, it has been resurrected with a greener façade—claiming better science—in an effort to win over skeptics. The plan, whose supporters include five water agencies, says it could supply 400,000 people with drinking water in thirsty Southern California communities within just a few years. According to conservationists, including the Sierra Club, however, mining the aquifer could harm the threatened Mojave Desert tortoise, bighorn sheep, as well as the nearby Mojave National Preserve.[81]

Consider further, how long the quick fix would last with the current wasteful use of water and the growing population its availability would allow, plus the fact that once drained, the aquifer will not recharge for centuries to come. What then? What will the next symptomatic quick fix be for a Southern California population that is even larger than today's because of the first symptomatic quick fix? Who will be responsible for it? What will the options be? Have you—today's decision maker—even thought about it?

As a decision maker, would you vote to curb growth in thirsty Southern California communities as part of a systemic solution or cast a symptomatic vote to alter the Mojave Desert irreversibly and forever by stealing its water, which, in unforeseen ways, will have permanent, negative affects for centuries to come, especially in the face of indisputable global warming and the potential for increasingly protracted droughts?

Rule 7—All Systems Have Cumulative Effects, Lag Periods, and Thresholds

A decision maker must help the public to understand that all systems have incremental, *cumulative effects* that occur below our level of awareness (*lag period*) until such a time that sufficient change has taken place to cross a *threshold of visibility*. In a human sense, this often means accepting the delayed gratification of a vision's outcome, despite wanting to see immediate results. Herein lies a major challenge for leaders of industrialized countries, where the citizenry insists on and uses technology that constantly ramps up the speed of transactions. The faster technological systems work, the faster people want them to work.

Unfortunately, we in the Western industrialized countries—and the United States in particular—are too often like the rich but foolish man in the Buddha's parable who wanted instant results:

> Once there was a wealthy but foolish man. When he saw the beautiful three-story house of another man, he envied it and made up his mind to have one built just like it, thinking himself just as wealthy. He called a carpenter and ordered him to build it. The carpenter consented and immediately began to construct the foundation, the first story, the second story, and then the third story. The wealthy man noticed this with irritation and said: "I don't want a foundation or a first story or a second story; I just want the beautiful third story. Build it quickly."[82]

Everything in nature has its own timing, and we would be wise to accept that timing with patience, including the patience of achieving consensus prior to committing an action that will affect all generations to come. Most people in the industrialized West seem to have the attitude that "time is money" and so are in a hurry to act, often without thinking through the potential consequences of their actions on either people or the environment.

Some years ago, while working with the Shinto priests in Japan, I learned a valuable lesson. The priests spent much time discussing the pros and cons of various decisions, as well as the potential outcomes of this choice or that. While the meetings seemed interminable and inefficient to me, when the priests came to consensus, they acted correctly with their collective wisdom the first time because they had winnowed the possibilities and probabilities through an informal, but very effective, risk analysis. This experience taught me that patience and delayed gratification are often more efficient than instant gratification because doing it right the first time is both effective *and* efficient since it precludes the necessity of having to do it over at the extra cost of time, labor, and capital.

We, too, would have a far greater probability of achieving our desired outcome, both efficiently *and* effectively, by doing whatever we do correctly the first time. But first, we have to properly understand the notion of *leisure*.

Whereas we tend to think of leisure as the privilege of the well-to-do, Brother David Steindl-Rast (a Benedictine monk) reminds us that leisure "is a virtue, not a luxury. Leisure is the virtue of those who take their time in order to give to each task as much time as it deserves. … Giving and taking, play and work, meaning and purpose are perfectly balanced in leisure. We learn to live fully in the measure in which we learn to live leisurely."[83] Therefore, decision makers must teach leisure and patience by example if their legacy is going to be a true benefit for all generations.

Rule 8—Change Is an Irreversible Process of Eternal Becoming

One of the more complex issues of decision making—especially social-environmental decisions—is dealing with the irreversibility of change as a constant process. Moreover, all systems are cyclical, but none are perfect circles, which often confronts decision makers with perplexing choices because most people seem to want everything to go perpetually their way. For example, Wall Street, the U.S. government, and the majority of the citizenry insist that our economic system be linear and ever growing. This myopic economic view has produced a symptomatic configuration based so completely on competition and continual growth that it is destroying the global environment's ability to serve humanity as a viable life-support system. This legacy, from many centuries of humanity's persistent form of decision making, cannot continue unabated if we are to avoid becoming the authors of our own demise as a viable society in the short term—or even as a species in the long run.

It is imperative that people become aware of the long-term effects of their decisions. I say this because children are one of the two silent parties in all social-environmental decisions; the land and its productive capacity is the other. Leaders, at whatever level of society, must understand the social, environmental, and economic circumstances to which they are committing future generations through their decisions of today. If the outcome of their decisions and actions is a deficit in terms of the future options for humanity, the productive capacity of ecosystems to serve human necessities, or both, it is analogous to "taxation without representation," and that goes against everything our American democracy stands for.[84]

In keeping with this statement, those decision makers who deny the measurable social-environmental changes wrought by a warming climate to protect economic interests serve as illustrative of the irreversible consequences of ill-informed decisions. This irreversibility is best understood within the historical context of humans and a warming climate.

Millions of years ago, when hominids became omnivores by including meat in their diet, they needed to range farther and wider than their herbivorous counterparts if they were to procure a sufficient amount of food. The animals they hunted were also moving targets, save for the windfall of an occasional carcass, which meant they had to expend a lot more energy to obtain their meals. In the case of early hominid hunters and scavengers, who still spent some time in the trees, natural selection transformed their apelike proportions into a long-legged body that was built for sustained striding and running. But, these elevated levels of activity came at a price—a greatly increased risk of overheating.

Keeping cool became problematic, especially for those early hominids who lived in hot places and thus generated abundant heat from prolonged walking or running. They had to regulate their core body temperature carefully because overheating damaged their tissues and organs, specifically their brain.

Consequently, we humans evolved an effective strategy to avoid overheating. In addition to lacking fur, we possess between 2 million and 5 million eccrine glands, which are the major sweat glands of the human body. These glands are found in virtually all skin, where they reside relatively close to the surface and discharge up to 12.7 quarts of thin, watery sweat a day through tiny pores. This combination of naked skin and watery sweat, which sits directly on top of the skin rather than collecting in fur, allows humans to eliminate excess heat effectively.

When did this metamorphosis occur? About 1.6 million years ago, an early hominid evolved body proportions that were essentially modern, which would have permitted prolonged walking and running. Moreover, details of the joint surfaces of the ankle, knee, and hip clearly demonstrate that these hominids actually exerted themselves in this way. Thus, according to the fossil evidence, the transition to naked skin and an eccrine-based sweating system must have been well under way by that time to offset the greater heat stress that accompanied our forebearers' increasingly strenuous way of life.

The evolutionary loss of fur was just a partial answer to the avoidance of overheating from a strenuous lifestyle. As with many other animals that live in hot climes, hominid evolution also increased the surface area of the naked skin to the volume of the body, thereby making it easier to expel deleterious heat, although this same evolutionary trait makes it more difficult for humans to retain the necessary body heat in cold climes.

Another clue to when hominids evolved naked skin has come from investigations into the genetics of skin color. A specific gene variant always found in Africans with dark skin pigmentation originated 1.2 million years ago. Early human ancestors are thought to have had pinkish skin covered with black fur, much like that of chimpanzees, so the transition to permanently dark skin was presumably an evolutionary requisite to the loss of our sun-shielding body hair.[85]

In addition to naked skins, maintaining a closely regulated, high body temperature of 98.6° Fahrenheit characterizes people and approximates that of other mammals. Although maintaining such a high inner temperature is costly in the amount of energy that must be garnered through the consumption of food, the trade-off is the prevention of pathogenic fungi from invading the body because they cannot tolerate the heat.[86]

Today, despite the uncertainty about the impacts of climate change, it is often assumed that we humans could adapt to any possible warming. This assumption is not necessarily true, however, because all life has a limited tolerance when it comes to temperature—both hot and cold. This said, the peak heat stress is today similar across diverse climates.

The wet-bulb temperature never exceeds 87.8° Fahrenheit. The wet-bulb temperature can be measured by using a thermometer with the bulb wrapped in wet muslin and exposed to the flow of air. It is a measure of the evaporation of water from the thermometer and the cooling effect the evaporation has. The rate of evaporation depends on the humidity of the air, which means the higher humidity, the more water vapor the air contains, the less effective evaporative cooling is. The second law of thermodynamics does not allow an object, such as a person, to lose heat to an environment when the wet-bulb temperature exceeds the person's or object's temperature, no matter how wet or well ventilated. In effect, the wet-bulb temperature is a measure of the lowest body temperature a person, or other mammal, could achieve solely by evaporative cooling in a given circumstance without the aid of technology. If the wet-bulb temperature were to exceed 95° Fahrenheit for extended periods, it would induce hyperthermia in humans and other mammals because the dissipation of body heat through evaporative cooling becomes impossible.

While the wet-bulb temperature does not currently exceed 95° Fahrenheit for extended periods, it would begin to occur if the world warmed an average of about 47° Fahrenheit—a temperature that would make habitability of some regions doubtful. Moreover, with a warming of 52–54° Fahrenheit, such regions would spread to encompass the majority of the human population as currently distributed. A global climate warmed by 54° Fahrenheit *is possible* from the burning of fossil fuel.

Heat stress is associated with not only warm nights and sleep deprivation but also hot days, which alter the lifestyles and productivity of people living at low latitudes. Although both impacts will worsen in warmer, more humid climates, most believe people will simply adapt because humans already tolerate a very wide range of climates. But, when measured in terms of peak heat stress—including humidity—this turns out to be a bogus assumption. Even modest global warming could expose large fractions of the human population to unprecedented heat stress, and severe warming would make such stress intolerable.

As mentioned, humans maintain a core body temperature near 98.6° Fahrenheit, which varies slightly among individuals but does not adapt to local climate. Human skin temperature is strongly regulated at 95° Fahrenheit or less under normal conditions, which is critical because the skin must be cooler than the core temperature of the body for the body's internal temperature to be regulated by conducting excess heat to the skin for evaporative cooling. Sustained skin temperatures above 95° Fahrenheit imply elevated core body temperatures (hyperthermia). If the temperature of the skin is sustained at 98.6–100.4° Fahrenheit, the core temperature can reach the lethal values of 107.6° to 109.4° Fahrenheit, even for acclimated and fit individuals. Thus, sufficiently long periods when the wet-bulb temperature would remain at 95° Fahrenheit would most likely prove to be intolerable.[87]

Although people with sufficient monetary means in the industrialized countries could, and already do, turn to technology for such things as air conditioning, it is a temporary, symptomatic fix. The biophysical costs extracted from the environment would ultimately catch up and not only be catastrophic ecologically but also affect everyone without regard to monetary riches. There would be no opportunity for a person to "buy" his or her way out of the consequences. Moreover, poor people in both industrialized and nonindustrialized countries would be the first to suffer greatly—to say nothing of large mammals, whose body mass is greater than the surface area of their skin, making them prone to fatal heat stress, as well as dramatic changes in the availability of their normal food. Finally, genetic adaptation to a warmer world would not be readily possible given the speed of the potential changes in temperatures.

What today's and tomorrow's decision makers must understand is that human-induced climate change depends not only on the magnitude of the change but also on the potential reversal of the process of change, which will take about 1,000 years after the emissions cease due to the slower loss of heat to the world's expanding and warming oceans.[88] Even with the potential for a reversal of global warming as a process, the current and ongoing biophysical consequences of climate change are already and forever irreversible.

Rule 9—Systemic Change Is Based on Self-Organized Criticality

According to the theory called *self-organized criticality*, the mechanism that leads to a minor event (analogous to the drop of a pin) is the same mechanism that leads to a major event (analogous to an earthquake).[89] Not understanding this, analysts have typically and erroneously blamed some rare set of circumstances (some exception to the rule) or some powerful combination of mechanisms when catastrophe strikes.

Self-organized criticality is an important concept for decision makers to understand because most current leaders base their decision making on linear, symptomatic thinking. They do not realize that virtually all social collapses are initiated from within, even if triggered from without. Consider that all civilizations—past and present—evolved by similar steps: growth of knowledge and technology through discoveries and inventions and through the ideas of government, family, and property, all of which are based on observation and the accumulation of experimental knowledge. As such, all civilizations have much in common, and their evolutionary stages are connected in somewhat similar sequences of cultural development.

The arts of subsistence and the achievements of technology can be used to distinguish the periods of human progress. People began their journey by gathering fruits and nuts. Then, they learned to hunt, fish, and use fire; invented the spear and atlatl, advancing to the bow and arrow. They developed the art of making pottery, learned to domesticate animals and cultivate plants, began using adobe and stone in building houses, and learned to smelt bronze and iron and use it in making tools. Finally, "civilization" began with agriculture[90] and the invention of the phonetic alphabet, culminating in all today's wonders of human accomplishment.

Earlier civilizations have been marked by birth, maturation, and demise, the last commonly caused by uncontrolled population growth that outstripped its sources of available energy, especially during times of climate stress, which initiated internal and external conflicts. In some cases, survivors moved on to less-populated, more fertile areas as their civilizations collapsed. Today, however, there is nowhere left on Earth to go.

Yet, having learned little or nothing from history, the rich, industrialized nations are engaged in the competitive race of destruction of the very environments from which their societies sprang and on which they—and all inhabitants of the biosphere—rely for continuance. The separation of work from the rest of life, which began many centuries ago with the inception of agriculture, is today fully manifested. Jobs are so fundamental to people's sense of identity and position within the social hierarchy that many become severely depressed when they lose their jobs. Moreover, today's competitive marketplace, where people are "bought" and "sold" at the economic convenience of businesses and corporations, is a breeding ground for stress-related illnesses due to the uncertainty and unpredictability of everyday life over which people feel increasingly out of control. Long-term stress not only wears down the body but also initiates the potential for high blood pressure, heart disease, mood disorders, chronic pain brought on by relentless muscle tension, and insomnia.[91]

But, when we couple cooperation and coordination with sharing and caring, it precludes the perceived need to compete for survival and social status, except in play—and perhaps storytelling. Linking individual well-being strictly to individual production is the road to competition, which in turn leads inevitably to social inequality, poverty, and environmental degradation. Self-centeredness and acquisitiveness are not inherent traits of our species but rather acquired traits based on a sense of fear and insecurity within our social setting, which fosters our ego's perceived need to impress others with our personal prowess, which means that *more* is always better, and *enough* does not exist.

Therefore, society, as we know it, cannot be the final evolutionary stage for human existence. But, what lies beyond our current notion of civilization? What is the next frontier for "civilized" people to conquer? Is it outer space, as so often stated? No, it is not outer space. What then? It is inner space—the conquest of oneself, which many assert is life's most arduous task. As the Buddha said, "Though he should conquer a thousand men in the battlefield a thousand times, yet he, indeed, who would conquer himself is the noblest victor."[92]

Sociologists, on the other hand, might think of self-organized criticality as the *butterfly effect*, so-called because it is analogous to the flapping wings of a butterfly, which represents a small change in the initial condition of the system that in turn causes a chain of events leading to large-scale alterations of events. When framed in the positive sense, it means that what a few influential, psychologically mature decision makers think and say can spread and bring about significant changes in the thinking and behavior of large numbers of people, which in turn would affect the social-environmental sustainability and inheritance of all generations.[93]

Rule 10—Dynamic Disequilibrium Rules All Systems

All social systems, like those of nature, are ruled by dynamic *dis-equilibrium*, which simply means that the universally constant process of change, with its ever-novel outcomes, precudes a state of lasting balance. Therefore, the notion of any kind of "steady-state economics," as it is known on the Internet, is impossible. Here, the abiding paradox of life is that we have a choice in everything we think and almost everything we do—except practicing relationships, experiencing ourselves as we experience relationships, choosing, changing the world, living without killing, and dying. In those we have *no* choice of what we do, but we *do* have a choice of how we do it—and we *must* choose. Moreover, we must understand that not choosing *is still a choice*. In addition, we make a new choice (even if it is doing nothing) each time a circumstance in our life changes, which, of course, is an ongoing process, be it the outworking of biophysical principles that govern life or how we view the life changes as we mature in years.

The constancy of change dictates the omnipresence of choice. Life can therefore be viewed as an eternal parade of decisions, each of which marks a fork in the path we follow. Each time a decision is made, others are foregone—one of the trade-offs. Nevertheless, each decision creates a kaleidoscope of additional choices. In turn, choice leads to both wisdom and folly manifested in the ensuing consequences. However, Israeli statesman Abba Eban observed that to date, "History teaches us that men and nations behave wisely once they have exhausted all other alternatives."[94] Now, the beginning of the 21st century, is the time let go of our resistance to change and find the humility to learn from nature by accepting and living within the inviolable constraints of nature's biophysical principles.

In this sense, everything we think and do has a trade-off of positive *and* negative consequences at the time a thought is formed, a decision is made, and a choice of action is taken. Hence, each choice—and the subsequent decision to act—involves the trade-offs of hoped-for outcomes amid the irreversibility of life's unknowns and uncertainties that are the essence of life's *dynamic dis-equilibrium*.

Rule 11—Success or Failure Lies in the Interpretation of an Event

All relationships are self-reinforcing feedback loops that, in nature, are neutral in valuation, as all values are intrinsic. However, these same feedback loops carry either a positive or a negative accent in human valuation because we want specific, predetermined outcomes to provide the illusion of being in control of circumstances. This human dynamic is the same as that driving the notion of success or failure, which lies in the interpretation of an event—but not the event itself, something every public decision maker would do well to learn.

To illustrate, I was once asked to mediate a conflict in Northern California over how to "restore" the Mattole River. During the process, some of the long-time, older residents began lamenting how "newcomers" into "their valley" had destroyed "their river" through years of overuse and abuse. Finally, a youth in his late teens, who had been in juvenile detention, spoke up and said, in effect: "I don't know what you're talking about. I've been working for 3 years with a crew to improve the river's condition. The river is so much better than when I started. What's your problem?"

Whereas the only thing the old-timers could see was the loss of what, to them, had been a "pristine condition," the boy perceived as a vast improvement in a short period of time. The success or failure of the efforts to heal the river was therefore perceived differently, depending on the life experiences of those involved. In this case, the old-timers constantly reinforced their collective grief over the *negative* changes for which they blamed others, although they had been complicit in rendering the changes they perceived through their own actions. The boy, on the other hand, was encouraged by his ability to produce a positive change in the river's condition—to which the old-timers were blind but for which he gladly took responsibility. In this case, to the old-timers the glass was half empty (failure), whereas the boy saw the same glass as half full (success), while the level of water in the proverbial glass had not changed in either case.

Rule 12—People Must Be Equally Informed If They Are to Function as a Truly Democratic Society

For a group of people to be socially functional, they must be given the opportunity to be equally informed about what is going on and how it is projected to affect them. In other words, there must be no secrets that are or can be potentially detrimental to any member. Inequality of any kind within the group related to gender or social class is merely the fear of inadequacy disguised as privilege by those who would impose their will.

Rule 13—We Must Consciously Limit Our "Wants"

We in the world today are so ensnared in the process of selling and buying things in the marketplace that we cannot imagine human life being otherwise. Even our notion of well-being and of despair are wedded to the flow and ebb of the markets. Why is this so much a part of our lives? It is largely because people have yet to understand the notion of conscious simplicity, which is based on the realization that there are two ways to riches: want less or work more. Put differently, riches lie in the scarcity of one's wants as opposed to the abundance of one's possessions.

Author James B. Twitchell puts it nicely: "Once we are fed and sheltered, our needs are and have always been cultural, not natural. Until there is some other system to codify and satisfy those needs and yearnings, commercialism [consumerism]—and the culture it carries with it—will continue not just to thrive but to triumph."[95]

By consciously limiting our wants, it is likely that we can have enough to provide comfortably for our necessities as well as some of our most ardent desires—and leave more for other people to do the same. Unfortunately, capitalistic systems are based on dissatisfaction and a continual stimulus to purchase superfluous items at the risk of personal debt, the long-term expense of the environment, and thus growing impoverishment for all future generations. Herein lies one of the greatest challenges of contemporary decision makers, a challenge that must be met if our society—and the environment that supports it—is to survive the 21st century with any kind of dignity and well-being for more than the top 1% of the monetarily wealthy.

Rule 14—Simplicity Is the Key to Contentment, Adaptability, and Survival

Any fool can complicate life, but it requires genius to keep things simple. Simplicity in living and dying depends on appreciating things simple, small, sublime, and sustainable. What is more, simplicity is the key to contentment, adaptability, and survival as a culture. Beyond some point, complexity becomes a decided disadvantage with respect to cultural longevity, just as it is to the evolutionary longevity of a species. As artist Hans Hoffman puts it, "The ability to simplify means to eliminate the unnecessary so that the necessary may speak."[96] To this, His Holiness the 17th Gyalwang Karmapa Ogyen Trinley Dorje would add, "A life full of material goods and barren of compassion is quite unsustainable from an ecological and karmic point of view. Of course, advertisements are always telling us that the path to happiness lies in purchasing the goods they sell. … We don't have to live a life that is sold to us—we can make the brave choice to live simply."[97]

Rule 15—Nature, Environmental/Cultural Wisdom, and Human Well-Being Are Paramount

Placing material riches, as symbolized by the money chase, above the wealth of nature, spirituality, and human well-being is the road to social impoverishment, environmental degradation, and the collapse of societies and their life-support systems. We must, therefore, rethink our priorities and place the viability of ecosystems at least on par with economics of "natural capital," which means reconnecting with nature, if our own well-being is to prevail.[98]

Here, it is instructive to consider the difference between money and the true wealth of nature's services and our personal well-being. Conventional money knows no loyalty to a sense of place, a person, a local community, a landscape, a region, or even a nation, so it flows toward a global economy in which traditional social bonds give way to a rootless quest for the highest monetary return at virtually any social-environmental cost. The real price we pay for money is the hold it has on our sense of what is possible—the prison it builds around our imaginations, which American journalist Sydney J. Harris captured in a few words: "Men make counterfeit money; [but] in many more cases, money makes counterfeit men."[99]

According to Bernard Lietaer, of the Center for Sustainable Resources at the University of California at Berkeley, "Money is like an iron ring we've put through our noses. We've forgotten that we designed it, and it's now leading us around. I think it's time to figure out where we want to go—in my opinion toward sustainability—and then design a money system to get us there."[100]

While textbooks on economics claim that people and corporations are competing for resources and markets, they are really competing for money by using markets to foster competition in the exploitation and commercialization of resources—both natural and human.[101] "A more fascinating aspect of money," notes author Caroline Myss, "is the fact that it can weave itself into the human psyche as a substitute for the life-force."[102] Through the way in which we spend money, according to Myss, we make our private beliefs into public declarations.

Today's money is either a numbered piece of paper or an electronic computer trace—both of which are merely symbolic representations of energy—that allows a person to obtain either the necessities of life or fulfill a desire that constitutes the *real wealth of life*. But, in our confusion over the true nature of real wealth (if we have thought about it at all), we chase the almighty dollar and neglect those things that actually sustain a life of quality—both spiritual *and* material. Money has only extrinsic value, only the potential to be converted into something else that may have the intrinsic value of real wealth.[103]

To illustrate, if you are marooned on a desert island without food or water, but with a trunk containing $1 million or even $1 trillion, what value would it have? Oh, you could burn it to stay warm for a while, but you could not drink it to quench your thirst. You could eat some, perhaps, but it would not sustain you.

Now, let us suppose you are a billionaire with inoperable, terminal cancer. Could all your money buy your way out of death? The cattle ranchers I used to work for as a young man were of the opinion that, "If you have your health, you have everything." That notion goes a long way in defining true wealth.

Nevertheless, our fixation with the "money chase" makes it difficult to express the basic difference between money and wealth through the symbology of language. Real wealth means the biophysical integrity of the diverse ecosystems on which we rely for social-environmental sustainability is intact. It means human equality and dignity, meaningful work, having a good home and adequate food, a good education, and so on. In a healthy economy, money serves the people in helping to create and protect the real wealth but is neither the dominant value nor the sole—or even the main—medium of exchange.

One of the most important indicators of economic health is social-environmental sustainability, which means not only quality interpersonal relationships but also quality reciprocal relationships between people and their environment. A truly viable economy is based on love and reciprocity, where people do kind and useful things for one another with no expectation of financial gain. Such mutual caring is the soft social capital that both creates and maintains the fabric of trust, which in turn is the glue of functional families, communities, and societies.

Pathology and its subsequent dysfunction enter the economic system when money, derived as a convenient means of exchange, becomes the factor that defines the purpose of life for individuals and their communities. For example, the majority of Americans in a November 28 to December 1, 2011, poll said they would need an annual income of $150,000 to feel rich. And, men had a higher median threshold, $150,000, than women—$100,000. Moreover, as people earn more, they want more. To wit, people earning $75,000 or more say they would need a median of $250,000 to feel rich. Such focus on dollar amounts by the majority of a community's or nation's citizens results in the social and biophysical capital on which the well-being of every human, community, and nation depends being sacrificed on the altar of competitive exploitation: Those who already have money prosper at a level of social influence above those who do not.[104]

The growing dominance of money as master is also revealed in the increasing commercialization of human relationships. Not long ago, even in such rich, industrialized countries as the United States, at least half of the adult population—predominantly women—worked without salary to create and maintain home and community, which are among the most fundamental values of a healthy, sustainable society and its social-environmental economy.

Today, financially supporting a household usually requires two adults holding two, and sometimes three or four, paying jobs between them, but at the expense of quality human relationships because, of necessity, they rarely see each other, and the care of children and the home is either neglected or hired out. In addition, the once-shared, mutual caring has morphed into "community service," which is the work of hired public employees—to the extent the public is willing to pay for it.[105]

As the soft social capital of mutual caring dwindles and the resulting quality of family life withers, a community's members typically become increasingly apathetic or competitive. As human relationships become more and more dysfunctional, a community's infrastructure crumbles into ever-greater disarray at an ever-increasing social-environmental cost to all generations.

Rule 16—Every Legal Citizen Deserves the Right to Vote

Every legal citizen of every country deserves the right to an equal vote of their conscience on how their country is to be governed because they and their children and their children's children must live with the consequences of the collective choices and actions.

Rule 17—This Present Moment, the Here and Now, Is All We Ever Have

This eternal, present moment is all we ever have in which to act. The past is a memory, and the future never comes. *Now* is the *eternal moment.* These inviolable rules of decision making are enshrined in every moment of everyday life, whether we recognize them or not. As such, the leader who honors them will move forward unafraid.

Perhaps the toughest decision a leader is confronted with is to bear, unflinchingly, all the abuses that the parties normally hurl at one another. In effect, a person, such as a mediator who serves the people, must pass the tests described in the eulogy that Senator William Pitt Fessenden of Maine delivered on the death of Senator Foot of Vermont in 1866:

> When, Mr. President, a man becomes a member of this body he cannot even dream of the ordeal to which he cannot fail to be exposed;
>
> of how much courage he must possess to resist the temptations which daily beset him;
>
> of that sensitive shrinking from undeserved censure which he must learn to control;
>
> of the ever-recurring contest between a natural desire for public approbation and a sense of public duty;
>
> of the load of injustice he must be content to bear, even from those who should be his friends;
>
> the imputations of his motives;
>
> the sneers and sarcasms of ignorance and malice;
>
> all the manifold injuries which partisan or private malignity, disappointed of its objects, may shower upon his unprotected head.
>
> All this, Mr. President, if he would retain his integrity, he must learn to bear unmoved, and walk steadily onward in the path of duty, sustained only by the reflection that time may do him justice, or if not, that after all his individual hopes and aspirations, and even his name among men, should be of little account to him when weighed in the balance against the welfare of a people of whose destiny he is a constituted guardian and defender.[106]

Such is the price of true social-environmental leadership—to be the keeper of everyone else's dignity by keeping one's own in the eternal, present moment—*leading by example.*

SUMMARY

Chapter 2 expanded the premise of Chapter 1 into the inviolable rules of decision making (= leadership), which are vital to understand and follow if social-environmental sustainability is to prevail. Section II examines why a command-and-control form of decision making tends to become dysfunctional and how to break the cycle of such behavior. Chapter 3 begins the discussion by examining the ways in which we carry our respective families with us throughout life—both the functional and dysfunctional aspects. Further, it points out that the governments, agencies (used here as a microcosm of society), and various organizations are as appropriately functional or dysfunctional as we are as individuals. Therefore, we—individuals acting as a single, collective body—are society's driving force and thus enable or limit the ability of both its social and environmental agencies to function as originally envisioned—or allow them to degrade into a military-style, command-and-control mode when dysfunction sets in.

NOTES

1. David Orr. 2020: a proposal. *Conservation Biology*, 14 (2000):338–341.
2. Russ Beaton and Chris Maser. *Economics and Ecology: United for a Sustainable World.* CRC Press, Boca Raton, FL. 2011.
3. Ibid.
4. (**1**) Robert McNamara, Krakatoa volcano eruption in 1883 was a worldwide weather and media event. http://history1800s.about.com/od/thegildedage/a/krakatoa.htm (accessed December 25, 2011); and (**2**) How volcanoes work. Krakatau, Indonesia, 1883. http://www.geology.sdsu.edu/how_volcanoes_work/Krakatau.html (accessed December 25, 2011).
5. (**1**) S. Zhang, S. Hou, Y. Wu, and D. Qin. Bacteria in Himalayan glacial ice and its relationship to dust. *Biogeosciences*, 5 (2008):1741–1750; and (**2**) S. Zhang, S. Hou, X. Ma, and others. Culturable bacteria in Himalayan glacial ice in response to atmospheric circulation. *Biogeosciences*, 4 (2007):1–9.
6. (**1**) Daniel A. Jaffe, Alexander Mahura, Jennifer Kelley, and others. Impact of Asian emissions on the remote North Pacific atmosphere: interpretation of CO data from Shemya, Guam, Mid-way, and Mauna Loa. *Journal of Geophysical Research*, 102 (1997):28627–28636; and (**2**) Dan Jaffe, Theodore Anderson, Dave Covert, and others. Transport of Asian air pollution to North America. *American Geophysical Research Letters*, 26 (1999):711–714.
7. I. Koren, Y. Kaufman, R. Washington, and others. The Bodélé Depression: a single spot in the Sahara that provides most of the mineral dust to the Amazon Forest. *Environmental Research Letters*, 1 (2006):1–5.
8. (**1**) S. Genualdi, S. Massey Simonich, T. Primbs, and others. Enantiomeric signatures of organochlorine pesticides in Asian trans-Pacific and western U.S. air masses. *Environmental Science and Technology*, 43 (2009):2806–2811; and (**2**) Ralph J. Delfino, T. Tjoa, D. Gillen, and others. Traffic-related air pollution and blood pressure in elderly subjects with coronary artery disease. *Epidemiology*, 21 (2010):396–404.

9. (**1**) B. Adhikary, G.R. Carmicheal, Y. Tang, and others. Characterization of seasonal cycle of South Asian aerosols: a regional-scale modeling analysis. *Journal of Geophysical Research–Atmospheres*, 112 (D22):Art. No. D22S22; and (**2**) Sid Perkins. What goes up. *Science News,* 172 (2007):152–153, 156.
10. Kathleen C. Weathers, Mary L. Cadenasso, and Steward T.A. Pickett. Forest edges as nutrient and pollutant concentrators: potential synergisms between fragmentation, forest canopies, and the atmosphere. *Conservation Biology,* 15 (2001):1506–1514.
11. Dieldrin. http://www.epa.gov/pbt/pubs/aldrin.htm (accessed March 23, 2011).
12. DDT. http://www.epa.gov/history/topics/ddt/01.htm (accessed March 23, 2011).
13. (**1**) Matthew Brown. Federal study: pesticides prevalent in national wilderness, Corvallis (OR) *Gazette-Times*, February 27, 2008; and (**2**) Roman Marks and Magdalena Bedowska. Air-sea exchange of mercury vapor over the Gulf of Gdansk and southern Baltic Sea. *Journal of Marine Systems,* 27 (2001):315–324.
14. The preceding two paragraphs are based on Don Phillips, Connie Burdick, Becky Merja, and Norm Brown. *The City,* 24 (Winter 2011–2012):5.
15. Chris Maser and James R. Sedell. *From the Forest to the Sea: The Ecology of Wood in Streams, Rivers, Estuaries, and Oceans.* St. Lucie Press, Delray Beach, FL, 1994.
16. (**1**) Chris Maser. The humble ditch. *Resurgence,* 172 (1995):38–40; (**2**) Chris Maser. Abnormal coloration in *Microtus montanus*. *Murrelet*, 50 (1969):39; and (**3**) Earl Bardsley. Conveying waste with water. *New Zealand Science Monthly*. http://nzsm.webcentre.co.nz/article449.htm (accessed April 16, 2010).
17. Willard S. Moore, Jorge L. Sarmiento, and Robert M. Key. Submarine groundwater discharge revealed by ^{228}Ra distribution in the upper Atlantic Ocean. *Nature Geoscience*, 1 (2008):309–311.
18. (**1**) Perrine Fleury, Michel Bakalowicz, and Ghislain de Marsily. Submarine springs and coastal karst aquifers: a review. *Journal of Hydrology,* 339 (2007):79–92; (**2**) D. Reide Corbett, William C. Burnett, and Jeffrey P. Chanton. Submarine groundwater discharge: an unseen yet potentially important coastal phenomenon. University of Florida IFAS Extension. http://edis.ifas.ufl.edu/SG060 (accessed April 16, 2011); (**3**) Sid Perkins. Invisible rivers. *Science News*, 168 (2005):248–249; and (**4**) Takeshi Uemura, Makoto Taniguchi, and Kazuo Shibuya. Submarine groundwater discharge in Lützow-Holm Bay, Antarctica. *Geophysical Research Letters*, 38 (2011):L08402, 6 pp., doi:10.1029/2010GL046394.
19. Carrie Halperin. How climate change may make killer diseases worse. http://abcnews.go.com/Politics/GlobalHealth/risk-malaria-diseases-rise-global-temperatures-climate/story?id=13277843 (accessed April 2, 2011).
20. Perkins. Invisible Rivers.
21. (**1**) Andrew Sugden, Richard Stone, and Caroline Ash. Ecology in the underworld. *Science*, 304 (2004):1613; (**2**) Jocelyn Kaiser. Wounding Earth's fragile skin. *Science*, 304 (2004):1616–1618; (**3**) Elizabeth Pennisi. The secret life of fungi. *Science,* 304 (2004):1620–1622; (**4**) David A. Wardle, Richard D. Bardgett, John N. Klironomos, and others. Ecological linkages between aboveground and belowground biota. *Science,* 304 (2004):1629–1633; and (**5**) I.M. Young and J.W. Crawford. Interactions and self-organization in the soil-microbe complex. *Science* 304 (2004):1634–1637.
22. (**1**) Mark Ferns. Geologic evolution of the Blue Mountains region, the role of geology in soil formation. *Natural Resource News*, 5 (1995):2–3, 17; (**2**) James L. Clayton. Processes of soil formation. *Natural Resource News*, 5 (1995):4–6; and (**3**) Elaine R. Ingham. Organisms in the soil: the functions of bacteria, fungi, protozoa, nematodes, and arthropods. *Natural Resource News*, 5 (1995):10–12, 16–17.
23. *The Holy Bible*, Authorized King James Version. World Bible, Iowa Falls, IA. Numbers Chapter 35, Verse 34.

24. Ancient Indian proverb. http://www.gardendigest.com/earth.htm (accessed March 23, 2011).
25. Franklin D. Roosevelt. http://www.gardendigest.com/earth.htm (accessed March 23, 2011).
26. J.R. McNeill and Verena Winiwarter. Breaking the sod: humankind, history, and soil. *Science,* 304 (2004):1627–1629.
27. **(1)** Louise H. Emmons. Tropical rain forests: why they have so many species, and how we may lose this biodiversity without cutting a single tree. *Orion,* 8 (1989):8–14; and **(2)** F. Stuart Chapin III, Brian H. Walker, Richard J. Hobbs, and others. Biotic control over the functioning of ecosystems. *Science,* 277 (1997):500–504.
28. **(1)** Michael Kiefer. Fall of the Garden of Eden. *International Wildlife* (July–August 1989):38–43; and **(2)** A.R.E. Sinclair, Simon A.R. Mduma, J. Grant, and others. Long-term ecosystem dynamics in the Serengeti: lessons for conservation. *Conservation Biology*, 21 (2007):580–590.
29. Aldo Leopold. http://www.nap.edu/openbook.php?record_id=2216&page=202 (accessed March 26, 2011).
30. Emmons. Tropical Rain Forests.
31. Wardle et al. Ecological Linkages.
32. Rym Momtaz. Female protesters targeted by Egyptian government, say activists. http://abcnews.go.com/Blotter/female-protestors-targeted-egyptian-government-activists/story?id=15206453#.TvTFhRw0i4A (accessed December 21, 2011).
33. Ibid.
34. Israeli Ultra-Orthodox rally for gender separation. http://abcnews.go.com/International/wireStory/israeli-ultra-orthodox-rally-gender-separation-15266088#.Tv93qBw0i4A (accessed December 31, 2011).
35. The Earth receives an incredible supply of solar energy. http://www.altenergy.org/renewables/solar.html (accessed November 6, 2011).
36. George Herbert Read. http://www.backyardnature.net/101/quotes.htm (accessed on February 4, 2011).
37. Richard Buckminster Fuller. http://www.skyscript.co.uk/brady.html (accessed March 24, 2001).
38. Winston Churchill. In: T.A. Warren. Leaders need followers. *The Rotarian*, (October 1945):10–12.
39. The foregoing discussion is based on Chris Maser. *Sustainable Community Development.*
40. Ibid.
41. Ibid.
42. Chris Maser. Do we owe anything to the future? In: *Multiple Use and Sustained Yield: Changing Philosophies for Federal Land Management?* Proceedings and summary of a workshop convened on March 5 and 6, 1992, Washington, DC. Congressional Research Service, Library of Congress. Committee Print No. 11. U.S. Government Printing Office, Washington, DC, 1992, pp. 195–213.
43. Maser. *Sustainable Community Development.*
44. **(1)** Oil Sands. http://en.wikipedia.org/wiki/Oil_sands (accessed November 7, 2011); and **(2)** Matthew Daly. U.S. delays massive oil pipeline from Canada. http://abcnews.go.com/Politics/wireStory/us-delays-massive-oil-pipeline-canada-14926053#.Trw4KRw0i4A (accessed November 7, 2011).
45. Mark Green. No unnecessary delays on Keystone XL. July 27, 2011. http://blog.energytomorrow.org/2011/07/no-unnecessary-delays-on-keystone-xl.html?gclid=CLvo7quCpawCFccbQgodIEn0Cg (accessed November 7, 2011).
46. Who we are. http://energytomorrow.org/who-we-are (accessed November 7, 2011).
47. Green. No Unnecessary Delays.

48. John Parkinson. Speaker's line in the tar sand: "guarantee" Keystone Pipeline will be in jobs bill. http://abcnews.go.com/blogs/politics/2011/12/speakers-line-in-the-tar-sand-guarantee-keystone-pipeline-will-be-in-jobs-bill/ (accessed December 16, 2011).
49. Matthew Daly. Interior Department to limit mining near Grand Canyon. http://abcnews.go.com/Politics/wireStory/interior-dept-ban-mining-grand-canyon-15317831#.TwoobBw0i4A (accessed January 9, 2012).
50. John Parkinson. Keystone pipeline delay sparks outrage on Capitol Hill. http://abcnews.go.com/blogs/politics/2012/01/boehner-says-obamas-jobs-policies-have-made-the-economy-worse/ (accessed January 18, 2012).
51. Robert Rodale. Big new ideas—where are they today? Unpublished speech given at the Third National Science, Technology, Society (STS) Conference, February 5–7, 1988, Arlington, VA.
52. Charles Petit. Cold panacea. *Science News*, 175 (2009):20–23.
53. **(1)** Georgi M. Daskalov, Alexander N. Grishin, Sergei Rodionov, and Vesselina Mihneva. Trophic cascades triggered by overfishing reveal possible mechanisms of ecosystem regime shifts. *Proceedings of the National Academy of Sciences of the United States of America,* 104 (2007):10518–10523; **(2)** Ransom A. Myers and Boris Worm. Extinction, survival or recovery of large predatory fishes. *Philosophical Transactions of the Royal Society of London: Biological Sciences,* 360 (2005):13–20; **(3)** Peter Ward and Ransom A. Myers. Shifts in open-ocean fish communities coinciding with the commencement of commercial fishing. *Ecology,* 86 (2005):835–847; and **(4)** Kenneth T. Frank, Brian Petrie, Jae S. Choi, and William C. Leggett. Trophic cascades in a formerly cod-dominated ecosystem. *Science,* 308 (2005):1621–1623.
54. **(1)** Chris L.J. Frid, S. Hansson, S.A. Rijnsdorp, and S.A. Steingrimsson. Changing levels of predation on benthos as a result of exploitation of fish populations. *Ambio,* 28 (1999):578–582; **(2)** Chris L.J. Frid, Odette A.L. Paramor, and Catherine L. Scott. Ecosystem-based management of fisheries: is science limiting? *Journal of Marine Science,* 63 (2006):1567–1572; **(3)** Shelley C. Clarke, Jennifer E. Magnussen, Debra L. Abercrombie, and others. Identification of shark species composition and proportion in the Hong Kong shark fin market based on molecular genetics and trade records. *Conservation Biology,* 20 (2006):201–211; and **(4)** R.Q. Grafton, T. Kompas, and R.W. Hilborn. Economics of overexploitation revisited. *Science,* 318 (2007):1601.
55. **(1)** Pauly and V. Christensen. Primary production required to sustain global fisheries. *Nature*, 374 (1995):255–257; **(2)** John A. Barth, Bruce A. Menge, Jane Lubchenco, and others. Delayed upwelling alters nearshore coastal ocean ecosystems in the northern California current. *Proceedings of the National Academy of Sciences of the United States of America*, 104 (2007):3719–3724; **(3)** F. Chan, J.A. Barth, J. Lubchenco, and others. Emergence of anoxia in the California current large marine ecosystem. *Science,* 319 (2008):920; and **(4)** Ryan R. Rykaczewski and David M. Checkley, Jr. Influence of ocean winds on the pelagic ecosystem in upwelling regions. *Proceedings of the National Academy of Sciences of the United States of America,* 105 (2008):1965–1970.
56. Michael Grunwald. An agency of unchecked clout. *The Washington Post*, September 10, 2000.
57. The discussion of the U.S. Army Corps of Engineers is based on **(1)** Jim Robbins. Engineers plan to send a river flowing back to nature. *The New York Times*, May 12, 1998; **(2)** Michael Grunwald. The corps' divided mission. LA Times–Washington Post Service. Portland (OR) *The Oregonian*, Portland, OR, February 17, 2000; **(3)** Michael Grunwald. More powerful than a river. *The Washington Post.* Albany (OR) *Democrat-Herald,* Corvallis (OR) *Gazette-Times*, November 23, 2000; **(4)** Frederic J. Frommer. Groups identify "wasteful" Corps water projects. Associated Press. Corvallis (OR) *Gazette-Times*, March 3, 2000; and **(5)** Amalie Young. Court: Snake dam operation violates Clean Water Act. Associated Press. Corvallis (OR) *Gazette-Times*, February 17, 2001.

58. John Howard. Crews struggle to save California levees. Associated Press. Corvallis (OR) *Gazette-Times*, January 7, 1997.
59. California crews bolster levees; drinking water faces peril. *The New York Times*, January 7, 1997.
60. The Associated Press. Levy failure causes new floods in California. Corvallis (OR) *Gazette-Times*, January 11, 1997.
61. John Howard. Receding floodwaters reveal fields of dead cows, horses. The Associated Press. Corvallis (OR) *Gazette-Times*, January 10, 1997.
62. The preceding discussion of flood damage and reservoirs is based on Matthew Yi. Brimming reservoirs keep California flood threats alive. The Associated Press. Corvallis (OR) *Gazette-Times*, January 10, 1997.
63. Scott Faber. Get people off of nation's floodplains. Corvallis (OR) *Gazette-Times*, January 21, 1997.
64. Ibid.
65. Ibid.
66. Luna B. Leopold. Ethos, equity, and the water resource. *Environment,* 2 (1990):16–42.
67. **(1)** B. Bates, Z.W. Kundzewicz, S. Wu, J.P. Palutikof (eds.). *Climate Change and Water*. Technical Paper of the Intergovernmental Panel on Climate Change, Geneva, Switzerland, 2008; **(2)** Jeffrey J. Opperman, Gerald E. Galloway, Joseph Fargione, and others. Sustainable floodplains through large-scale reconnection to rivers. *Science*, 326 (2009):1487–1488.
68. **(1)** Roger A. Pielke, Jr., Mary W. Downton, and J. Zoe Barnard Miller. Flood damage in the United States, 1926–2000: a reanalysis of national weather service estimates. University Corporation for Atmospheric Research, Boulder, CO, 2002. http://www.flooddamagedata.org/full_report.html (accessed December 18, 2011); **(2)** Costs are normalized to 2007 dollars using a gross national product inflation index from NOAA National Climatic Data Center, National Oceanic and Atmospheric Administration. http://www.ncdc.noaa.gov/oa/reports/billionz.html (accessed December 18, 2011); **(3)** Nicholas Pinter. One step forward, two steps back on U.S. Floodplains. *Science*, 308 (2005):207–208; **(4)** Recommendations for a national levee safety program. National Committee on Levee Safety, 2009. http://leveesafety.org/docs/NCLS-Recommendation-Report_012009_DRAFT.pdf (accessed November 13, 2011); and **(5)** Opperman et al. Sustainable floodplains.
69. **(1)** National Committee on Levee Safety. Recommendations for a national levee safety program. 2009. http://leveesafety.org/docs/NCLS-Recommendation-Report_012009_DRAFT.pdf (accessed November 13, 2011); **(2)** Robert Costanza, Ralph D'arge, Rudolf De Groot, and others. The value of the world's ecosystem services and natural capital. *Nature*, 387 (1997):253–260; and **(3)** Opperman et al. Sustainable floodplains.
70. **(1)** Costanza et al. The value of the world's ecosystem services; and **(2)** Opperman et al. Sustainable floodplains.
71. Mangrove swamp. http://water.epa.gov/type/wetlands/mangrove.cfm (accessed December 20, 2011).
72. Janet N. Abramovitz. Learning to value nature's free services. *Futurist,* 31 (1997):39–42.
73. The preceding two paragraphs are based on **(1)** National Committee on Levee Safety. Recommendations; **(2)** Costanza et al. The value of the world's ecosystem services; **(3)** K. Tockner and J.A. Stanford. Riverine flood plains: present state and future trends. *Environmental Conservation*, 29 (2002):308–330; **(4)** T.A. Volk, T. Verwijst, P.J. Tharakan, and others. Growing fuel: a sustainability assessment of willow biomass crops. *Frontiers in Ecology and the Environment*, 2 (2004):411–418; and **(5)** David Tilman, Jason Hill, and Clarence Lehman. Carbon-negative biofuels from low-input high-diversity grassland biomass. *Science*, 314 (2006):1598–1600; and **(6)** Opperman et al. Sustainable floodplains.

74. Martha J. Groom, Elizabeth M. Gray, and Patricia A. Townsend. Biofuels and biodiversity: principles for creating better policies for biofuel production. *Conservation Biology*, 22 (2008):602–609.
75. **(1)** Tad W. Patzek. Thermodynamics of the corn-ethanol biofuel cycle. *Critical Reviews in Plant Science,* 23 (2004):519–567; **(2)** David Pimentel and Tad W. Patzek, Ethanol production using corn, switchgrass, and wood; biodiesel production using soybean and sunflower. *Natural Resources Research,* 14 (2005):1, 65–76; **(3)** Jason Hill, Erik Nelson, David Tilman, and others, Environmental, economic, and energetic costs and benefits of biodiesel and ethanol biofuels. *Proceedings of the National Academy of Sciences of the United States of America,* 103 (2006):11206–11210; **(4)** Manfred Kroger. Forum: corn is food, not fuel. *Pittsburgh Post-Gazette*, April 8, 2007; **(5)** Alice Friedemann. Peak Soil. http://www.culturechange.org/cms/index.php?option=com_content&task=view&id=107&itemid=1; **(6)** Lian Pin Koh. Potential habitat and biodiversity losses from intensified biodiesel feedstock production. *Conservation Biology,* 21 (2007):1373–1375; and **(7)** Sid Perkins, Groundwater use adds CO_2 to the air. *Science News,* 172 (2007):301.
76. Robert F. Service. Another biofuels drawback: the demand for irrigation. *Science*, 326 (2009):516–517
77. Discussion of the dead zone is based on Thomas O'Connor and David Whitall, Linking hypoxia to shrimp catch in the northern Gulf of Mexico. *Marine Pollution Bulletin,* 54 (2007):460–463; Donald Scavia and Kristina A. Donnelly. Reassessing hypoxia forecasts for the Gulf of Mexico. *Environmental Science and Technology,* 41 (2007):8111–8117; and Sarah C. Williams, Dead serious. *Science News,* 172 (2007):395–396.
78. Perkins, Groundwater use.
79. Ibid.
80. The preceding discussion of ocean acidification is based on **(1)** Helmuth Thomas and Venugopalan Ittekkot. Determination of anthropogenic CO_2 in the North Atlantic Ocean using water mass ages and CO_2 equilibrium chemistry. *Journal of Marine Systems,* 27 (2001):325–336; **(2)** Jonathan Shaw. The great global experiment: as climate change accelerates, how will we adapt to a changed Earth? *Harvard Magazine,* 105 (2002):34–43, 87–90; **(3)** Kathy Tedesco, Richard A. Feely, Christopher L. Sabine, and Cathrine E. Cosca. Impacts of anthropogenic CO_2 on ocean chemistry and biology. NOAA Archive of Spotlight Feature Articles. http://www.oar.noaa.gov/spotlite/spot_gcc.html (accessed June 1, 2005); **(4)** Lisa Stiffler. Research in Pacific shows ocean trouble. *Seattle Post Intelligencer*, March 31, 2006; **(5)** Ruth Bibby, Polly Cleall-Harding, Simon Rundle, and others. Ocean acidification disrupts induced defences in the intertidal gastropod *Littorina littorea. Biology Letters,* 3 (2007): 699–701; **(6)** Scott C. Doney, Natalie Mahowald, Ivan Lima, and others. Impact of anthropogenic atmospheric nitrogen and sulfur deposition on ocean acidification and the inorganic carbon system. *Proceedings of the National Academy of Sciences of the United States of America,* 104 (2007):14580–14585; **(7)** Rosane Gonçalves Ito, Bernd Schneider, and Helmuth Thomas. Distribution of surface fCO_2 and air-sea fluxes in the southwestern subtropical Atlantic and adjacent continental shelf. *Journal of Marine Systems,* 56 (2005):227–242; **(8)** J.C. Blackford and F.J. Gilbert. pH variability and CO_2 induced acidification in the North Sea. *Journal of Marine Systems,* 64 (2007):229–241; **(9)** Igor P. Semiletov, Irina I. Pipko, Irina Repina, and Natalia E. Shakhova. Carbonate chemistry dynamics and carbon dioxide fluxes across the atmosphere-ice-water interfaces in the Arctic Ocean: Pacific sector of the Arctic. *Journal of Marine Systems,* 66 (2007):204–226; and **(10)** O. Hoegh-Guldberg, P.J. Mumby, A.J. Hooten, and others. Coral reefs under rapid climate change and ocean acidification. *Science,* 318 (2007):1737–1742.

81. Oasis or mirage? Company wants to tap Mojave water. http://abcnews.go.com/US/wireStory/oasis-mirage-company-tap-mojave-water-14892164#.Traplxw0i4A (accessed November 6, 2011).
82. *The Teaching of Buddha.* Bukkyo Dendo Kyokai, Tokyo, 1985.
83. Brother David Steindl-Rast. *Gratefulness and the Heart of Prayer: An Approach to Life in Fullness.* Paulist Press, Ransey, NJ, 1984.
84. Chris Maser. *Resolving Environmental Conflict: Towards Sustainable Community Development.* St. Lucie Press, Delray Beach, FL, 1996.
85. Why humans became naked is based on **(1)** Shannon P. McPherron, Zeresenay Alemseged, Curtis W. Marean, and others. Evidence for stone-tool-assisted consumption of animal tissues before 3.39 million years ago at Dikika, Ethiopia. *Nature*, 466 (2010):857–860; **(2)** Nina G. Jablonski. The naked truth. *Scientific American*, 302 (2010):42–49; **(3)** Alan R. Rogers, David Iltis, and Stephen Wooding. Genetic variation at the MC1R locus and the time since loss of human body hair. *Current Anthropology*, 45 (2004):105–108; and **(4)** Nina G. Jablonski and George Chaplin. Skin deep. *Scientific American*, 287 (2002):74–81.
86. Aviv Bergmanand and Arturo Casadevall. Mammalian endothermy optimally restricts fungi and metabolic costs. *mBio* 1(5):e00212–10. doi:10.1128/mBio.00212–10.
87. The preceding discussion of heat stress is based on **(1)** Steven C. Sherwood and Matthew Huber. An adaptability limit to climate change due to heat stress. *Proceedings of the National Academy of Sciences of the United States of America*, 107 (2010):9552–9555; **(2)** A.B. Alley, J. Marotzke, W.D. Nordhaus, and others. Abrupt climate change. *Science*, 299 (2003):2005–2010; **(3)** R. Sair Kovats and Shakoor Hajat. Heat stress and public health: a critical review. *Annual Review of Public Health*, 29 (2008):41–55; **(4)** Thomas R. Karl, Richard W. Knight, Kevin P. Gallo, and Thomas C. Peterson. A new perspective on recent global warming-asymmetric trends of daily maximum and minimum temperatures. *Bulletin of the American Meteorological Society*, 74 (1993):1007–1023; **(5)** A.P. Sokolov, P.H. Stone, C.E. Forest, and others. Probabilistic forecast for twenty-first-century climate based on uncertainties in emissions (without policy) and climate parameters. *Journal of Climate*, 22 (2009):5175–5204; and **(6)** Anthony J. McMichael and Keith B.G. Dear. Climate change: heat, health, and longer horizons. *Proceedings of the National Academy of Sciences of the United States of America*, 107 (2010):9483–9484.
88. Susan Solomon, Gian-Kasper Plattner, Reto Knutti, and Pierre Friedlingstein. Irreversible climate change due to carbon dioxide emissions. *Proceedings of the National Academy of Sciences of the United States of America*, 106 (2009):1704–1709.
89. Per Bak and Kan Chen. Self-organizing criticality. *Scientific American,* January (1991):46–53.
90. **(1)** Stacey Y. Abrams. The land between two rivers: the astronomy of ancient Mesopotamia. *The Electronic Journal of the Astronomical Society of the Atlantic,* 3 (no 2). Georgia State University Press. 1991; and **(2)** The fertile crescent. http://visav.phys.uvic.ca/~babul/AstroCourses/P303/mesopotamia.html (accessed January 7, 2011).
91. Radha Chitale. Job loss can make you sick. http://abcnews.go.com/Business/WellnessNews/story?id=7530730&page=1 (accessed June 1, 2009).
92. *The Teaching of Buddha.*
93. Butterfly effect. **(1)** http://en.wikipedia.org/wiki/Butterfly_effect; and **(2)** From butterfly wings to single e-mail, if one action can cause a torrent, Cornell researchers find the best place to start. *Cornell News.* http://www.news.cornell.edu/releases/Feb04/AAAS.Kleinberg.ws.html (accessed February 4, 2011).
94. Abba Eban. The quotations page. http://www.quotationspage.com/quote/298.html (accessed February 2, 2011).
95. Christopher Lehmann-Haupt. Sales pitches that put the M (for mega) in Madison Ave. *The New York Times*, January 3, 2001.

96. Hans Hoffman. http://quotationsbook.com/author/3495/ (accessed January 7, 2011).
97. [His Holiness] 17th Gyalwang Karmapa Ogyen Trinley Dorje. Walking the path of environmental Buddhism through compassion and emptiness. *Conservation Biology*, 25 (2011):1094–1097.
98. Russ Beaton and Chris Maser. *Economics and Ecology: United for a Sustainable World*. CRC Press, Boca Raton, FL, 2011.
99. Sydney J. Harris. http://www.google.com/#pq="men+make+counterfeit+money%3B+%5Bbut%5D+in+many+more+cases%2C+money+makes+counterfeit+men."&hl=en&cp=28&gs_id=m&xhr=t&q="Men+make+counterfeit+money%3B+in+many+more+cases%2C+money+makes+counterfeit+men."&pf=p&sclient=psy-ab&source=hp&pbx=1&oq="Men+make+counterfeit+money%3B+in+many+more+cases,+money+makes+counterfeit+men."&aq=f&aqi=&aql=&gs_sm=&gs_upl=&bav=on.2,or.r_gc.r_pw.,cf.osb&fp=8eeef67bd8da93e5&biw=983&bih=866&bs=1 (accessed May 5, 2011).
100. Sarah van Gelder. An interview with Bernard Lietaer. *YES! A Journal of Positive Futures*, Spring (1997):10–12.
101. Beaton and Maser. *Economics and Ecology*.
102. Caroline Myss. *Anatomy of the Spirit: The Seven Stages of Power and Healing*. Three Rivers Press, New York, 1966.
103. David C. Korten. Money versus wealth. *YES! A Journal of Positive Futures*, Spring (1997):14–18.
104. Susanna Kim. Poll: Americans need $150k income to feel rich. http://abcnews.go.com/blogs/business/2011/12/poll-americans-need-150k-income-to-feel-rich/ (accessed December 9, 2011).
105. The foregoing discussion of the difference between money and wealth is based in part on Korten. Money versus wealth.
106. John F. Kennedy. *Profiles in Courage*. Harper & Row, New York, 1961.

Section II

When the Decision Making Becomes Self-Serving

While our citizens cry out for government action to make their cities and lives livable, for an imaginative assault on the new social ills that are eroding the quality of their lives, the national leadership continues to over-emphasize economic growth at the expense of both the environment and a more natural pace of life, increasing productivity at the expense of job satisfaction, technology at the expense of human spirit, bureaucracy at the expense of more imagination and more public participation in government, and individualism at the expense of a sense of community.

—New Zealand Values Party*

Only when the last tree has died, and the last river has been poisoned, and the last fish been caught, will we realize we cannot eat money.

—Cree Indian proverb†

* New Zealand Values Party. *Beyond Tomorrow 1975 Values Party Manifesto*. Wellington, New Zealand. 1975.

† North American Indian wisdom. http://www.unitedearth.com.au/tipiwisdom.html (accessed February 4, 2011).

3 The Stages of an Environmental Agency

INTRODUCTION

Please keep in mind as you read the text that what we are doing to our environment in the name of "management" is really a peek into our human inner workings, the cause behind our environmental crises brought about largely by our institutionalized resistance to change. It can be summed up as "the need to know and the fear of knowing." Whether we like it or not, we now have the technological capability to disarrange and disarticulate the entire global ecosystem. Thus, for the survival of human society as we know it, we must face ourselves, as uncomfortable as it may be, and we must help one another to confront our human failings and our blind spots. We must be willing to risk changing our thinking and our behavior and get back in touch with our repressed feelings, our exiled intuition, and our lost wisdom.

Although I, as an individual, cannot change what is written in history, I can change myself and thereby influence what may be written in history when the present becomes the past. In all likelihood, I cannot change overnight. But, like the man who eventually moved the mountain by carrying away small stones, I can begin.

Change, an immutable law of the universe, is an ongoing process that is always in present tense. I am, therefore, writing about the power of the individual human being; the power of choice; the power of decision making; the power to effect change in any arena of life, be it personal, a community, agency, corporation, or government. Each and every one of us takes our self with us wherever we go. We are therefore the common denominators—the threads running throughout the entire tapestry of our social structure. The irony is that, as a species, we must actually study ourselves so that we can learn how to survive with and—at times—in spite of ourselves.

If my interpretation of what I observe is correct, that an individual is an extension of the family and, through a collection of individuals, the community, agency, corporation, and government are also familial extensions, then society is an extension of the community, agency, corporation, and government. As such, an individual who is willing to change can change society. If this is true, then the dynamics in a family are similar to the dynamics in the community, agency, corporation, and government. An understanding of one may help us to understand the other because each is composed of individuals like you and me—each blessed or hampered by the wisdom of our choices.

Here, it is necessary to keep in mind that a negative word is but the manifestation of a negative thought, and once spoken, it can never be withdrawn, despite an apology, because words are but the public extensions of our private thoughts. By our thoughts, we privately define and by our actions we publicly declare who and what we are.

According to author George Crane, "Language is the apparel in which your thoughts parade before the public. Never clothe your thoughts in vulgar or shoddy attire."[1]

I write these pages with the hard-won realization that, of all species on Earth, we humans are both blessed and cursed with the greatest of powers—the power to consciously change ourselves, to struggle toward an ideal of being, and to frequently fall short of that ideal. In struggling, however, we must understand and remember that anything worth doing well is worth doing poorly for a while. That said, our human ability to create is powerful because we create and re-create ourselves through our thoughts, and then, with those thoughts, we create the society and the environment in which we live in relative comfort, just survive, or become extinct.

We have control over what we choose to think and do. The outcome of our choice and decision to act is therefore our responsibility. Thus, it is within our creative power to change ourselves from collectors of the society's negative, psychological garbage into trustees of one another's dignity, and it is within our power to transform the world from its growing degradation into a viable, biological living trust for all generations.

Today, we are at a cultural/ecological crossroads of social survival, a threshold at which we have never stood before. We must now come to grips with what we face. Fortunately, we, as individuals, can change and move our community, agency, corporation, government, and human society in a new direction.

For the sake of simplicity, I use federal agencies as the focus of this discussion, rather than repeating community, agency, corporation, government, society, and nation over and over. My critique is purposefully focused on government agencies, which in one way or another came into being through directives to safeguard the nexus between society and the natural environment, such as the U.S. Forest Service, Bureau of Land Management, National Park Service, U.S. Fish and Wildlife Service, Environmental Protection Agency, and others. As such, how they function affects every person in the United States—and ultimately the world—for better or worse. In addition, I worked as an agency scientist in the Bureau of Land Management for over a decade and the Environmental Protection Agency for a year, which allowed me to observe the agencies' appropriate and dysfunctional behavior firsthand, both within and without.

An agency is a business or service authorized to act for others. When speaking of an agency, we often use the possessive, such as "the Forest Service's point of view," as though it was an individual, and in a way it is. An agency is a collection of people, much like an extended family. Therefore, its benefits and deficits are the collective contributions of both its past and present members. Making changes in the decisions and actions of an agency, however, requires understanding something about agencies in general.

AN INTRODUCTION TO COMMAND-AND-CONTROL DECISION MAKING

Each agency is an aggregate of individuals, which, through its official doctrine, is a reinforcement of society's perception of itself, just as a family, through its behavioral patterns, is a reinforcement of a community's perception of itself. I never thought about the meaning of an "agency" until I went into the Army when I was 17 years old. My first week was at Fort Ord, California. Having nothing to do, we were given something with which to look busy. I remember a struggling piece of lawn, which should have been granted the Congressional Medal of Honor for valor above and beyond the Almighty's intent for grass. I saw that poor piece of lawn sheared with an old, push lawn mower 10 times in one day; I mowed it twice! In addition, I learned about the chain of command—in other words, command-and-control decision making.

A man with one stripe on his sleeve told me to mow the lawn, and while I was so engaged, a man with three stripes on his sleeve came up to me and told me to do something else. "Sir," I replied, "I've been told to mow the grass."

The man narrowed his eyes and asked through tight lips, "How many stripes did he have? How many do I have? Who are you going to obey?"

My eyes bugged out, and I gulped, "Sir, I guess I'd better do as you say. You have three, and he had only one."

"A damn wise choice, son!" he said as he walked away.

After I got to Fort Lewis, Washington, and started basic training, learning how to kill a potential enemy for the sake of "God and country," I found myself confronted with the greatest proverbial wisdom of mindless obedience the military has to offer: "Yours is not to question why; yours is but to do or die," and "If it moves, salute it; if it doesn't, paint it." And then there was the perpetual hurry-up-and-wait syndrome.

I was also told that, "You're in the Army 24 hours a day, 7 days a week, remember that. We pay you 11 cents an hour 24 hours a day, and you'd better damn well be grateful! Where could you get a better deal than that: 11 cents an hour 24 hours a day, three hot meals, and a bed? But your ass is ours!" I remember thinking: I know a lot of places where I could get a better deal than that, and I'd prefer 22 cents an hour with 12 unpaid hours off the base.

The Army was my first introduction into the collective thinking of a dysfunctional agency. Of course, I did not recognize my dysfunctional family as belonging to the same category. Nor did I think about the school system, which tried to make me into a right-handed person, which tried for 12 years to make me conform to some preconceived notion of acceptability by attempting to steal my individuality and my imagination.

And yet, an agency can serve a useful purpose in society. The problem arises when the agency "forgets" what its purpose is; the purpose becomes obsolete, confused, or confounded; or the agency becomes self-serving to its own survival through command-and-control decision making.

THE INCEPTION OF AN AGENCY

An agency not only gives form and substance to a mutually held ideal during its inception, which is projected into the future, but also gives the individual a sense of participation, achievement, and personal and "family" pride in the furtherance of that ideal, something no one can accomplish alone. Another opportunity an agency holds out to its employees and volunteers is the potential for a "team effort," wherein everyone's talent counts, which makes the whole greater than the sum of its parts.

Although many people tend to be cynical about both public and private agencies, each began with noble ideals of service worthy of their time and place in history and worthy of the money and effort invested in them. Somewhere along the way, however, people can lose sight of themselves and the collective continuance of their ideals, which allows the corporate/political elite to take over, much like the theme of Ayn Rand's book *The Fountainhead*.[2]

I came to this conclusion after working in the Bureau of Land Management (commonly referred to as the "BLM") for 13 years, while housed with the U.S. Forest Service (once the preeminent agency in the world), and thereafter worked for a year in the Environmental Protection Agency. In many ways, they were wonderful years, and I learned much about myself and about society. I also learned that these agencies had become largely dysfunctional monoliths, each of which seemed to believe it had the only "true" answer.

WE ARE THE AGENCY

Before we can see how to change an agency, such as the Environmental Protection Agency, U.S. Forest Service, or Bureau of Land Management, we must understand that nothing about an agency is sacred. An agency is merely a collection of people—as good as the most functional, as bad as the most dysfunctional, and as mediocre as the average. Further, our individual motives and conduct—the carryover from life in our families of origin and whatever experiences we have gleaned from living—give the agency its meaning, purpose, integrity, foresight, and the credibility ascribed to it. In essence, an agency is an extension of our families of origin in that our reactions to our coworkers and authority figures are often symbolic not only of whatever stage of personal development we are in but also of the dynamics of our relationships with significant members of our own families, especially in relation to unresolved issues.

Whoever works in an agency sets the agency's standard, determines the agency's budget, oversees the agency, and *those citizens who vote* all constitute the agency through the selection of decision makers. Because we are the agency, we also share in its destiny, and because it is within our power to be authentic individuals, it is also within our power to heal dysfunctional behavior.

STAGES IN THE CYCLE OF AN AGENCY

I am not aware of any reference that describes the developmental stages of an agency as has been done for an individual and a family, and I have not taken part in the formation of an agency. But, I can think of four generalized stages of development.

STAGE ONE

The inception of an agency is based on a perceived need that is in the public interest. This perception revolves around one person or a small nucleus of people with a vision, as clearly stated by Gifford Pinchot in his book, *Breaking New Ground*. He saw the "Conservation policy," which he helped to forge, as the guiding principle of the U.S. Forest Service, of which he was the first chief from 1898 through 1910:

> The Conservation policy … has three great purposes.
>
> First: wisely to use, protect, preserve, and renew the natural resources of the earth.
>
> Second: to control the use of the natural resources and their products in the common interest, and to secure their distribution to the people at fair and reasonable charges for goods and services.
>
> Third: to see to it that the rights of the people to govern themselves shall not be controlled by great monopolies through their power over natural resources.[3]

On February 1, 1905, the same day the Transfer Act was signed, Pinchot received a letter written by Secretary of Agriculture James Wilson. This letter, based on the Organic Act, authorized the secretary of the interior to formulate such rules and regulations as necessary for the proper management of the forest reserves. The letter came not as surprise, however, because Pinchot had drafted it; he stated:

> In the administration of the forest reserves it must be clearly borne in mind that all land is to be devoted to its most productive use for the permanent good of the whole people and not for the temporary benefit of individuals or companies. All the resources of the reserves are for use, and this use must be brought about in a thoroughly prompt and businesslike manner, under such restrictions only as will insure the permanence of these resources … and where conflicting interests must be reconciled, the question will always be decided from the standpoint of the greatest good for the greatest number in the long run.[4]

Yet, even with the clearly perceived need for a guardian of the public interest, the formation of the Forest Service was no easy task. There were many bitter, political battles to be fought with men who wanted all the land put in private ownership for their own personal gain, the inevitable birth pains of that which we today take for granted as our national heritage—our *national forests.*

Although the ideals, which are seminal in the inception of an agency, may be clearly defined for their time and place in history, such as those of the Forest Service, we now look back and wonder exactly what was meant. In this sense, I have heard the following question raised several times in recent years, as we struggle to meet today's perceived "needs" from our national forests. What exactly did Pinchot meant by "the greatest good for the greatest number in the long run"?

We the people may not agree with what we perceive his motives to have been, as attested by the political actions taken by contemporary decision makers and leaders, but we must remember that whatever the ideal with which the Forest Service was founded, whatever actions were taken to implement that ideal, it was new and daring in its time, and it was meant to be a service held in trust for all the people—both present *and* future as part of our *national commons.*

We can today easily snipe at the "simplistic vision" of the pioneers of the past because we have greater biological knowledge about our forests, a different perception of desires and necessities from our forests, and a different perception of ourselves as a society. But, it is important to remember that those pioneers did the best they could with the knowledge and the vision they had, and that today we are the pioneers of the future. Are we doing the best we can with what we know? Will the agencies in which we now serve or which we now create fare any better than those of the past?

Stage Two

After its inception, an agency goes through a period of false starts and apparent fumbling, often with much political infighting. Such behavior is much the same as an adolescent searching for an identity. Out of this fumbling can come growth, a coming together to fulfill the vision—provided the vision is clearly stated and firmly agreed on.

Pinchot was a great and electric leader by any standard. Stewart Udall called the Forest Service's Washington headquarters during Pinchot's regime "the most exciting place in town." And it was, as Pinchot and President Theodore Roosevelt successfully conspired against private western ranching, timber, and water interests to set aside 148 million acres, three-fourths of today's system, as national forests.

This is the excitement of growth not only in a project with a clearly articulated vision, goals, and objectives but also in an agency where the people are clearly committed to and empowered to follow the vision of public service for which the agency is being created. "Ideals." said journalist Carl Schurz, "are like the stars, we never reach them, but like the mariners of the sea, we chart our course by them."[5]

Stage Three

Growth, with the stimulation of its unknowns, its gropings, its many false starts, and surprising insights eventually gives way to maturity in which the outcome seems ensured. The first few years are commonly marked by a struggle to put the ideas coherently on paper. The pieces then begin coming together, slowly at first and then faster and faster. People from every necessary discipline seem to materialize "out of thin air" just when they are needed most, which keeps the interest charged and the enthusiasm crackling.

At some point, people know where the agency is going and what the product will be. At that point, the agency comes into its maturity, and everyone seems to sense it. There are functions to perform, and the agency begins to accomplish its stated mission (fulfill its function). Once this has been accomplished, a turning point has been reached. Thus, the beginning of Stage 4 is pivotal to the life of an agency for here its direction is either clearly determined or becomes more and more confused.

Stage Four

Having fulfilled its original charge, it must be continuously reenvisioned, reoriented, rechartered, reinvented, and revitalized or *disbanded* as times and perceptions change. If left solely to its status quo, senescence creeps in. At that point, the agency begins a slide toward dysfunctional behavior. In other words, for some reason or other, the agency becomes increasingly unable to achieve its stated mission.

The now-declining agency tries to "hang on," to live caught by a clearly defined mission, as in the past, which is becoming less viable or more confounded over time. In reality, however, it has become a self-perpetuating machine, which begins to flounder because its original function has been dropped or deemphasized and not replaced by a newly defined mission.

WHEN DYSFUNCTIONAL DECISION MAKING CREEPS IN

We, as individuals, represent the behavioral strengths and weaknesses of our upbringing and experiences, and we tend to repeat our behavioral patterns—be they functionally appropriate or dysfunctional—over and over again unless we choose consciously to break the cycle.

There are three basic ways in which an agency becomes dysfunctional. The first is when a person takes over as a command-and-control dictator and pays mind solely to vested interests at the expense of the agency's mission, its people, and the public at large. If you have never experienced real command-and-control decision making, here is a graphic example:

> Daily NK [North Korean online news] learned from a source from North Hamkyung Province on January 10th [2012], "The authorities are handing down at least six months in a labor-training camp to anybody who didn't participate in the organized gatherings during the mourning period, or who did participate but didn't cry and didn't seem genuine."
>
> Furthermore, the source added that people who are accused of circulating rumors criticizing the country's 3rd generation dynastic system are also being sent to re-education camps or being banished with their families to remote rural areas.[6]

The late Hwang Jang Yop, a North Korean defector and former official, described a similar response after the death of Kim Il Sung, Kim Jong Il's father and the founder of North Korea, in 1994.

> "The party conducted surveys to see who displayed the most grief, and made this an important criterion in assessing party members' loyalty," Yop wrote. "Patients who remained in hospitals and people who drank and made merry even after hearing news of their leader's death were all singled out for punishment."[7]

The second an agency becomes dysfunctional is when people in an agency, created for one purpose, try to make the agency into something it was not designed to be (i.e., change its legally defined function). The third is when an agency has outlived its original purpose (function) or has simply lost sight of it and becomes a self-serving, institutionalized machine committed to the perpetuation of its own survival.

In any case, the agency no longer serves the public it was designed and empowered to serve but instead compels the public to serve the survivability of the agency. Here, it must be remembered that individuals are the agency. Therefore, the more dysfunctional the leadership's decision making is, the more dysfunctional the agency becomes; the more dysfunctional the behavior of rank-and-file members is, the more dysfunctional the agency becomes. Although the functionality of an agency is a reflection of the functionality of its individual members in the collective—especially the leadership—there is more to it than this statement suggests.

Complex organizations, such as the U.S. Forest Service, the Bureau of Land Management, or the Environmental Protection Agency, are typically composed of people that fall into three major categories, each prone to becoming subcultures:

1. Politicians—who expect their futures to lie in appointed or even elected office and therefore concentrate on maintaining and expanding their career opportunities outside the agency, but these days they are often appointed as heads of agencies about which they know little or nothing.
2. Careerists—whose rewards come largely from inside the agency of employment. They align their primary goals and loyalties within the agency's hierarchical career ladder in a way that maintains the stability of the agency and elevates their own position within it.
3. Professionals—who relate to norms set by others in the same profession, say hydrology and wildlife, norms that generally cut across organizational boundaries. A professional's rewards come from peers within a professional society, not from within the agency of employment. The professional role is seldom the most powerful in an agency and is often misused by those in power.

Although an individual often plays two or more roles in daily working life, one is dominant. So, it is not surprising that people in these three categories within a single agency often come into conflict when they are brought together to deal with a complex, integrated issue because they each belong to a different subculture with different supporting structures and different systems of ideals, rewards, incentives, and thus hierarchical, decision-making authority. This is especially true if an agency tries to shift its identity in a way that also shifts the perceived basis of power from one subculture to another.

Adding to the confusion is the fact that, while an agency's formal goal may be sound forestry, range "management," or social work, its informal, operating goals may differ substantially among the three subcultures based on differing senses of idealism, integrity, and self-interest. Thus, each subculture not only perceives a given situation differently but also defines the problem and envisions its solutions differently.

All of the many professionals I have known over the years, including myself, want to keep two professional norms inviolate: *personal integrity*, which is a sense of authenticity based on sound, sincere, personal and professional principles; and *professional integrity*, which means dealing with biophysical issues with as little political bias or prejudice as humanly possible. The best of these professionals learned how to operate effectively within an agency directed by those of other subcultures. The way an agency is organized and the attitude of agency managers (= decision makers) can either encourage and nurture professionalism or abuse and stifle it.

It takes more than cooperative behavior and a sense of goodwill to protect integrity. Such cooperation and goodwill must be nurtured by the appropriate organizational design, one that is "safe" for dissenting views and interpretations. In addition, the organizational design must support and nurture integrity and its creative source, as well as differing ideologies, through positive rewards and incentives. However, when final decisions are rendered, it is the responsibility of professionals to support the decision, remain silent, speak out and put their job on the line, or resign. "Leaks," sabotage, and sniping after retirement are not part of professional behavior.

Finally, the U.S. Forest Service, now over a century old, is an example of a public agency that has been deprived of its original charge. It has become increasingly dysfunctional due to a flood of new, often-conflicting laws, the rise of the environmental community, lack of clear direction from Congresses and administrations, and more and more involvement by the courts. It did not have as much to do with "exploitive capitalism" as it had to do with the growing environmental consciousness of the American people concerning the long-term consequences of exploitive capitalism, brought to the fore through the democratic form of government.

If exploitive capitalism were the driving force, the Forest Service would still be cutting 13 billion board feet of timber per year instead of less than 3 billion board feet of timber per year. The "driving force" for today's Forest Service is—oddly enough—fire suppression, recreation, and wildlife (especially threatened and endangered species, which led to the age of so-called ecosystem management). This shift points out that our present form of exploitive capitalism is not working for the good of the American people either, especially those who will need healthy forests in their future. When a dysfunctional agency serves the special interests, who is it that serves the biophysical sustainability of the forests and social-environmental sustainability of the people of the United States, people who by and large have no vested profit motive in exploitation but rather are dependent on the potable water supplied by forested water catchments?

The same question can be asked about a dysfunctional Bureau of Land Management, which, by Congressional directive, serves the special interests of a small portion of the livestock, mining, and oil and gas industries with attendant impacts on public rangeland ecosystems. Again, the American people as a whole have no vested profit motive in grazing livestock, mining, or drilling for oil and gas on public lands, but they do have an increasing need for open space and non-polluted water catchments, both of which are shrinking resources.

Dysfunctional behavior tends to creep into an agency when the production of a product, coveted by a politically strong special-interest group, becomes more important than the process of human interactions that fulfill the agency's original charge. In other words, the product is more important than people, human dignity, or trusteeship of the land itself. The sole focus becomes timber, forage, minerals, oil, gas—or whatever else—at any cost, regardless of the long-term, environmental degradation such myopia causes future generations to increasingly pay for.

What happened? Rather than recognizing that the people of the nation are both the agency's bosses and its customers and therefore treating them accordingly, the agency bows to moneyed interests through political pressures and serves the corporate/political elite—for better or worse:

> Money and politics go hand-in-hand. Politicians need money to run campaigns. The rich have ideological beliefs just like other Americans. But their fortunes afford them an oversized influence in the political process. Once they acquire their billions, these heavy hitters tend to focus on changing the U.S. political climate and influencing global issues, either by using their clout to steer policy debates or using their money to back parties and candidates.[8]

In fact, 47% or 249 current members of Congress in both the U.S. House and Senate are millionaires, yet the American taxpayer funds their salaries, according to a new study by the Center for Responsive Politics. The study is based on the center's analysis of the lawmaker's median values as shown in financial disclosure forms covering their personal assets and liabilities for calendar year 2010. Thirty-six Senate Democrats, 30 Senate Republicans, 110 House Republicans, and 73 House Democrats reported an average net worth in excess of $1 million in 2010. Clearly, they are part of the 1% of American citizens who enjoy this financial status.

"The vast majority of members of Congress are quite comfortable, financially, while many of their own constituents suffer from economic hardships," said Sheila Krumholz, executive director of the Center for Responsive Politics. "It's no surprise," Krumholz continued, "that so many people grumble about lawmakers being out-of-touch. Few Americans enjoy the same financial cushion maintained by most members of Congress—or the same access to market-altering information that could yield personal financial gains."[9]

Again, our political system—as it currently functions—serves the private power base, not the people. Thus, enemy number one to a dysfunctional agency is the very public it is meant to serve—a public that requires a healthy, sustainable ecosystem to produce nature's free services (such as clean air; good-quality water; and clean, fertile soil) if the people are to enjoy their birthright of a quality of life worth living.

When such dysfunctional behavior is pointed out in news stories and court cases, the "views and policies" of an agency become the backbone of the agency's homeostatic defense, a defense that is usually upheld by the courts of the land if legal procedure has been followed—regardless of any negative, long-term environmental consequences.

THE HOMEOSTATIC COMMAND-AND-CONTROL OF DECISION MAKING

Homeostasis (from the Greek *hómoios*, "similar"; and *stásis*, "standing still") is not only the attempt to maintain a desired equilibrium of control within a system, such as a family, but also a mechanism through which familial dysfunction is hidden and the family's name kept pure. A family is a system governed by a set of rules, which determine and control the interaction of the family members in organized, established patterns. The family rules are directives of what shall and shall not occur within and without the family. In turn, homeostatic mechanisms maintain the ongoing arrangement among members of the family by activating the rules defining each member's relationship to the whole. As with a family, homeostasis in an agency is an attempt to maintain a desired equilibrium through top-down, command-and-control decision making to skirt inconvenient legalities.

Once an agency's function has become inappropriate, it begins to go through institutionalized rituals to ensure its survival and to hide its dysfunctional behavior. And, like a family, an agency is a system governed by a set of rules that determine and control the interaction of its members in organized, established patterns. As Andrew Carnegie once said, "The organization is the shadow of the man."

In today's agencies (from the office of mayor to the presidency of the United States), those ideas contrary to the established view are termed *heresy*, and those ideas implicitly supporting the established view are termed *policy*. Policy is therefore the prescription for directives of what shall and shall not occur within and without the agency, and homeostatic mechanisms help to maintain the ongoing arrangement among members of the agency by activating the rules defining each member's relationship to the whole. Homeostasis is thus a mechanism through which an agency's inner workings are kept hidden from "outsiders." This is a very important concept.

To illustrate, the Environmental Protection Agency, like any other agency, is but a collection of people with special talents who have been brought together to carry out the functions assigned to them by law and budgetary direction, thereby theoretically serving the public—not to make the public serve them. For a blindly "loyal," unthinking, unquestioning agency person, however, this is an intolerable point of view because the homeostasis of the agency is always in jeopardy from the unknown, impure, questioning, and therefore potentially dangerous actions of a thinking "outsider" who is "within" the agency.

Such concern is reasonable, to some extent, in that to function, even within its original charge, an agency must work as a number of teams and subteams that are in constant flux. One uncaring maverick can disrupt the whole effort. On the other hand, homeostasis within an agency can hide inappropriate behavior.

Like members of a dysfunctional family who have their roles assigned so the family's "good name" will be safeguarded, an agency that functions inappropriately becomes a self-serving machine that can define more and more narrowly and rigidly the job descriptions of its employees—and thereby control its employees. In this case, the description of an employee's job, which is available to the public, directs the employee to do something as an employed professional, but the agency, through its homeostatic equilibrium, is such that the employee must be prevented from carrying out an assignment in a 100% professional, questioning manner if the agency's deception of the public is to be kept under wraps.

In the Bureau of Land Management, for example, Mary, a hydrologist, and Joe, a forester, each has been given work assignments. Mary is to determine how to balance the water requirements of young salmon and a local community that rely on the same water catchment, whereas Joe is to determine how to maximize the amount of timber that can be cut and removed from the water catchment as part of the agency's timber removal target.

Both Mary's and Joe's reports must be done independently. They are not allowed to work together or craft any kind of compromise that would best serve both the salmon *and* the community. In other words, Mary and Joe are to limit their performances strictly within the narrow confines of their job assignments within their job descriptions—rigid, mental walls that prevent them from either pondering or looking beyond their assignments—if they want to survive with a good job rating.

With the completion of their assignments, both Mary and Joe are instructed to give their individually accumulated information to their superiors in the form of written reports. Once this is done, it will, in all probability, be their last connection with their data. They will have no idea if or how their data are used in the decision of when, where, or how much timber will be cut in that particular water catchment until it is done. By then, however, both Mary and Joe will have been immersed in several other assignments. Thus, instead of finding their jobs, through their job descriptions and their day-to-day working relationships, to be professionally enriching and ennobling, both Mary and Joe have fulfilled their assignments in dehumanizing isolation not only of each other but also of the outcome and all the generations it will affect.

Few people who have the courage to stand up and speak out for what they feel is right are rewarded for their honesty and their loyalty to their professional ideals because preventing these kinds of "leaks" is what homeostasis is all about. In some cases, the homeostasis fails, and change is affected. When this happens, the truly loyal people die a slow, unknown "professional death" for their integrity and courage as human beings and professional employees.

Moreover, I was told in the Army, in the Bureau of Land Management, and in the Environmental Protection Agency to beware of my behavior because I was serving them 24 hours a day, 365 days a year. Unknowingly, with the stroke of a pen, I ceased to be an individual and became an expendable, numerical item of government "property" (as I was once told) to be controlled and used as deemed necessary to further the agency's purposes. Where is the line, the boundary between a human being as an environmental trustee and the broken machine?

Thus, knowing nothing consciously about either functionary employees or homeostasis, I found out the hard way that not everyone in the agencies wanted new or better scientific data with which to care for or monitor our forests or other public lands. I was once told by a chief of information and education in the Bureau of Land Management that he did not want me to make a particular videotape or to publish a particular book about our scientific findings because it would make the agency look bad. I did not comply. And the forestry staff in the Washington, D.C., office of the Bureau of Land Management tried for 3 years to quash the publication of one of our reports because our data were perceived to be adverse to the bureau's unstated liquidation policy for old-growth forests in western Oregon and western Washington. Again, the report not only got published after a 3-year political battle but also has been reprinted several times and is today in libraries in more than 70 countries worldwide.

Even at the highest, national level, one must be aware that unconscious, narrowly focused people in the U.S. Congress, lacking a sense of personal or professional boundaries, often become the role models for a dysfunctional agency. When, for example, a senator or representative, trained as a lawyer, makes ecologically uninformed decisions about forests, rangelands, national parks, or oceans, it is about the same as a forester, range conservationist, park ranger, or oceanographer making equally uninformed legal decisions about courtroom etiquette. The difference is that the lawyer-senator will not even consider the forester's comments, but the forester is supposed to obey the lawyer-senator unquestioningly.

When everyone in an agency performs this way, the homeostatic control is transformed into command and control by the person with the most to gain from compliance with the wishes of the corporate/political bosses who speak through the senators and representatives in the U.S. Congress. In this way, the Bureau of Land Management, Forest Service, Environmental Protection Agency, to name a few, no longer serve as guardians and trustees protecting the nexus between the environment and human society as they were originally designed to do but rather are made to serve the corporate/political bosses through certain senators and representatives who, in turn, have forgotten whom they are supposed to serve in our democracy.

The use of job descriptions as an isolating mechanism is also one of the ways an agency protects its members from "unfavorable" information. Institutionalized, internal policies of the agency shape perceptions and beliefs of the uncritical employee in ways that protect the agency's dysfunctional behavior—even when catastrophic outcomes are involved.

A distortion of information is not limited to willful deceit on the part of an individual, who is perceived as loyal to the agency. Even honesty within an inappropriately functioning agency is insufficient to prevent the widespread distortion of information. The weakness lies with an individual's acceptance of a life that involves completing one's assignments without thinking critically about or questioning the consequences—especially the long-term, social-environmental consequences of narrowly focused decision making.

Granted, this behavior hardly sounds untrustworthy, much less dangerous. Nevertheless, it is just this "functionary" behavior that allows systematic, homeostatic distortions to occur. And, this is why decisions are often difficult to deal with in public land management agencies, regulatory agencies, or even local, governmental institutions—because one is seldom sure exactly who makes the decisions. Why? This is because, afraid of taking a risk, almost everyone makes sure that more than one person is involved in a decision, especially when the responsibility for a miscalculation must be assigned to someone. Therefore, inappropriate decisions "seem to just happen" when an agency distorts unfavorable data affecting a decision to further its own, self-serving ends.

Moreover, performing an assignment that has social-environmental consequences (simply taking orders without questioning them) is personally safe and environmentally and socially risky. On the other hand, it is often personally risky—if you want to keep your job—to question orders, but questioning the orders that do not *feel right* is both environmentally and socially responsible.

Most professionals in land-management and regulatory agencies, as well as many other governmental institutions, are told what level of professionalism they will practice if they want to stay employed. Therefore, people often—*but not always*—trade their dignity and professional ethics for the security and "safeness" of their jobs.

A person thus becomes an unconscious, closed-minded functionary by limiting his or her inquiries only to questions of how best to accomplish an immediate assignment, thereby turning his or her mind over to the agency and allowing the agency to shape personal perceptions according to its homeostatic needs. The fault, therefore, lies neither in the job description nor in the assignment. The fault lies in not accepting personal responsibility for the outcome of the assignment. At this point, the average employee under a command-and-control agency head becomes afraid of losing his or her job and thus pleads impotence to decision making in one form or another, which is the mythology of abnegating personal responsibility. This is the germinated seed of dysfunctional human behavior, which is the birth of the machine through the loss of identity, individuality, courage, and human dignity.

THE MYTHOLOGY OF ABNEGATING PERSONAL RESPONSIBILITY

Although mythology as a fundamental frame of reference for how to lead one's life is variously construed, it is here meant as an intellectual fabrication used to justify existing in one's fear of change, rather than fully engaging life, which entails a measure of risk. The mythology of abnegating personal responsibility, like a chameleon, assumes a number of guises, of which seven are discussed: (1) I can't change because I'm locked in; (2) I can't commit future leaders to a course of action; (3) it's not my responsibility because it's not in my job description; (4) I can't make a decision because I lack definitive data; (5) it can't be true, so I won't believe it; (6) yes, but I have to face reality; and (7) what you are asking can't be done—it's impossible.

Myth One—I Can't Change Because I'm Locked In

A common myth of being stuck, which has been around for many years, is the notion of being locked into a certain position or circumstance in life, of being out of control, and thus unable to change one's current existence. The truth, however, is that we, each and every one of us, always have a choice, that no one is "locked" into anything, that change is always an option. Therefore, a person of psychological maturity will examine first and foremost the opportunities presented by an impending change, be they personal growth or material gain, and will weigh the associated risk accordingly, whereas a psychologically immature person will focus first and foremost on the perceived risk of losing what they already have and so decline the opportunity presented, no matter how good or important it is intuitively known to be. Thus, as British philosopher James Allen observed: "Circumstance does not make the man; it reveals him to himself."[10] Or, as French-Cuban author Anaïs Nin wrote: "We don't see things as they are. We see things as we are."[11]

Some years ago, for example, a man, I will call him "Bob," worked for the Environmental Protection Agency, as did I. Bob, who originally worked in a different state with a different government agency, had been sold on the job with the Environmental Protection Agency by a friend, and he took the job only to find out that it was neither ethically planned nor ethically administered. Although Bob could have gone back to work for the agency he left, where he had felt good about what he did, he said that he was "locked in" to his new job despite his better judgment, and that, when he allowed himself to think about it, he felt betrayed, miserable, depressed, and dishonest.

When I asked why he did not go back to his original agency, which had gladly offered him his old job, Bob said it was too expensive to move again, that he had just gotten his family settled, that he was just learning the ropes of his new job, which he hoped might get better, but he did not see how it could. Finally, he said it was not fair to let his friend down, while steadfastly refusing to acknowledge that his friend had duped him. He had a litany of reasons that sounded good, but when really pressed, it turned out that Bob found it easier to stay, where he knew in his heart he did not belong, rather than risk the potential ridicule of changing his mind and admitting, by going back to his old job, that he had made a terrible mistake.

True, his return would undoubtedly have caused short-term hardships, but it would have earned him his self-respect and the inner peace of feeling good about what he did to earn a living. Instead, Bob prostituted his fundamental beliefs to avoid the short-term pain of taking personal responsibility for what turned out to have been an unwise decision. In so doing, he paid a much higher personal cost over a much longer period of time than had he returned to his old job.

Another spin on this myth is that the "terrible known" is more comfortable than the unknown, even when one can clearly see that it promises to be better. How often I have heard someone say, "I can't change jobs, even though I know I must do something else because I'm no longer fulfilled by my job or doing it justice, but I only have a few years left until I can retire." The real question is: How many years of misery are you willing to accept rather than experiencing personal growth, joy, and fulfillment by risking change?

Myth Two—I Can't Commit Future Leaders to a Course of Action

When I worked as a research scientist in the Bureau of Land Management and later served as an advisor to county government in my home county, I was told by more than one person faced with an uncomfortable decision that he could neither speak for nor commit future leaders to a particular course of action, that it was not fair to "lock them in." In this way, they sought to avoid the risk of making a decision that would be unpopular with the people they feared might have the political power to turn them out of office.

Despite one's personal trepidations, some decisions (which in fact act as a "lock and key" to protect the future) must be made in the present moment, such as the legal acts that authorized the national parks and wilderness areas, created the national forests, and currently protect endangered species. If the majority of the people responsible for the passage of these acts had not had the individual courage embodied in psychological maturity to act for the benefit of all generations, despite fierce opposition, our nation and all its people would indeed be culturally poorer today, while a very few individuals would have made substantial amounts of money.

Myth Three—It's Not My Responsibility Because It's Not in My Job Description

It is quite common, I have found, for people afraid to make a decision to rationalize that the letter of the law, the letter of their job description, must be followed at any cost, rather than embrace the heart or intent of either and risk making a conscious choice based on their moral judgment of extenuating circumstances, social responsibility, or the greater good of humanity. And, make no mistake, all decisions are based on a sense of cultural morality because we humans are subjective creatures; we cannot be otherwise. If you think you are or can be objective, try holding a neutral thought in your mind for 1 minute.

Myth Four—I Can't Make a Decision Because I Lack Definitive Data

We will never have enough data, let alone definitive data, that translate into all the data we desire to make an entirely safe decision. But, not to make a socially responsible decision is still to make a decision, albeit one that usually proves to be unwise for all generations. For those who suffer interminable labor pains while giving birth to a decision, I point out that, in the end, there are but two choices—too soon or too late—because virtually all data are inconclusive. Generally speaking, however, I find that *too soon is better* than too late.

On the other hand, claims of not having definitive data or enough data to warrant a change has long been used by the timber industry, among others, to justify business as usual. As a research ecologist in forestry, an expert witness in both the U.S. and Canadian courts of law, as well as testifying before the Canadian House of Commons, I have encountered this argument endlessly from industrialists. The argument went something like this: We don't have enough data to prove conclusively that we need to change the way we do business; therefore, we won't change because it would introduce economic uncertainty into our business and cost us too much. If, however, *you can prove definitively* that change is necessary, we will consider it.

Although the statement sounds reasonable, conclusive proof is, of course, impossible, especially if one does not accept scientific data that go counter to one's desired outcome. This refrain is therefore played like a broken record, regardless of the quantity of data on hand that demonstrate the ecological necessity of change to ensure, as much as humanly possible, a sustainable future for all generations—witness the denial of global warming based on a lack of definitive evidence, despite measurable melting glaciers and rising sea levels worldwide.

But, not all people plead ignorance because of a lack of data to avoid making a responsible decision. I once sat next to a contractor on a flight from Alaska to Oregon. Knowing nothing about building a house, I asked him how he did it.

"Well," he replied, "I build each house at least a hundred times in my mind before I purchase the first nail or board. That way, I see the problems before they arise and have figured out how to fix them with the least cost and loss of time. I'm the only person who really 'knows' the house. I know it better than the owners ever will, even if they live in it for 50 years!"

Although the contractor did not have perfect data, he did the level best he could with the data on hand, and he took responsibility for his work. It was, after all, his identity as a person and an artisan that went into the construction of each house, and his integrity meant more to him than the money he could make by using cheap materials and cutting hidden corners.

Myth Five—It Can't Be True, So I Won't Believe It

When one refuses to accept data, no matter how clearly valid they are, one is steeped in an interesting dichotomy—the need to know and the fear of knowing, which can be thought of as "informed denial." In this instance, a person gathers all the data possible, always hoping the data will affirm the cherished point of view, while simultaneously rejecting out of hand any unfavorable data by denying or refusing to believe the data's scientific validity. To give this notion a human face, I know a man whose refrain to anything that threatens his point of view is, "I'm skeptical." With this statement, he summarily dismisses whatever he finds to be uncomfortable.

Another colleague of mine used a slightly different approach. He simply went through his professional life ignoring whatever he did not want to deal with on the theory that if something was ignored long enough, it would go away, including people.

In addition, I once knew a wildlife biologist who worked for the Oregon Department of Fish and Game (as it was known at that time). He was perhaps the most extreme example of informed denial. His professional responsibilities included looking out for the welfare of a herd of elk that used parts of two counties as its habitat. Scientists within the same department studied this particular herd of elk across its geographical range, but the biologist would *not* accept any data as valid from the neighboring county if it posed for him an uncomfortable decision. This is known as the "NIH factor," which means *not invented here, thus invalid.*

Informed denial is perhaps the most rampant myth when it comes to avoiding the personal risk of making an unpopular but socially responsible decision. I have found this myth in every conceivable bureaucratic closet in every level of government in the United States. If you doubt the accuracy of this statement, read the newspaper with an open mind and a discerning eye or listen to the news with an open mind and a discerning ear.

Myth Six—Yes, But I Have to Face Reality

"What you say is all well and good, but I have to face reality," or "It's fine to be idealistic, and it would indeed be nice if things could be that way, but the reality is … ." Note how the foregoing statements summarily dismiss the other person's point of view.

Facing reality, as it is put forth to avoid making a socially responsible decision one feels is risky, boils down to staying within the limits of someone else's intellectual, political, or economic "bottom line," which amounts to a mental fence to keep fear and discomfort from trespassing. Reality, however, is what we each make it to be based on the philosophical underpinnings of our individual worldviews. Such views are founded either on the fear of potential loss or on the faith of potential opportunities. Although the choice is ours, the vast majority of people unfortunately elect the former.

With the foregoing in mind, what do you think about this December 30, 2011, news story?

> A federal judge moved Thursday to block California from enforcing its first-in-the-nation mandate for cleaner, low-carbon fuels, saying the rules favor biofuels produced in the state.
>
> The lawsuit challenging the state regulations, which were adopted as part of the state's landmark 2006 global warming law, was filed in federal court last year by a coalition including the National Petrochemical & Refiners Association and the Consumer Energy Alliance.
>
> Fresno-based U.S. District Court Judge Lawrence O'Neill's written ruling Thursday said the low-carbon fuel rules violated the U.S. Constitution's commerce clause by discriminating against crude oil and biofuels producers located outside California.
>
> Out-of-state fuels producers hailed the decision as a win for California drivers.

"Today's decision … struck down a misguided policy that would have resulted in even higher fuel costs for Californian consumers while increasing the cost of business throughout the state," said Consumer Energy Alliance Executive Vice President Michael Whatley.

The board plans to ask a judge to stay the ruling, and appeal if necessary to the San Francisco-based 9th U.S. Circuit Court of Appeals, spokesman Dave Clegern said Thursday.

The rule is "an evenhanded standard that encourages the use of cleaner low carbon fuels by regulating fuel-providers in California," Clegern said, adding that it "does not discriminate against any fuels on the basis of geography."

The nonprofit legal organization Earthjustice, which was not party to the suit but works on climate-related issues, said the ruling was a major setback for the state's aggressive greenhouse gas emission regulations.

"California is leading the way on cleaner fuels and a cleaner power grid, and the state's programs are consistent with federal law," Earthjustice President Trip Van Noppen said. "It is not surprising that the oil industry is attacking these programs, but like previous attacks in the courts and at the ballot box, we expect this one ultimately to fail."[12]

Myth Seven—What You Are Asking Can't Be Done—It's Impossible

While I was still working as a scientist for the Bureau of Land Management, I wanted to hire an extremely well-qualified woman as a plant ecologist to help with some work. I went carefully through all the necessary hoops the personnel department put in front of me. After 6 months, however, the head of personnel told me I could not hire the woman, that it was impossible. When I asked him why, he simply repeated that it was impossible. Finding his answer unacceptable, I went to the state director and explained the situation.

"Ridiculous!" he exploded.

With that, he picked up the telephone, called the head of personnel, and the woman was hired within 15 minutes.

As it turned out, the head of personnel had used inappropriate judgment a few months earlier and had been reprimanded. So, when my request reached him, he was taking no chances. His problem was that, by not acting appropriately this time out of fear of criticism, he would once again get himself in trouble.

By suggesting that the required decision is impossible, one is pleading impotence from a position of authority, thereby seeking to avoid personal responsibility. Yet, when Napoléon Bonaparte was confronted with such a situation, he said, "You write to me that it's impossible; the word is not French."[13]

So, in the end, what are these myths protecting? They are protecting our resistance to change by attempting to hide behind the abnegation of personal responsibility for our decisions. In the process, however, we are perpetuating life in the maw of our fears.

Unfortunately, some frightened employees are judged harshly for being functionaries, judged by people outside the "system" who do not understand the employee's fear and the insidious, pervasive tentacles of the agency's dysfunctional decision makers. There are no enemies—only frightened people who may have shirked their responsibilities as professional employees and have thus forfeited control of their lives to a dysfunctional agency that uses homeostatic, command-and-control strategies to further dehumanize them to maintain itself.

THE UNPREDICTABILITY OF ENVIRONMENTAL LAWS

Rachel Carson lamented, "Contemporary society seldom evaluates the risk of a new technology before it is embedded in a vast economic and political commitment, becoming virtually impossible to alter."[14] And in *Silent Spring*, she says: "This is an era of specialists [symptomatic thinkers], each of whom sees his own problem and is unaware of, or intolerant of, the larger frame into which it fits. It is also an era dominated by industry, in which the right to make a dollar at whatever cost is seldom challenged. When the public protests … it is fed little tranquilizing pills of half truth"[15] because outside control is the terror of any agency—especially a dysfunctional one—never a cure.

Carson's admonishment points to yet another aspect of the homeostasis of our public land management agencies, which has to do with such laws as the National Environmental Policy Act of 1969[16] and the Endangered Species Act of 1973.[17] Both of these laws were passed not only to control abuses of ethics by dysfunctional public land management agencies but also to correct the inappropriate behavior and to protect nature's free ecological services for all generations. (For an in-depth discussion, see *Earth in Our Care: Ecology, Economy, and Sustainability.*[18])

The problem with such laws as the Endangered Species Act is that, to a dysfunctional agency, they are like an unpredictable disease, which can strike at any time and totally disrupt the homeostasis. Take, for example, the ongoing battle over the spotted owl and its habitat—the ancient forest. First, the U.S. Fish and Wildlife Service dallied in listing the owl as "threatened." Once listed, however, both the U.S. Forest Service and the Bureau of Land Management had to pay legal attention to the spotted owl and its habitat requirements under the Endangered Species Act. Both agencies, at the order of the administration in power, did as much as legally possible to ignore this responsibility with respect to their timber sales[19]—especially the Bureau of Land Management. To ignore the law, however, the Bureau of Land Management set up a "spotted owl SWAT team" (a special weapons and tactical team) whose sole charge was to maintain the agency's homeostasis—level of clear-cutting the ancient forest—at any cost, a tactic that took place while I was still employed by the agency.

Unfortunately, not even Congress is immune to such dysfunctional, SWAT team mentality.[20] For example, Republican Senator Mark Hatfield, who served constituents in Oregon, is quoted in a 1988 United Press International article as saying (with respect to listing the spotted owl as a threatened species), "It is an attack on all Oregonians, because they threaten our jobs, our roads, our schools, our ability to meet Oregon's diversification needs for the future." Despite Hatfield's rhetoric, the effort to ignore the owl's welfare failed and was replaced with the far-reaching Northwest Forest Plan[21] developed by a team of university and agency experts headed by Jack Ward Thomas, the Forest Service's chief research wildlife biologist.

Western senators and representatives often forget that these are *public forests*—belonging to every citizen of the United States—not their own private, industrial forests.[22] Further, court appeals of government actions are the legal right of American citizens on their own public lands, thanks to such farsighted laws as the Equal Access to Justice Reform Act of 2005.[23]

The SWAT team mentality and congressional tinkering with ecologically wise (as opposed to politically correct) laws is founded solely on trading political support for short-term, economic favors based on abysmal ecological ignorance. During my 30 some years as a research scientist, I have repeatedly noticed that at least 90% of what we know ecologically about biophysical capabilities of our forests, ranges, deserts, arctic, and oceans is ignored in their management. In fact, during my tenure as a government scientist, the vast majority of our hard-won ecological knowledge about the way in which our home planet functions was summarily dismissed not only by agency decision makers but also in court, where I was summoned as an "expert witness." Each time, the data were unilaterally dismissed by the judge—data that would have interfered with someone's private interest in short-term profits, political support, or both and that included congressional tinkering.

According to biologists Robert F. Baldwin and Stephen C. Trombulak, "For all of the advancements in our understanding of biological systems and in our tools for predicting the consequences of society's choices, the state of affairs for nature on private lands, where we live and work, has become steadily worse."[24] This situation is not surprising when you consider that, while postwar Europe was strengthening its controls of how land was used, the United States was attempting to stimulate land development through the "free-market," process.[25] Already in 2007, the tendrils of outdoor residential lights along the East Coast, which are of lower wattage than conventional streetlights, festooned myriad areas beneath the tree canopy of heretofore forestland and farmland, which a mere a decade ago were undeveloped. Today (2011), these dimmer lights of housing subdivisions increasingly illuminate the heretofore-undeveloped areas across the rest of North America.[26]

Why, you might ask? The answer lies in our fixation with the "money chase," which makes it difficult to express the basic difference between money and wealth through the symbology of language. Real wealth means the biophysical integrity of the diverse ecosystems on which we rely for social-environmental sustainability is intact. Either not understanding or denying this truth, the monetary interests have prevented the formation of a nationally funded land conservation policy to protect the biophysical integrity of forestlands and farmlands for future generations, a failure that will become increasingly burdensome to those who follow us.[27]

Like all industrialized nations, the population of the United States, which was predominantly rural in the early 20th century, has moved to exurban and urban settings. As of 2008, over half of humanity lived in cities. The number of "megacities" (engines of growing economies with populations over 10 million) grew from 3 in 1975 to 19 in 2007 and is projected by the United Nations to increase to 27 in 2025.[28]

This decentralization of community life has been accompanied by and facilitated by the decentralization of amenities typical of cities. In keeping with this migration and consequent economic boom, transportation systems have been designed around economic efficiency in moving people from here to there rather than being accountable for landscape integrity, which has caused not only widespread fragmentation of the landscapes but also widespread illumination of areas that were formerly dark at night.[29]

Countries, like France and Germany, which have for centuries faced a progressively limited availability of land, have developed wiser land use policies than has the United States. European nations have been making progress since the 1960s with increasingly strict zoning and land use laws aimed at an international "spatial policy." For example, the European Spatial Development Perspective of 1999 has many specific environmental goals, such as reducing fragmentation by transportation networks, as well as accelerated spatial planning across all jurisdictions for development, agriculture, and environmental protection. An early draft states, "The profound changes brought about in the economic and social structures of the countries of Europe and their relations with other parts of the world demand a critical review of the principles governing the organization of space, to avoid their being wholly determined by short-term economic objectives without taking into consideration social, cultural and environmental aspects."[30]

In the United States, however, Congress might pass a law to create a legal principle of reasonable governance of social behavior for the long-term good of society, but then, while the lawmakers want others to obey that principle in their own parts of the country, certain members want to exempt themselves from that self-same edict in their home districts or states. They each want to be a "special case," above and beyond their own law, to protect their own special interests, as noted with the failure to protect ecologically critical lands from development. The danger of such irresponsible shortsightedness is that we are directing the social-environmental course from a position of ignorance, illegality, and irresponsibility. Today, more than ever, the wisdom of Henry Ford is sorely needed: "The highest use of capital is not to make more money, but to make money do more for the betterment of life."[31]

Leonard Abess, Jr., a banker, exemplifies Ford's philosophy—as did Leonard's father before him. The senior Abess started City National, a community bank in Miami Beach, Florida, in 1946, which grew into 18 branches and one of the top-rated banks in the United States.

The junior Leonard remembers, "I grew up at the dinner table with tales of the Depression and tales of what a banker meant to a community, about responsibility, handling other people's money. And that these are your neighbors, the people you live with, you see all the time and you have to protect their money." Guided by these thoughts, he sold the bank at age 60 but remains as chair and chief executive officer. As for the profit he made in selling the bank, he said, "I never ever once thought that this success was, you know, about me. I know what it's about; it's about 400 wonderful, decent people. When I sold the bank, I didn't want the money so I gave $60 million of it to my employees [399 current employees and 72 former ones]. And I feel really good about it."

The values Leonard's father instilled at the family dinner table have led him, albeit reluctantly, into the national spotlight, as a guest of the White House, where President Obama applauded his generosity. "I'm amazed by the attention that this has gotten," he confessed. "I'd prefer to live in a world where this is ordinary and didn't need to be mentioned to anybody."[32]

That said, it is important to keep in mind that the main thing produced in a university, an agency, or Congress, for that matter, is information, which is translated into laws, policies, directives, management plans, and public relations—all with the intent of protecting the integrity of our social-environmental sustainability through decision making in the public interest. Therefore, all "products" are really translations of the information and the system that produces it, as dictated by the relative psychological maturity and personal integrity of the decision makers. The way in which that information is used determines whether an agency retains its appropriate function as a trustee of the public good or becomes a dysfunctional, self-serving entity dependent on human coping mechanisms to maintain its protective homeostasis.

SETTING THE STAGE FOR THE ENGAGEMENT OF COPING MECHANISMS

There is a vast array of coping mechanisms, first deciphered and named *defense mechanisms* by psychologist Sigmund Freud. Coping mechanisms begin as thoughts devised in our families of origin to protect us from that which we deemed dangerous to our well-being. What begins as a thought manifests over time into a behavior when we are confronted with perceived life-threatening (or ego-threatening) circumstances. If the combination of thought and action is successful, we have devised a mechanism for survival, a "coping mechanism," which is reinforced by a psychological feedback loop every time it works as expected. As we become experienced with the mechanism, the thought process behind it fades into the subconscious until the behavioral pattern alone manifests itself. Coping mechanisms thus become the unconscious behavioral devices we use to help us retain, or regain, control in uncomfortable circumstances, as we attempt to cope with a constantly changing environment.

Coping mechanisms, as a strategy for survival, may be functional, positive, and entirely appropriate for a given circumstance when first developed, but they eventually can, and often do, become outmoded and dysfunctional as circumstances change, such as sucking a pacifier to calm yourself when you are 40 years old. Clinging to dysfunctional coping mechanisms in meeting uncomfortable situations can lead to a hardening of attitudes.

Robert T. Lackey, an employee of the National Health and Environmental Effects Research Laboratory, U.S. Environmental Protection Agency, in Corvallis, Oregon, has some profound insights into the politics of issues surrounding social-environmental policies, which he says "are politically contentious, socially wrenching, and replete with scientific uncertainty,"[33] all of which bring out the various coping mechanisms of those involved. Before discussing specific coping mechanisms, however, a summary of Lackey's insights will help to set the stage, as it were.

All problems concerning social-environmental policies share the same general qualities:

1. Their complexity ensures myriad choices and trade-offs.
2. Each engenders clashes among competing values, which guarantees political polarization.
3. Because every choice we humans make is beneficial for some people, detrimental to others, and uncertain for still others, there are winners and losers for every policy choice.

4. Every immediate, symptomatic fix and the benefits of often-painful concessions are seldom evident for decades. Moreover, today's symptomatic fixes almost always become problems of tomorrow.
5. Advocates of cherished positions often appeal to strongly held values and distort or hide the real policy choices and their consequences.
6. The priorities of a given policy often differ substantially among local, regional, and national levels, inciting conflicts.
7. Science is seldom pivotal in evaluating policy options; nevertheless, it frequently ends up serving inappropriately as a surrogate for debates over values.[34]

Lackey goes on to list what he terms "policy axioms," which I shall list as simple statements, each of which clearly elicits the coping mechanisms of those engaged in a policy debate:

1. Every proposed policy choice results in winners and losers, despite the fact that a search for a "win-win" choice, while tantalizing to decision makers, is a non sequitur—even though people running for office may attempt to convince the voters otherwise.
2. Complicating a policy analysis is the fact that one person's benefits may be measured solely in monetary terms but more broadly encompass all desirable things likely to happen. Conversely, the antithesis represents another person's costs.
3. Theoretically, a democracy operates on delegated compromise, which is validated periodically by public voting. To gain sufficient political support (votes) for a proposed policy, it is prudent for a decision maker to advertise its benefits sufficiently to garner majority support. The corollary is that the minority who bear the costs should be as small as possible—yet include all future generations.
4. With many social-environmental policy questions, those bearing the costs, the extant losers, exert a disproportionately greater influence on the decision-making process. Although the analysis of a policy tends to evaluate the rationality of competing arguments, the political process tends to weigh the breadth and vigor in support of competing options. Here, the major drawback is that issues of perceived fairness, although important in the political process, are difficult to assess quantitatively in policy analysis.
5. People with scientific training who are responsible for implementing law or social-environmental policy often have an impulse to insert what they think is, or should be, the appropriate outcome of a public policy—in other words, a value judgment. However, a policy choice is a political judgment—not a scientifically derived decision. The charge of ecologists and other scientists must be to assess the feasibility or the social-environmental consequences from among the choices, which nevertheless is a societal enterprise.
6. The notion that current scientific knowledge is inadequate to resolve social-environmental policy issues is common among both scientists and decision makers. Despite the ubiquitous calls for more research pertaining to debates of social-environmental policy, definitive scientific knowledge

about any or all aspects of an issue is an ever-receding horizon. And, even if definitive data were available, the same debates would emerge in defense of entrenched political positions because policy differences are invariably about personal values and preferences, not scientific knowledge or even observable facts, such as global warming, as demonstrated and quantified by melting glaciers and rising sea levels over the past few decades.

7. In fractious debates over social-environmental policy, proponents often spend more energy demonizing their opponents than sticking to rational policy analysis. Negative images often seem more effective in swaying people than positive ones, much to the frustration of scientists and policy analysts.
8. Although most scientists prefer to talk about things they can measure with some degree of confidence, to policy makers the most important factors cannot be quantified in a credible way. This disconnect, between what is important to policy makers and what scientists can measure, is a reality scientists must recognize—reality not likely to change in the foreseeable future. In a pluralistic society with a wide array of personal values each competing for dominance, the social-environmental policy debate is usually centered around whose values will win in the end rather than over rationality of scientific information.
9. Advocates of various competing choices often argue over semantic nuances rather than actually making decisions. Although the precise meaning ascribed to a word is important, a given word may have different meanings to various people based on their backgrounds, life experiences, and resulting worldview. Therefore, what might appear to be semantic nuances, which many think will lead to a particular policy option, is really a surrogate debate over values and policy preferences.[35]

HOW COPING MECHANISMS INFLUENCE THE EFFECTIVENESS OF DECISION MAKING

A few of the more common, dysfunctional coping mechanisms that infect agency personnel—and so the agency—are discussed, and the personal pronouns I, me, and my are used in a generic sense as illustrative. These are coping mechanisms I have observed while employed by the Bureau of Land Management and the Environmental Protection Agency as part and parcel of the homeostatic defense shield whereby an agency attempts to conceal its inappropriately oriented and top-down, command-and-control decision making to protect its chosen policies.

Anger and Aggression

Anger and aggression are coupled because anger is the emotion, a feeling of hostility, that triggers aggression as an act. Anger is extreme fear that is violently projected outward and is today fed extensively by the media.[36] It is a temporary insanity that isolates us not only from the facts but also from ourselves and from one another.

We are never angry for the reason or at the person or thing we think we are. We are always angry for being afraid of circumstances and therefore feeling out of control, which has nothing to do with the person or thing at which we level our anger.

Unless I fully understand the dynamic, I really think I am angry for the reason and with the person or thing at which I level my anger. I therefore use my anger as a means of not having to deal with the circumstance I am afraid of. Let us suppose, for example, that I am at a public meeting dealing with people from an agency who are presenting a land use issue, and that I am hearing things over which I have no control and with which I disagree. I get angry and start yelling: "You selfish, greedy bastards are out to rape the land, any idiot can see that! All you care about is money!"

In the intensity of the emotion, I feel that I am right in projecting my anger to those who seem to be in control, those who have "taken" control away from me. And, in the grip of my anger, I am unable to see beyond my own, narrow self-interest. Under such circumstances, I do not perceive that I have a choice because I feel out of control, and I am terrified of being out of control. So, I am really angry at feeling out of control in the first place and being terrified in the second place.

Anger often translates into aggression, which translates into the habit of launching attacks, of being hostile. If I show enough aggression toward the person or people I think I am angry at, I am coping with my fear by causing them to back away from me. Through aggression, I can avoid having to deal with "their kind," which really means that I will not have to deal with the circumstance in which I am out of control and of which I am afraid. But, all I have really accomplished is to isolate myself from any understanding of the facts and from the people who are presenting them. If, on the other hand, I had been patient and open-minded and asked questions kindly, then I may have been able to overcome my fear because I would have found that there are no enemies out there—only other frightened people like me, people who may or may not have been able to answer my questions.

Appraisal

Appraisal is the act of evaluating something; of estimating its quality, amount, size, and other features; of judging its merits. As such, appraisal, as a coping mechanism, can effectively prevent forward motion into an uncomfortable situation. It is like being on the platform at the train station and being so afraid of missing the train that I spend my whole time checking and rechecking the schedule. I am so engrossed in appraising the schedule I do not even see my train come and go.

Another example of an appraiser is the shopper who goes to the grocery store to buy three items and has to read every comparative label in minute detail and then weigh and reweigh the data and the price before making a choice. Thus, what would take one person 5 minutes to buy takes another 45 minutes.

Like the shopper, many scientists and policy makers are so afraid of making a mistake and of being criticized or held accountable for an unforeseen error that they get bogged down appraising the details in an attempt to cover all of the bases and thus never make a commitment. This coping mechanism is no stranger to government employees, both state and federal.

Appraisers cope with their self-doubt and fear of criticism and responsibility by checking, rechecking, and further rechecking the data, seldom willing to make a decision for which they are accountable. In other words, when in doubt, make another study, but refrain, at any cost, from saying or doing anything until *all* the data are in and carefully and "properly analyzed," which of course will never happen, because even if one could get definitive data, the appraiser would still define *properly analyzed*. And, the mantra is: Appraise, always appraise but avoid making a decision at any cost.

Defensiveness

To be defensive means to protect that which already is, to resist a new view, to resist the possibility of change, and to resist the truth about myself. Defensiveness is a limitation to my growth in that I am arguing for my old self rather than taking a new look and embracing a new possibility. I am defending the rut in which my old belief, my old behavioral pattern, is stuck. I become defensive because, at some level, I know that what is being said is at least partly true. And, I am afraid to listen to the truth or even a different point of view if I will have to act on it, which means I will have to change my stance, something I am afraid to do. I thus feel obliged to defend my old groove. After all, it is like home. It is a comfortable, known entity.

Defensiveness is a coping mechanism that comes rapidly to the fore when we feel unsafe, when we are losing control of a circumstance. It not only robs an individual of dignity but also isolates the individual from peers and fellow workers. Defensiveness creates a sense of distrust that can spread like an epidemic throughout an agency.

Denial

Denial is a refusal to recognize the validity of a statement. It is a contradiction, a rejection of what is. Although denial as a coping mechanism is part and parcel of almost all other coping mechanisms, it is also an entity unto itself. Think, for example, of your mind as the honeycomb in a beehive and visualize the feelings you do not want to deal with as the honey put into the comb and sealed—out of conscious awareness, out of mind.

If you stuff your feelings into an empty comb and seal it shut with mental wax so you do not have to deal with them, you are now effectively in denial of the uncomfortable thoughts. The rest of your mind seems to be cleared of your suffering. You are free to live, but only as long as you continually mend the already-full comb and continually create more empty comb cells to accommodate future discomforts.

Denial is probably one of the most pervasive coping mechanisms in the world. A simple example is refusal by many—*but not all*—industrialists and their political allies to admit to their role in global warming. Nevertheless, according to the U.S. Environmental Protection Agency:

> The most detailed data yet on emissions of heat-trapping gases show that U.S. power plants are responsible for the bulk of the pollution blamed for global warming.
>
> Power plants released 72 percent of the greenhouse gases reported to the Environmental Protection Agency for 2010. The data include more than 6,700 of the largest sources of greenhouse gases, or about 80 percent of total U.S. emissions.

> According to an Associated Press analysis of the data, 20 power plants in 14 states account for the top-releasing sites.
>
> The largest greenhouse gas polluter is the Scherer power plant in Juliette, Ga., which is owned by Southern Company.
>
> That coal-fired power plant reported releasing nearly 23 million metric tons [507 million U.S. tons] of carbon dioxide, the chief greenhouse gas, in 2010.[37]

A complex example can be found in the highlands of Kenya, Africa, where temperatures have increased significantly since the 1980s, and, due to the rising temperatures, the mosquitoes that carry malaria have better conditions in which to breed, reproduce, and transmit the disease and its death tolls, according to Dr. Maria Neira, director of public health and environment for the World Health Organization. This observation is corroborated by Madeleine Thomson, a senior research scientist for the past 25 years with the International Research Institute for Climate and Society, working with the Kenya Meteorological Department. When Thomson visited Tigre, a mountainous region in Ethiopia, she found people with malaria in the highlands, who were trying to understand what was going on because heretofore malaria had not been an issue. In addition to malaria, contends Dr. Neira, global warming will touch the very pillars of our health, food, water, and shelter. In fact, she says, a child already dies every 21 seconds from a lack of access to clean water.[38] Yet, there are people who steadfastly deny that global climate change is an issue.

We isolate ourselves when we do not accept change. We become defensive, fearful, and increasingly rigid in our thinking; we harden our attitudes and close our minds. If I become defensive about anything, if I start to form a rebuttal before someone is finished speaking, if I filter what is said to hear only what I want to hear, I am in denial of what I do not want to deal with.

More to the point, however, if we become defensive and deny observable, measurable facts, such as the global meltdown of the world's glaciers and the resultant rise in sea levels,[39] we are courting "informed denial" of *what is*. Informed denial means data that conflict with a desired outcome based on a cherished belief are summarily dismissed as inaccurate, inappropriate, or inconclusive. For example, some years ago a neighbor of mine was stirring the rotting vegetation in the bottom of his composter when he slipped and fell headfirst into the container. His wife and I laughed so hard that our knees grew weak, whereupon my friend extracted himself from the composter, looked at us, and denied that the incident had even happened. When I pointed to some compost debris clinging to him, the intensity of his denial merely grew. Consequently, his wife and I just about laughed ourselves into oblivion.

There is also uninformed denial by those with closed minds: "Whatever it is, I'm against it."

The oldest argument of denial I encountered while working for the government, as both a scientist and an expert witness, was informed denial in which the timber industry and a government agency attempted repeatedly to "hide" from an uncomfortable circumstance—the passage of a restrictive, environmental law that would negatively affect their allowable cut of old-growth timber and so their profits. To wit, despite all the scientific data at hand, the industry's and the government's mantra was: "It's inappropriate to do anything until we have conclusive data. What we have to date is inconclusive." Of course, data are always inconclusive because knowledge is—and only can be—*some version of the truth, but* ***never The Truth***. In other words, deny there is a problem until a catastrophe of some sort happens, and then it is too late—but not for pointing fingers and placing blame.

Consider the following illustration, which I use instead of a natural resource issue because it is classic case of informed denial: The tobacco industry usually denies the validity of the data, even when confronted by 50,000 published articles on the dangers of smoking. The president of Reynolds Tobacco Company, for example, is reported to have said, "Honestly, I have not seen one piece of medical evidence presented by anybody, anywhere that absolutely, totally said that smoking caused disease or created it."[40]

Why would anyone preface a sentence with "honestly," "to tell you the absolute truth," or "to tell you the honest truth" if the person habitually told the truth? It always comes across to me as if the person is doing me a special favor by assuring me that, because I am special, the person is willing to be honest with me. The question is, who is being convinced? Moreover, why would I need convincing if the person were indeed truthful?

Displacement

Displacement is used to shift the focus from that which is uncomfortable to that which is safe—a "smoke screen," if you will. I have even had attorneys for the federal government try to distract me with this tactic while I was under oath as an expert witness. They did not want me to complete my answer to a question they had asked because they were afraid of what I was saying, so they interrupted and asked a totally unrelated question.

Another handy way to cope with the fear of being out of control is to displace the uncomfortable situation onto *time*, such as people who have their lives so tightly scheduled that they have not "a second to waste." In this way, they control what they do, whom they see, and for how long without ever having to take the responsibility for confronting an uncomfortable situation.

The use of time in this sense is another form of displacement, of avoidance embodied in the dynamic of the word *can't*. I use time to control those circumstances I wish to deal with and to see those people whom I wish to see for as long as I want to see them. At the same time, I am pleading a case for being innocently out of time, out of control—by saying, "I can't"—to those circumstances or those people with whom I do not want to deal.

Displacement is a common coping mechanism in any dysfunctional agency, not only on the personal level but also on the agency level. The focus on "jobs lost" and "community stability" by both the timber industry and most of the old cadre of foresters in the Bureau of Land Management and the U.S. Forest Service has been used as a displacement for having to deal with the rapidly increasing scientific evidence that ancient forests have value for something besides wood fiber—or even as habitat for spotted owls. In fairness, the conservationists have often used the spotted owl as the hammer for their arguments to save the ancient forest instead of directly facing the issue of the forest itself. Both sides are not only employing displacement behavior but also misleading themselves and the public at large because, while each side strives to save the ancient trees for its own purposes as a product of the political struggle, society is losing the forest in terms of the processes embodied in the biophysical sustainability of the forest itself.

"Higher officials" in an agency often used time as a displacement to achieve their own outcomes, as the following three examples illustrate: The first way time is used to avoid undesirable data or change is to be very stringent with the allotted time for a particular job. An example is an agency's taking 2 or more years with the full-time commitment of an entire planning team to create a land management plan and then limiting the public to a 30-day comment period of review in which they are supposed to receive the plan; understand it and all its alternatives (the proposed methods of operation, their respective outcomes, the potential sustainability of the timber sales or grazing allotments, the ecological and economic consequences of the proposed array of actions on all resources); reply, usually in writing; and submit their reply—all within the agency's chosen timetable.

A second way in which "not enough time" is used is by making time a moving target. I once conducted a weekend workshop with the Forest Service in which the district ranger and a lady from a conservation group were determined to work out their differences and keep the proposed management plan out of court. The ranger invited 30 people from the Forest Service, the timber industry, and several conservation groups to meet for a weekend and begin to resolve their differences over the management plans. At the close of the 2-day workshop, conservationists were talking with local industrialists and Forest Service personnel and vice versa, and there was a general agreement that the process they had established was working and needed to be continued.

All parties worked diligently for some time and came to more agreement than ever before. As the parties got closer to working out their differences, which would have altered the management plan, the upper echelon in the Forest Service, prodded by the upper echelon of the timber industry, suddenly shortened the promised time for the plan to be completed, which broke faith with everyone involved, including their own employees, who had worked so long and hard to build the bond of trust.

To see how time is used a third way, it is necessary to present the following view of Senator Mark Hatfield's (R-OR) attempt to use the allocation of time as a displacement, which is clearly pointed out in the last paragraph:

> The premise behind Hatfield's legislative tactic [trying to prevent the public from challenging timber sales in court] is that frivolous lawsuits by environmentalists hamstring federal agencies because courts don't make timely, responsible decisions. Were that actually the case, the answer for Congress would be to enact clearer laws or provide for faster judicial action.
>
> …
>
> The real threat of lawsuits, of course, is not the time it takes to resolve them. Federal judges can act quickly when time is of the essence. Rather, the threat is that the suits will have merit, that judges will find agencies have acted improperly. That is the very reason that Americans should not be denied their day in court, whatever the perceived urgency of a particular issue.[41]

Filters

A filter is a device through which a substance, such as light, water, or thoughts, is passed to remove what we define as "unwanted impurities." In the sense of a coping mechanism, we filter out unwanted information. That way, we can accept and understand whatever we want. Have you, for example, ever tried to explain something to someone and had him or her hear only part of it, the part he or she wanted to hear?

When speaking to a group of people comprised of the timber industry, environmental organizations, land management agencies, city councilors, county commissioners, or personnel in city planning, it is often the case that each hears what he or she wants to hear in what is said. Not surprisingly, they each address these disparate aspects of the presentation during the question-and-answer period. The more polarized the audience is, the more predictable the questions they are likely to ask.

At times, we live as though we are in a giant "safe" with filters to control what we see, what we hear, and what we feel. In other words, we hear only what we want to hear, see only what we want to see, and feel only what we want to feel. That way, we can accept and understand that which we choose, and we do not have to get out of our comfort zone and be accountable in the world, which is the meaning of "looking at life through rose-colored glasses." Filtering is a common coping mechanism of selective hearing and seeing, as exemplified in two of the three monkeys—hear no evil and see no evil.

Although filters can be very frustrating for the person who is trying to communicate with someone who does not want to hear what is being said, we all filter information simply because we have different frames of reference. Whereas filters in an agency are almost always fully engaged, they serve a different purpose.

By filtering the information and hearing only what fits into the homeostatic pattern, agency personnel can dub it "So and So's unproven hypothesis," discount it, and be "safe" from having to confront the agency's dysfunctional dynamics by changing policy to accommodate new data. This type of filtering system to ward off new, *imaginative* data cannot last indefinitely, however, as observed by the English poet William Blake in 1793: "What is now proved was once only imagined."[42]

Projection

> And Aaron shall lay both his hands upon the head of the live goat, and confess over him all the iniquities of the people of Israel, and all their transgressions, and all their sins; and he shall put them upon the head of the goat, and send him away into the wilderness. … The goat shall bear all their iniquities upon him to a solitary land; and he shall let the goat go in the wilderness.[43]

In biblical times, on the Jewish day of atonement, Yom Kippur, all the transgressions of the Jewish people were heaped—projected—onto the back of a "scapegoat," which was then driven away into the wilderness, taking all the people's transgressions with it. Thus, projection always has a scapegoat.

Projection is the "externalization" of blame, guilt, or responsibility as a defense against anxiety and is so used as a coping mechanism. Just as an empty movie projector casts only light until a specific film is put in it, we can project onto other people only what we think about ourselves because without the thought, there is nothing to project. Like a movie screen, which reflects the projected image, people are mirrors that reflect the projected image. Thus, we see in others what we both consciously and unconsciously see in ourselves—nothing more, nothing less. As such, judgment, that which we see in ourselves, is the projectile we cast at the mirror in which we see our own reflection in the form of another person or thing.

Rationalization

To rationalize, in the sense of a coping mechanism, is to devise self-satisfying, but potentially incorrect, reasons for one's behavior. For example, I have been told to do something in my job with which I ethically disagree, but if I do not comply, I will lose my job—a real possibility in days of corporate/political administrations. So, I rationalize that I can do more good by keeping my job and working unobtrusively for change on the inside of the agency than I can by getting myself fired for sticking openly to my ethical principles. In so doing, I intellectually rationalize the "okayness" of the order and comply with it, but I simultaneously have to repress the honesty of my feelings. I have thus murdered a vital, creative part of myself.

Some years ago, I put on a workshop for the Forest Service fuels managers—folks who help control and prevent forest fires. I spent an entire day with 74 men and 1 woman. We talked about forest ecology, fire ecology, Forest Service policy with respect to the management of ecosystems and fire, and human dynamics.

Finally, toward the end of the day, I said, "Now, turn off your minds. I don't care about what you think. I want to know how you feel about what you're doing."

Silence.

Then, an older gentleman got up and said, "I'm going to retire in about 6 months, and damn it, I don't feel good about what I'm doing. I haven't for some time now."

It turned out that was the general feeling of the audience. So, then I asked: "If you don't feel good about what you're doing, why are you doing it?" The answers had the tenor: "I do it because it's Forest Service policy. I can rationalize its being okay in terms of policy and politics if I don't examine it too much. It may sound good, but now that you've asked, it doesn't feel good."

Our truth is how we feel about something, not what we think about it. Those 75 folks could no longer just think—rationalize—that something was okay when in fact it was not. From that day, they would have to check in with their feelings at some level, and if they were untrue to their feelings, they would have to deal with an ethical crisis.

Finally, *can't* is a great word in the arsenal of rationalization, so it is not surprising that it is one of the favorite words in the federal government. What it means is the following: Just because your request is explicitly permissible in the personnel manual has nothing to do with the fact that I perceive your request as a risk, and there is no way I am taking a chance doing something for you, whatever it is, for which I might be held accountable.

Repression

Repression can be thought of as a one-way, spring-loaded valve into the unconscious. Any thought or emotion causing us anxiety passes through this one-way valve, building tension in the coiled spring as it does so. Once trapped in the unconscious, neither the thought nor the emotion is allowed to reappear in our consciousness.

It might be expressed as follows: John really wanted to slug his brother for having made fun of him in front of Alice, but that was not acceptable behavior at the party. So, he tamped down his anger, put a lid on it, as it were, and left the room. In other words, he repressed his feelings. Without a release, however, energy continually builds in the spring over time because repression allows no acceptable "safety valve" for growing tension. How does this translate into an agency?

I have always believed that if I disagreed with how an agency was being run and I was unwilling to compromise my integrity, I had an obligation to voice my feelings while I was still employed and vulnerable and, if necessary, lay my job on the line for what I believed in, which I did. I ultimately resigned from the Bureau of Land Management.

I felt that I had an ethical commitment to myself and to the folks in the agency to let them know where I stood at all times. I deserved that and so did they. After all, integrity is a gift one can only give oneself but simultaneously owes to others. In the words of Thomas Paine, "Character is much easier kept than recovered."[44]

However, few people seem to express their true feelings. They repress their emotions until they build to unhealthy proportions, all too often carrying an incredible bitterness within them to their graves.

Repressed feelings take their toll in government employees, many of whom drink and eat too much, have high blood pressure and heart attacks, miss days of work due to malingering illness or chronic, low-level burnout. Many a government employee is dissatisfied, unfulfilled, and unappreciated and ends up capitulating to political decisions, all the while putting in the required time until retirement. This situation is a terrible squandering of human potential, dignity, and life's essence.

Yet, is it either courageous or ethical for agency employees—who cope with "the machine" by repressing how they truly feel while reaping the benefits of medical insurance, paid vacations, perceived security, and so on, for 20 or more years—on retirement to let all the repressed bitterness boil out by publicly attacking the agency for which they worked and whose benefits they willingly accepted?

In addition to repressing personal feelings, someone in the chain of command often represses, omits, quashes, or conveniently "loses" information for a variety of reasons. Once, while I was in Washington, D.C., on detail with the Bureau of Land Management, I heard some of the director's staff talking about "protecting him," the director, from things they did not think he would want to hear about or know. Here was the dog's tail deciding what information the head needed.

Well, as it so happened, I met this particular director a year after he was out of office, and I told him that I had tried in vain to reach him while he was still the director of the bureau to tell him about an impending crisis and how it could be defused, but that I could never get through his layers of protection. "I sure wish you'd been able to reach me," he said. "You'd have saved me a lot of grief. If I had to do it over again, I wouldn't allow myself to become so isolated and inaccessible."

Resistance

Why do we fear change so much? We fear change and thus resist it because we are committed to protecting our existing belief system. Even if it is no longer valid, it represents the safety of past knowledge in which there are no unwelcome surprises. We try to take our safe past and project it into an unknown future by skipping the present, which represents change and holds both uncertainty and accountability. Thus, when confronted with change, we try to control the thoughts of others by accepting what to us are "approved" thoughts and rejecting "unapproved" thoughts. Such control is a defense against change, which, after all, is in the mind. As British playwright George Bernard Shaw said, "My own education operated by a succession of eye-openers each involving the repudiation of some previously held belief."[45]

To resist is to work against, to actively oppose or fight off. Resistance is not bad in and of itself. It is simply a conservative, stabilizing tendency, which keeps us from overstepping limitations too quickly and rashly. Problems arise, however, when our resistance becomes overreactive, obsolete, maladaptive, or in other words, dysfunctional. Then, we become stuck and unable to realize our potentials. Resistance, in the dysfunctional sense, is one of the most commonly used coping mechanisms to ward off change, to avoid the responsibility of moving forward, of participating in life.

Resistance is like swimming directly against the current of a swift river. The swimmer in such a circumstance, despite maximum effort, gets worn out, is carried downriver by the overwhelming strength of the current, and sometimes drowns. If, perchance, he or she is strong enough and determined enough just to stay even with the current, it soon becomes apparent that, while the current does not tire from its effortless flowing, the swimmer weakens from the effort of swimming and is thus carried away by the current—the tired borne away by the tireless.

Circumstances are the river of life, and change is its current. The individual swimmer can choose to resist the current, become fatigued and perhaps drown, or can choose to flow with the current and, with patience, learn the skill necessary to cross the river easily. Herein lies the secret: To be in control, we must give up the desire to control. Only when we give up trying to control life can we master navigating its currents.

That which we resist persists in the degree to which we resist it, and we become like that which we resist. It cannot be otherwise. What we resist is a lesson in life not learned, and life seems to persist in its lessons until we learn them. Then and only then are we free to go on to life's next lesson. Resistance as a coping mechanism is a subtle, inner device that urges us to back away from the difficulties and demands of living.

Resistance serves two purposes in life, one positive and one negative. My feeling of resistance is positive when it is my intuition telling me that what I have been asked to do really goes against my deepest sense of principles. In this case, I honor my feelings by resisting.

On the other hand, there are times when I simply do not want to do something that I know I need to do. Then, resistance works against me because I am stepping on the gas pedal of guilt and the brakes of resistance at the same time with equal pressure and therefore spin my wheels in place.

According to British author George Monbiot, people resist acting because struggle for change outweighs the potential material benefits. On the flip side of the coin, Monbiot points out that coercive power is sustained by an unassailable belief in its validity. Therefore, "those who possess power will surrender it only when they see that the costs—physical or psychological—of retaining it are higher than the cost of losing it."[46]

Resistance to things also takes on the disguise of confusion. If I am confused, I have a legitimate reason for not participating in something that makes me uncomfortable.

It always amazed me while in the Bureau of Land Management that whenever something "new" or "unexpected" came down "from on high," there was a rush of "applied" confusion and resistance. The immediate reaction was to try to find a way around whatever it was. When that failed, as it often did, the next line of homeostatic defense was to overreact and blow whatever it was out of proportion.

For instance, when an unexpected order or mandate, especially a new law that might have an effect on business as usual, came down "from on high," there was an immediate scurrying at every level to decode it—read between the lines—and see what the words "really meant." Nothing was taken at face value, although the laws and mandates seemed perfectly clear and unambiguous. Change was always suspect because it disrupted the status quo and required adjustments in operations.

The reason for all this deciphering was to avoid having to change anything any more than was absolutely necessary because even the bare minimum was seen as a threat to business as usual. The heart of the law or mandate during my tenure was not considered as a working option, only as an obstacle to get around. When that failed, we would comply, but the resistance continued.

We complied through bare minimums in that we never seemed to have the time or money to do it right the first time, but we always had the time and money to do it over—and sometimes over and over again. Thus, we inevitably dealt with minimum standards of land trusteeship while sustaining maximum opportunities to extract products, such as timber, pounds of beef cattle, or natural gas. Anything else was stalwartly resisted.

The Upshot

All personal coping mechanisms, in one way or another, become the collective coping mechanisms of the agency as a whole. I remember, for example, sitting in one high-level meeting in the Oregon State Office of the Bureau of Land Management while some of the top decision makers tried to figure out how they could lie to the public about the spotted owl on bureau lands without appearing to do so. The charade was for the sole purpose of keeping up the allowable cut of old-growth timber—the espoused goal of which was total liquidation of the "standing inventory" on bureau lands in western Oregon within 30 years.

I also remember an excellent internal report on the spotted owl made by the bureau's own biologists. It was professional by any standard, but the folks in Washington, D.C., did not like the data, so they quashed it. Someone leaked it, however, and the bureau got caught.

But even then, some of the director's staff in Washington, D.C., openly lied about it through a dazzling array of coping mechanisms. All the while, a feverish, sweeping hunt was being conducted for the agency traitor, who would be aggressively reprimanded for leaking the truth. I was on detail in Washington, D.C., when this episode took place.

It seems that the industrial/political elite have so completely and so insidiously taken over public land management and regulatory agencies (to say nothing of local agencies and organizations) that many people have lost touch with themselves and no longer know the difference between the truth and their coping mechanisms. More and more frightened people are being chewed up and spat out by the dehumanizing elitism of self-interested decision makers, who seek short-term profits at the tremendous cost of truth, trust, human dignity, and public service, which merits an observation by American author and engineer Richard Buckminster Fuller, "Great nations [and some of their governing factions] are simply the operating fronts of behind-the-scenes, vastly ambitious individuals who had become so effectively powerful because of their ability to remain invisible while operating behind the national scenery."[47]

SUMMARY

Chapter 3 examined the ways in which we carry our respective families with us throughout life—both the functional and dysfunctional aspects. Further, it pointed out that the governments, agencies, and various organizations are as functional or dysfunctional as their individual employees are. Therefore, we—individuals acting as a single, collective body—are an agency's driving force and thus enable or limit its ability to function as originally envisioned, the latter often in the military style of command-and-control decision making. Chapter 4 is a case study depicting the duplicity with which the sustainability programs were controlled in one agency—the Southern Nevada Water Authority.

NOTES

1. George Crane. http://www.great-quotes.com/quote/213051 (accessed March 27, 2011).
2. Ayn Rand. *The Fountainhead.* Bobbs-Merrill, Indianapolis, IN, 1943.
3. Gifford Pinchot. *Breaking New Ground.* Harcourt, Brace, New York, 1947.
4. **(1)** Christopher McGrory Klyza. *Who Controls Public Lands*? University of North Carolina Press, Chapel Hill, NC, 1996; and **(2)** U.S. Forest Service history. http://www.foresthistory.org/ASPNET/policy/Agency_Organization/Wilson_Letter.aspx (accessed March 27, 2011).
5. Carl Schurz quotes. http://www.brainyquote.com/quotes/authors/c/carl_schurz.html (accessed November 15, 2001).
6. Choi Song Min. Harsh punishments for poor mourning. January 12, 2012. http://www.dailynk.com/english/read.php?cataId=nk01500&num=8668 (accessed January 12, 2012).
7. Ben Forer. North Korea reportedly punishing insincere mourners. January 12, 2012. http://abcnews.go.com/blogs/headlines/2012/01/north-korea-reportedly-punishing-insincere-mourners/ (accessed January 12, 2012).

8. David Besnainou. Top 8 most powerful businessmen influencing politics. November 14, 2011. http://abcnews.go.com/Politics/top-powerful-businessmen-influencing-politics/story?id=14932475#1 (accessed November 14, 2011).
9. The preceding two paragraphs are based on a study by the Center for Responsive Politics. Most members of Congress enjoy robust financial status, despite nation's sluggish economic recovery. November 16, 2011. http://www.opensecrets.org/news/ (accessed November 16, 2011).
10. James Allen. *As a Man Thinketh*. Wilder, Radford, VA, 2007.
11. Anaïs Nin. http://www.quotationspage.com/quote/27655.html (accessed November 16, 2011).
12. Garance Burke. Federal judge blocks Calif. low-carbon fuels rule. http://abcnews.go.com/US/wireStory/federal-judge-blocks-calif-low-carbon-fuels-rule-15256588#.Tv0GqRw0gpc (accessed December 30, 2011).
13. Napoleon Bonaparte quotes. http://www.qotd.org/search/search.html?aid=5114&page=11 (accessed November 16, 2011).
14. Sophie Poklewski Koziell. Two women of the soil. *Resurgence,* 195 (1999):36–38.
15. Rachel Carson. *Silent Spring*. Haughton Mifflin, New York, 1962.
16. The National Environmental Policy Act of 1969. http://ceq.hss.doe.gov/nepa/regs/nepa/nepaeqia.htm (accessed December 29,2011).
17. Endangered Species Act of 1973. http://www.fws.gov/laws/lawsdigest/ESACT.html (accessed December 29, 2011).
18. Chris Maser. *Earth in Our Care: Ecology, Economy, and Sustainability*. Rutgers University Press, New Brunswick, NJ, 2009.
19. Miles Burnett and Charles Davis. Getting out the cut: politics and national forest timber harvests, 1960–1995. *Administration & Society*, 34 (2002):202–228.
20. Ibid.
21. Northwest Forest Plan. http://www.google.com/#sclient=psy-ab&hl=en&source=hp&q=northwest+forest+plan+standards+and+guidelines&pbx=1&oq=Northwest+Forest+Plan&aq=2&aqi=g4&aql=&gs_sm=c&gs_upl=0l0l1l157681l0l0l0l0l0l0l0l0l0l0l0l0&bav=on.2,or.r_gc.r_pw.,cf.osb&fp=8eeef67bd8da93e5&biw=1020&bih=878 (accessed December 29, 2011).
22. Burnett and Davis. Getting out the cut.
23. Equal access to Justice Reform Act of 2005. http://thomas.loc.gov/home/gpoxmlc109/h435_ih.xml (accessed March 27, 2011).
24. Robert F. Baldwin and Stephen C. Trombulak. Losing the dark: a case for a national policy on land conservation. *Conservation Biology*, 21 (2007):1133–1134.
25. Graham Hallet (ed.). *Land and Housing Policies in Europe and the USA: A Comparative Analysis.* Routledge, London, 1988.
26. Baldwin and Trombulak, Losing the dark.
27. Elena G. Irwin, Kathleen P. Bell, and Jacqueline Geoghegan. Modeling and managing urban growth at the rural-urban fringe: A parcel-level model of residential land use change. *Agricultural and Resource Economics Review,* 32 (2003):83–102.
28. **(1)** David D. Parrish and Tong Zhu. Clean air for megacities. *Science*, 326 (2009):674–675; and **(2)** United Nations. *World Urbanization Prospects: The 2007 Revision Highlights.* United Nations, New York, 2008.
29. Baldwin and Trombulak, Losing the dark.
30. O.B. Jenson and T. Richardson. *Making European Space: Mobility, Power and Territorial Identity*. Routledge, London, 2004.
31. Henry Ford. http://www.brainyquote.com/quotes/authors/h/henry_ford_3.html (accessed December 15, 2011).

32. Alice Maggin. Banker gives $60 million profits to employees. http://abcnews.go.com/WN/PersonOfWeek/story?id=6973727&page=1#.Tuu00xw0i4A (accessed February 27, 2009).
33. Robert T. Lackey. Axioms of ecological policy. *Fisheries*, 31 (2006):286–290.
34. Ibid.
35. Ibid.
36. **(1)** Craig A. Anderson and Brad J. Bushman. The effects of media violence on society. *Science* 295 (2002):2377–2379; and **(2)** Jeffrey G. Johnson, Patricia Cohen, Elizabeth M. Smailes, and others. Television viewing and aggressive behavior during adolescence and adulthood. *Science* 295 (2002):2468–2471.
37. Dina Cappiello. EPA: power plants main global warming culprits. http://abcnews.go.com/Politics/wireStory/epa-power-plants-main-global-warming-culprits-15340558#.Tw3Ychw0i4A (accessed January 11, 2012).
38. **(1)** Carrie Halperin. How climate change may make killer diseases worse. http://abcnews.go.com/Politics/GlobalHealth/risk-malaria-diseases-rise-global-temperatures-climate/story?id=13277843#.Trbmbhw0i4A (accessed May 5, 2011); and **(2)** Richard A. Kerr. Northern India's groundwater is going, going, going ... *Science*, 325 (2009):325–798.
39. **(1)** Richard A. Kerr. A worrying trend of less ice, higher seas. *Science*, 311 (2006):1698–1701; and **(2)** Ian Joughin. Greenland rumbles louder as glaciers accelerate. *Science*, 311 (2006):1719–1720.
40. Larry C. White. *Merchants of Death: The American Tobacco Industry*. Beech Tree Books, Morrow, New York, 1988.
41. Editorial. Portland (OR) *The Oregonian*, June 27, 1988.
42. William Blake. http://books.google.com/books?id=qQzi_NzxKUgC&pg=PA79&lpg=PA79&dq=William+Blake+in+1793:+"What+is+now+proved+was+once+only+imagined."&source=bl&ots=BxeOUDk9dt&sig=xCqnRuyl8MvTN9BfdRMEofezEyI&hl=en&ei=zwRCTbPCIMvrOa3xneQB&sa=X&oi=book_result&ct=result&resnum=3&sqi=2&ved=0CCIQ6AEwAg#v=onepage&q=William%20Blake%20in%201793%3A%20"What%20is%20now%20proved%20was%20once%20only%20imagined."&f=false (accessed January 27, 2011).
43. *The Holy Bible*, Authorized King James Version. World Bible, Iowa Falls, IA, Leviticus 16:21–22.
44. Thomas Paine. http://www.quotedb.com/quotes/1206 (accessed January 27, 2011).
45. George Bernard Shaw http://wist.info/shaw-george-bernard/9385/ (accessed December 24, 2011).
46. George Monbiot. Why we conform. *Resurgence*, 221 (2003):16–17.
47. Richard Buckminster Fuller. http://www.great-quotes.com/quote/1392994 (accessed February 4, 2011).

4 Dissonance within Duplicity

Sustainability Programs at the Southern Nevada Water Authority, a Case Study

Jessica K. La Porte, PhD

It was a particularly cold night in Vermont. Snuggled in my yurt while snow piled up on the branches of a seemingly magical forest outside, I felt incredible gratitude for the self-sustaining life I was enjoying. Reading by candlelight to conserve the electricity I had generated from the sun during the day, I stayed up later than usual reading about the latest developments in the Southern Nevada Water Authority's pipeline, referred to internally as the Northern Resources Project. The project is designed to pump 200,000 acre-feet a year, or 65 billion gallons, 300 miles south to Las Vegas from groundwater basins in northeastern Nevada.[1]

This pipeline, if approved and constructed, will essentially eliminate the ability of small, agricultural communities to live in the desert, while augmenting the fast-paced shimmer of the Las Vegas metropolis. To the residents of rural eastern Nevada communities, who are at risk of having their water stolen, it is known as "the water grab."[2] This is a war for water in the West and one with irreversible, social-environmental consequences. The outcome will determine the future of the northern communities and of the nearly 2 million people living in the largest metropolitan area in the Mojave Desert.[3] The Mojave Desert is the arid region of southeastern California and portions of Nevada, Arizona, and Utah and encompasses more than 25,000 square miles. As such, it embodies the transition from the hot Sonoran Desert to the south and the cooler, higher Great Basin to the north.[4]

The availability of water has shaped the history of Las Vegas and seems a major limiting factor to the explosion of urban expansion in the landscape that is a dry and dusty expanse without this precious liquid. Keeping the water supply apace with the look and feel of the sparkling energy of development in the fabulous oasis, which sees 37 million visitors each year,[5] is becoming rapidly more difficult due to drought, the impacts of climate change, and a finite allocation of water from the Colorado River.

Keeping the water flowing into the Las Vegas area is the sole purpose of the Southern Nevada Water Authority—a quasi-governmental agency created in 1971 by seven local water agencies to address regional water issues as a powerful, united front rather than rely on the area's individual water purveyors. In 2001, the federal government officially transferred the Southern Nevada Water Authority to the Water Authority.[6] Patricia Mulroy has been the general manager of the Southern Nevada Water Authority in Las Vegas since its inception and continues to oversee the operations of both the authority and the Las Vegas Valley Water District, which is a member agency of the Southern Nevada Water Authority and of which she has been the general manager since 1989.[7] Mulroy is known in the bureaucracy of Clark County (which contains Las Vegas) for her outspoken and often politically savvy oversight of water operations; she is known as a "water warlord in a lawless region" and as the "water empress of Vegas."[8] As a former employee of the Southern Nevada Water Authority and from my limited experiences being in the same room with Patricia Mulroy, she commands respect, regardless of one's opinion of her.

It could be expected that sustaining an available supply of water, as well as reducing wasteful water consumption, would be a primary focus of the Southern Nevada Water Authority, especially when daily business is conducted in the intense heat of what was just the driest December (2011) in the past 130 years.[9] However, I have found the purported efforts to make the supply of water in Las Vegas sustainable to be a facade worth billions of dollars and political primacy and just valuable enough to keep audiences focused on the smoke and mirrors.

The following case study is a story of dissonance and duplicity within my experience as the senior policy analyst of sustainability programs while employed by the Southern Nevada Water Authority. Beginning with my initial interviews, I chronicle my experience of being hired in early January 2009 to lead sustainability initiatives within the Southern Nevada Water Authority and with local community partners. My experience concluded with the sudden elimination of my position in early 2010 by senior management. Nevertheless, the wars for water rages on as the fate of billions of gallons of life-sustaining groundwater rests with the decision of the state engineer.

I use prominent categories of the job description I was hired to work within as a tool to navigate the dissonance and duplicity that resulted from my direct participation in activities related to my employment with the Southern Nevada Water Authority. It is from this job description that I was directed to participate in the sustainability programs of the Southern Nevada Water Authority, and it was from this description that I first noticed the duplicity and then began to create dissonance. These categories include responsibilities, scope or participation, and required abilities. It was precisely these categories that caught my attention when I first considered applying for the position, and it was these categories that would compel me to uphold my personal and professional integrity within the purview of sustainability—and that ultimately cost my job.

Within the experiences presented, *dissonance* refers to the increasing state of conflict within the opinions, actions, and character of my relationship with the Southern Nevada Water Authority; I had to protect my personal and professional integrity while attempting to fulfill my terms of employment. In essence, I was cast as an "outsider" to the organizational culture of the authority. *Duplicity* refers to acting in bad faith or deception by pretending to entertain one set of intentions while acting out the hidden agenda of another. Overall, my direct participation with the Southern

Nevada Water Authority was both disheartening and awakening with respect to the inherent integrity that must accompany all decisions regarding social-environmental sustainability. I ultimately came to the realization that, while I cannot change the Southern Nevada Water Authority, I am able to change myself.

I recall the way the job description stood out from myriad other jobs related to sustainability that I had reviewed. The description for the position of "senior policy analyst—sustainability programs with the Southern Nevada Water Authority" boasted a leadership role that advised and participated with the senior management team in the development of an organization-wide sustainability program. This was in contrast to the many agencies hiring a "sustainability person" that embedded the role within their hierarchies and potentially stripped the incumbent of the ability to fulfill the prescribed duties and responsibilities effectively. At the time, I had never heard of the Southern Nevada Water Authority, but the glimmer of the job description, the six-figure salary, and the desperate need for sustainability within the social-environmental context of water in the largest desert metropolis was a challenge I was willing to accept.

I had, at the time, invested nearly 10 years in the field of sustainability and was in my second year of a doctoral program focused specifically on education in the arena of sustainability. I knew if there was ever the chance to create positive change this could be it. I was living roughly 120 miles east of Las Vegas in St. George, Utah, tenaciously promoting an education center for sustainability that I had founded the year before. The move to Las Vegas would not be nearly as extreme as that for some of the jobs I had applied for in other areas of the country, and it certainly would be more sustainable to move just 120 miles versus 1,200, or at least I thought that was the case.

I applied for the position and went on with my education center. Three months later, I received a call for an interview. Driving into Las Vegas, the importance of social-environmental sustainability gleamed like the thousands of marquees. While waiting for someone from the interview team to greet me in the high-security lobby of the Southern Nevada Water Authority, I had the opportunity to read a few informational plaques describing the many sustainable features of the building. I learned that the home of the Southern Nevada Water Authority was in the Molasky Corporate Center—the first LEED™ (Leadership in Energy and Environmental Design) Certified Gold Class A office building in downtown Las Vegas[10]—complete with organic deli and fitness center and specifically constructed to minimize the use of water.[11] At the time, this felt affirming, and I was excited by the potential of having my office in this space.

As the interview concluded, I had the opportunity to ask the interview team questions, one of which was extremely meaningful to me; it concerned what ways the team was sustainable within their own lives, outside their positions with the Southern Nevada Water Authority. After a look of surprise, the first, a trained biologist and current manager of the department I was applying to join, replied with recycling and water conservation. The second of the interview duo, Richard Holmes, the director of the same department, head of the sustainability program, and who I would later learn was former vice president of the real estate development agency Focus Property Group,[12] responded meagerly with "recycling." I had hoped for more but rationalized their responses as indicators of the need for sustainability in their organization.

During the second round of interviews, I was told I would be taken to the Springs Preserve for a tour of the property. I was, however, unfamiliar with the multimillion-dollar sustainability theme park prior to the interviewing process. It was touted as a 180-acre living example of sustainable living, design, and architecture and was awarded the status of LEED Platinum for the design and techniques of construction.[13] With a focus on education for sustainability, sustainable living, and an experiential Desert Living Center, I expected this to be the context for my employment and efforts toward creating an organization-wide sustainability plan. Instead of a tour of the Springs Preserve, however, I was taken to lunch at a restaurant just a block from the Freemont Street Experience, a pedestrian mall and attraction, occupying the westernmost five blocks of Freemont Street.[14]

It was not until much later into my employment with the Southern Nevada Water Authority that I actually set foot on the campus of the Springs Preserve. Instead of an office within the serene setting of the Springs Preserve, I would spend the next year trying to establish a cohesive and comprehensive sustainability plan, employee education programs, and community partnerships from the Southern Nevada Water Authority offices in the Molasky Corporate Center.

On January 12, 2009, I reported to the 10th floor of the Molasky Corporate Center for my first day of work. After being introduced to the environmental monitoring and management team and learning the layout of the office space, I was in a meeting with my direct supervisor, the manager of the department. During this meeting, I was given a list of top priorities for my job responsibilities. The top three priorities included developing a comprehensive sustainability plan, engaging community partners, and employee education. As we discussed the details of these responsibilities, it became clear that my position had become more deeply embedded within the hierarchy of the organization than previously agreed. I was now to report directly to this department manager.

According to the job description, I was to participate directly with the senior management team (the directors), exercising significant administrative discretion in the execution of assigned responsibilities, and overall leading organization-wide efforts to build awareness and support for sustainable development, both internally and among other governmental, business, and community organizations. I had originally accepted the position due to the leadership role and the place of the sustainability policy analyst within their chain of command. This was also discussed during the interview process and was heralded as an estimate of the organization's commitment to sustainability. Sustainability was touted as the highest focus.

Without any discussion, my position was embedded more deeply within the organization, and I was then considered a supervisor under the manager of the department and within the administrative power of the director. This was in stark contrast to coming in as an organization-wide leader who worked side by side with the team of directors. At the time, it was explained that I was within the probationary period of my early employment with the authority, and the scope of my position would be reevaluated within 90 days.

I later learned that, in the week prior to beginning my job, the manager of the department had gone to the director and asked that I report directly to her. The manager had voiced concerns about having her leadership of the department be undermined by my position due to my original place within the hierarchy. In other words, she wanted full command and control of my responsibilities and my scope of participation instead of having to work with me as an equal in authority.

I accepted the new scope of my position and expected to be regarded similarly as the two other supervisors of the department; both had offices and were given a level of respect that was appropriate for a supervisor. The main difference was that I was the only employee focusing on sustainability programs. The other supervisors had teams of 5 to 10 employees to oversee. But, instead of an office—and there were plenty available because the south side of the 10th floor was barely a third full—I was assigned to a cubicle among the interns. Seeing the larger vision of why I was there (to write a sustainability plan, engage with community partners, and develop employee education programs), I accepted my cubicle. After all, it was within a LEED Gold-certified building, with an expansive view of Las Vegas.

The next 6 months, from January to June, were a blur. As the sustainability person for the authority, I assumed the responsibility of attending any and all meetings related to sustainability within the greater Las Vegas area and internally within what were called "cross-departmental teams." This equated to a minimum of three meetings each day. There was a tangible buzz of energy within the Las Vegas community to establish a future vision of sustainability, and every organization, from casinos to regional planning coalitions, was in on the frenzy. It did not take long to realize that all the other agencies in the Las Vegas Valley had their eyes on the Southern Nevada Water Authority. As the perceived leader in the community for sustainability, the authority had successful programs that lent themselves well to the notion of leadership within the community. Most of these programs, however, were set in motion prior to my employment.[15]

From where I stood, overlooking the metropolis, this was a dream job. I quickly earned respect in the community as the Southern Nevada Water Authority's "sustainability guru," and my expertise was applauded. Internally, I was slowly fitting into the culture of the Southern Nevada Water Authority. When not traveling to and from community meetings, I was at my desk working tenaciously at creating a comprehensive sustainability plan, which was due by early May 2009. The endeavor of organizing past sustainability efforts into current programs and future visions engulfed my attention. I vividly recall walking into the office in the morning and the next thing I knew it was time to go home. I loved my job.

During this time, I also had regularly scheduled meetings with my director to continue learning the history and culture of the organization. I later learned that my director was a former land developer who, through various endeavors with Clark County, had earned the name "Teflon" because everything seemed to roll off of him. Teflon implied that he was unshakable, and nothing seemed to rock his boat or spark his impassioned efforts. But, Patricia Mulroy would later refer him to as "Mr. Sustainability" during an organization-wide event.

These one-on-one meetings with the director often turned into hour-long storytelling sessions; he would recount various water projects and future aspirations for water supplies. It was during one of these meetings that I first heard of the In-State Groundwater Resources Project.[16] It was explained to me as a necessary endeavor for the sustainability of the Las Vegas Valley; water rights, already owned by the authority, would be accessed in a remote and desolate northern county. As it was presented, I did not question the project. Little did I know this seemingly benign pipeline was the heart of a water war that had flared up after more than 10 years of silence. Being coined a "stimulus package for developers," as recently as October 2011, the original impetus for the Southern Nevada Water Authority pipeline originated in 1989, when the Las Vegas Valley Water District filed applications for water rights.[17] This initial action would become a wrench in the gears for the development of the pipeline and has left the current status of the project in a legal battle with the decision to be made by the state engineer.

I continued working on the sustainability plan and cultivating community partnerships while flashing the sustainable achievements of the Southern Nevada Water Authority. According to my job description, I was fulfilling the essential duties and responsibilities as directed, despite the embedding of my position deeper within the pecking order. The job continued to quench my impassioned thirst to establish the water authority as a model organization for sustainability. I was living the dream, enjoying a good salary, and actually applying many years of academia. The involvement of the Southern Nevada Water Authority with the World Wildlife Foundations' Earth Hour event[18] was the first accolade of my employment. A viewing party was coordinated at the Wolfgang Puck Café, located in the Springs Preserve campus, to celebrate the dimming or shutting off of decorative lighting on the Las Vegas Strip to show solidarity for the environment and raise awareness for climate change.[19]

Prior to my employment, the Southern Nevada Water Authority had hired external consultants from the Sustainability Partners group[20] to energize the organization and its employees about sustainability and to kick off the commitment to becoming leaders in the community. Three days after I started with the authority, the consultants had given another presentation to the Southern Nevada Regional Planning Coalition to report the latest issues and trends concerning sustainability and to recommend action.[21] They were scheduled to return in 6 months to check in with the Southern Nevada Water Authority on its own sustainability initiatives and potential requirements for consultation. On their return, I had finished the first draft of the organization-wide sustainability plan. I had just a few weeks before the plan was to be completed and submitted to the director of the department in which I worked. The director would then be presenting the plan to the entire management team for review and initiate implementation.

After months of effort, the plan was in review by the external consultants. As I eagerly awaited their input, I felt very optimistic about the continued growth of sustainability within the organization. Also during this time, the organization added "sustainability" to its set of values. All organizational communications were modified to include this new value, and the pressure to start to build employee educational programs began with their promotion. The Sustainability Partners responded to the plan with the recommendation that it be presented to the senior management team and even went as far as to call the volume a "comprehensive compendium" of sustainability initiatives. After a round of proofing by the public information department, a department I would later learn is referred to as the "Southern Nevada Water Authority's propaganda machine,"[22] the plan had gone to print and was to be delivered at the directors' meeting.

Weeks went by as I eagerly awaited a response from the initial meeting. The director of my department, the previously mentioned Teflon, was apathetic about the outcome of the meeting. During this hurry-up-and-wait period, the manager of my department, my direct supervisor, suggested I schedule a meeting with Kay Brothers, deputy general manager at that time, and try to garner some support from the executive level of management for the sustainability plan. I was, however, unaware of Brother's past associations with oil and gas companies and her track record of getting projects approved regardless of environmental regulations.[23] I met with Brothers for 30 minutes and left with a sense that she was supportive of the plan, despite her lack of actually verbalizing a willingness to champion the endeavor. I remember not understanding why someone in an upper management position was not stepping up to be an advocate for sustainability.

Fast forward a year, and in the early spring of 2010, Brothers would retire in the wake of a water-grab cover-up gone bad[24] after the state water engineer tried to slip the water rights issue associated with the 1989 groundwater applications by the Southern Nevada Water Authority into a special session of the Nevada Supreme Court in February 2010. This special session was to be focused on the budget. The engineer's maneuver was coupled with the surfacing issues related to the lack of due process for protestors that would have placed the northeastern counties at risk of having their water stolen. This illegal coupling caused the Nevada Supreme Court to order the state engineer to reevaluate the impacts of withdrawing water. This sparked a new round of protests and new hearings.[25] The special session, which included proposed amendments to state water law, resulted in a stunning upset to Southern Nevada Water Authority's pipeline proposal, delaying development and creating additional legal battles.[26]

These legal issues still lay on the horizon for the Southern Nevada Water Authority during my tenure as I worked to get approval for an organization-wide sustainability plan. Without a champion, the plan and all of its programs, partnerships, and expressions of collective visions for the Southern Nevada Water Authority and its community were nowhere close to approval—let alone implementation. A meeting was finally scheduled to discuss the future of the plan; the meeting included only a few select directors, the manager of the department where I worked, and me.

My manager and I had been told this meeting would be to discuss the implementation of the plan. Optimistically, I walked in and took the only open seat. It was located at the opposite end of the table from my manager and our director. It did not take long to realize that the meeting was not about the plan's implementation. Instead, it was a well-orchestrated highjacking, during which my director was silent as senior managers tag teamed me with criticism—eventually stripping the plan of its collaborative, cross-departmental nature. Each of the directors commandeered whichever program they were interested in taking over and modifying to best suit their own agendas.

During the entire meeting, I spoke out about the need for a collaborative, organization-wide endeavor that would establish a sense of community around the concepts in the plan. Through a shared vision, I asserted, unforeseen solutions could arise, as well as opportunities to be bona fide leaders of sustainability within the water agency. I analyzed opportunities within the plan while trying to shed light on potential issues associated with deconstructing it. I tried to engage the directors in a discussion about my recommendations and outlined a process to aid in the coordination of a comprehensive plan that would truly honor their recently added "corporate value of sustainability."

Based on my education, training, and professional experience, I exercised the initiative to make what I believed and knew in my heart to be sound judgments and recommendations for the success of sustainability programs within the Southern Nevada Water Authority. I was fighting for the plan that I had just spent the last 6 months following orders in creating and promoting within the community. Every local partner agency was scrambling to draft something comparable to the Southern Nevada Water Authority sustainability plan. From Clark County to the city of Las Vegas, everyone was desperately seeking to reach the bar that had just been set by the authority even though no one had ever seen it.

The more I talked, the more hostile the meeting became. I finally realized that my words were falling on deaf ears as one of the directors slammed her hand down on her papers and expressed that I was not "getting it." I fell silent. Throughout the entire meeting, my director, Teflon, sat silently staring down at the papers in front of him—never once speaking in support of the plan or me. The meeting adjourned, and my director, still silent, remained at the table. My manager rushed out of the meeting in tears.

For the next few months, I existed in a state of solitary confinement. My desk phone stopped ringing, colleagues kept their distance, and I ate lunch at my desk, alone. Moreover, I was no longer invited to internal meetings. Rumors began to circulate that I had "gone rogue" and had a "blowup." I began to understand that I had been blacklisted, and that the true intent of the meeting had been geared toward an agenda other than approving and implementing the plan.

Although I continued to attend meetings related to sustainability within the community, I was fiercely instructed that I was merely to take notes for the sake of my director and manager—but under no circumstance was I to express any opinion or comment. Everything I wrote became subject not only to review but also to censorship. My once-dream job had turned into a nightmare.

Nevertheless, I still could not understand why the sustainability plan I had written was the source of such conflict. After all, it had received praise from the external consultants and was the high-water mark all other community organizations in Las Vegas sought to reach. This plan, according to my job description, was my primary responsibility and a necessity to the organization; in addition, the ability to make sound, appropriate judgments was a criterion defined clearly in my job description. With my level of engagement reduced literally to clipping articles about topics related to sustainability from newspapers for the reading pleasure of my manager, I was no longer functioning within the agreed-to parameters of my job description.

On a seemingly typical morning in August, I made my way into work and noticed people with signs gathering outside the Molasky Corporate Center. On entering the building, armed security officers escorted me to my desk, where I learned of the Southern Nevada Water Authority board meeting taking place that morning and how it also entailed the voices of those who opposed the pipeline project—the residents of the rural communities in northeastern Nevada.[27] During this meeting, board members would hear concerns and opinions prior to a vote about continuing with the application and permitting processes necessary to develop the pipeline.

Because all Southern Nevada Water Authority board meetings are broadcast live on the Southern Nevada Water Authority Web site,[28] there was considerable buzz about the meeting from coworkers. Consequently, nearly everyone was tuning in. I turned on my computer and began to watch the meeting. Finally, it was time for the residents of eastern Nevada to speak. Cup of organic coffee in hand, I was glued to the computer screen. Their stories were genuine and resonated a level of authenticity that I had not heard in a long time. A woman approached the microphone, visibly nervous, and held up a large photo of a coyote. She expressed to the Southern Nevada Water Authority Board that she was there on behalf of the animals that depend on the water and yet have no voice. Tears started streaming down my cheeks.

Suddenly, the sound of laughter in the vicinity of my cubicle became louder. I looked up from my desk and noticed a group of coworkers huddled around a computer screen laughing at the woman who was just speaking on behalf of the animals. I felt sick to my stomach. Getting up from my desk, I walked over to the huge windows, where instead of looking out at the vista of Las Vegas, I looked down at the street below. Groups of protesters lined the street holding signs, united in their intentions and desperate for their voices to be heard. In that moment, I knew in my heart that I was on the wrong side of the fence.

In the months following the board meeting, something began to awaken within me. Intuitively, I started looking deeper into the issue of the sustainability plan and why it was not being honored. I was missing something. I began reading alternative opinions on the In-State Groundwater Resources Project and found there was quite a chorus of opposition. Seeking perspective, I submitted a travel request to my manager to attend a workshop on barriers to sustainable behavior[29] in San Francisco. Once at the weekend-long workshop, I felt at home in the group of dedicated and impassioned advocates for sustainability, each of whom had a cause and organization he or she was supporting. When my turn came to introduce myself, I said I worked for the Southern Nevada Water Authority of Las Vegas on sustainability programs. Following the introduction, there was an awkward moment of silence, and then the entire group began to laugh. The person sitting next to me asked if I was joking. Suddenly, it all became clear. Sustainability within the Southern Nevada Water Authority was ludicrous.

I had gone to San Francisco seeking perspective, and I most definitely found it. Once back at my desk, the wheels in my head began turning. I started to question the intention behind my position. Reviewing my job description with a highlighter, I made note of the ideals I was hired to uphold and the stark contrast to which these ideals actually became reality. It soon became the "800-pound gorilla" in my cubicle. I compared the reduction in the scope of my position, the pruning of my responsibilities, and the blatant dismissal of the required abilities to fulfill the obligations of the position of senior policy analyst of sustainability programs with what I was actually doing—attending meetings and promoting the success and valor of the Southern Nevada Water Authority as a leader in sustainability. Then, I added the disconnection between my position and the Springs Preserve, and it became even clearer. Having an intelligent, censored, public face for sustainability created a façade of admirable leadership during the application and permitting process for the pipeline.

I took it a step further and started to analyze the advancement of sustainability programs within the organization from the time I was hired, considering I was approaching my 1-year anniversary within the Southern Nevada Water Authority. Also during this time, the Southern Nevada Water Authority had been dealt an unfavorable ruling—the groundwater applications they had applied for in 1989 were no longer being upheld.[30] With the organization preparing for a continued legal battle for the rights to pump groundwater, I started to see my paychecks as a sort of "blood money" taken from future generations.

The results of my analysis of the advancement of sustainability in 1 year of my employment revealed that nothing had changed. Despite the plan I had written, the daily barrage of community partnership meetings, the efforts directed toward employee education and involvement, all that had changed was the unfounded presumption of the Southern Nevada Water Authority as a leader in sustainability. The community was looking to the Southern Nevada Water Authority as a leader and modeling efforts after the highly touted sustainability plan, yet the agency maintained its status quo. The highly advertised sustainability plan was "dead on arrival."

With my job description in front of me, I composed an e-mail to the general manager of the Southern Nevada Water Authority, Patricia Mulroy. Internally, there was always a highly regarded, open communication policy with the general manager; suggestions and comments were welcomed. I saw this as my Hail Mary pass.[31] I used my job description as a framework to convey my message. I spoke to the lack of progress during my time of employment, despite my participation, leadership, and expertise. I identified the need for sustainability to be repositioned within the organization to reflect the commitment and values of the agency. I presented opportunities and alternatives to create tangible programs that would support the Southern Nevada Water Authority in becoming an authentic leader in sustainability at a local, regional, and national scale. I exercised initiative and expressed the sound judgment that, if the Southern Nevada Water Authority were to be truly committed to sustainability, it would be prudent to actually create a credible sustainability program. I went so far as to suggest that my annual salary was an unsustainable use of organization funds due to the lack of support from senior management to truly commit to sustainability programs and establish a collective vision. I sent the e-mail.

Two very quiet, eerie weeks passed. I continued to look busy clipping newspaper articles related to sustainability for my manager. The morning of January 19, 2010, I was at my desk, looking out into the endless blue sky when my manager appeared suddenly at the entrance of my cubicle. She informed me of a meeting with our director that would take place in 5 minutes. I grabbed a notebook and a pen and followed her to his office, where it took but a moment to realize that the meeting was about my e-mail to Pat Mulroy. A rush of adrenaline raced through my body. I remained calm and stone-faced. The director began informing me that my e-mail created quite a stir with the general manager and her executive team. I made eye contact with him as he talked. He then expressed that the Southern Nevada Water Authority had decided to eliminate my position, effective immediately.

Consequently, I cleaned out my desk and was escorted out of the building by security. The dissonance I had been feeling was real. I believe the duplicity of sustainability had been confirmed by the elimination of my position.

I spent the next month with my family, retrofitting a retired school bus with solar panels and living space. We rented our home, gave away all unnecessary belongings, and moved into our school bus. We left Las Vegas in search of a more self-sustaining life. In the early summer of 2010, we stopped in Vermont and found it to be a welcoming and affirming place to root our lives. We currently live in an off-grid yurt, support the local food economy, and are in the design phase of developing self-sustaining systems to support our local community. These systems, a passive solar barn for collaborative use, an alternative wastewater infrastructure, a solar array capable of meeting our requirements for electricity, and collaborative food cultivation are intended to create resonance within the local community and continue to inspire creativity in reestablishing harmony with the environment. While my current experiences are dramatically different from my time with the Southern Nevada Water Authority, the level of personal and professional integrity, which I continue to cultivate for sustainability, remains deeply rooted.

Since my departure from Las Vegas, the Southern Nevada Water Authority's pipeline project has stumbled along a rocky road of legal trouble and associated hearings[32] surrounding the violation of due process with regard to the protestors and the estimated volumes of groundwater available for the proposed pumping. For more details on the timeline of the Southern Nevada Water Authority pipeline, refer to the timelines produced by the *Las Vegas Sun*[33] and the Protect Snake Valley organization[34] in addition to the many articles by water journalist Emily Green.[35]

The final decision on the development of the 300-mile pipeline from northeastern Nevada to Las Vegas has been made by the Nevada state engineer. On March 22, 2012, the Southern Nevada Water Authority won the approval of Nevada state engineer Jason King to pump 83,988 acre-feet of water from eastern rural Nevada.[36] The Southern Nevada Water Authority had originally asked for permits to pump roughly 126,000 acre-feet of water, but reduced their request to approximately 105,000 acre-feet during the decision making period. The approval brings the potential to meet the water needs of close to a quarter of a million homes in Las Vegas.[37] In essence, the permits will increase available water to Las Vegas by 25%. The approval indicates a staged implementation of withdraw over several years, required monitoring of biological effects and the flow of groundwater over the first 2 years and initiatives for mitigation and management to be in place prior to the exportation of any water.

The approval does not, however, afford the Southern Nevada Water Authority instant access to the water awarded by the Nevada state engineer. Within the approval comes a withdraw plan that spreads pumping over various stages during a period of eight years, across the various basins approved for pumping, and all contingent on monitoring. Additional permits from the Department of the Interior have yet to be granted because the environmental impact statement for the development is under review and final comments.[38] The final document is expected in the summer of 2012 and, once approved, will give the green light for the Southern Nevada Water Authority to begin development.

Opponents of the project have already vowed continued legal engagement. The socio-environmental consequences of constructing the 300-mile pipeline are also being widely discussed. Many fear economic and environmental devastation to the surrounding areas because the pipeline will take more than mountain snows can replenish, resulting in a dramatic decline of available water in the interconnected valley aquifers.

In a statement released by the Center for Biological Diversity, ecologist Rob Mrowka calls the permit approval "a disaster for rural communities, native plants and water needs." Mrowka goes on to state, "The winner in today's ruling is mindless Las Vegas growth, while biodiversity, rural residents and future generations are the clear losers. … There are other, better options for addressing southern Nevada's long-term water needs."

Mrowka raises many concerns associated with the Southern Nevada Water Authority's pipeline. Some of the impending socio-environmental consequences include the availability of water for the relatively rare Bonneville cutthroat trout; shrub cover for the potentially endangered sage grouse; food for foraging animals, such as elk; and drying plant roots.

Drying plant roots is of particular concern. Estimates of 35,000 tons of new dust swirling toward the Wasatch Front in Utah have been suggested. As warned by Mrowka, some of this dust would include radionuclides from old, nuclear-weapons tests, thus setting the stage for Salt Lake City to become a downwind community again.

Science and political pressure have been two very powerful forces in the decision made by the state engineer. However, many opponents believe political pressure stemming from Las Vegas contributed to the final approval. "The state engineer is ignoring science and bending to the winds of political pressure," said Mrowka. "Robbing Nevada's wildlife and rural communities to quench the insatiable thirst of Las Vegas is profoundly shortsighted."[39] Mrowka is not the only expert to be voicing these concerns.

Emily Green adds a deeper dimension to the political pressure surrounding approval of the water rights. She asserts that this water war was designed to secure approval with the weight of all the money and influence from Las Vegas behind Pat Mulroy, beginning 22 years ago when the initial applications were filed. Her article, "Nevada State Engineer refills Vegas pipeline," reveals additional details and specifics of the political associations that may have lead to this final approval.

In the article, Green describes a remark made by Pat Mulroy during a 2007 *Las Vegas Sun* editorial board meeting that points to the odds of betting on the approval of the pipeline. In this meeting, explains Green, Pat Mulroy was asked what she would do if the state engineer denied her claims for pumping water from eastern Nevada. Mulroy allegedly replied that she would have the Governor of Nevada replace him.[40]

Coming from a general manager who just had her first performance review since being hired by the Southern Nevada Water Authority over 20 years ago and just a week before the state engineer's decision, the point is not whether her alleged remarks regarding political power are accurate but rather that she has been leading the Southern Nevada Water Authority for two decades with a decision-making power that warranted no review or critique by the Authority.

Mulroy continues to be at the helm of the massive city of Las Vegas in the heat of the Mojave Desert on a never-ending quest for water that bets on environmental consequences over stringent water conservation. The stakes are high in Nevada-tourist dollars and glamorous development against the future of rural communities and the sustainability of thousands of species. Social-environmental sustainability has become a chip worth playing when the politics favor climate change as a guise to puppeteer the need for more water.

The duplicity within the decision-making in this war for water will only become more palpable as the socio-environmental consequences of these decisions are revealed. I maintain a great hope for the dissonance that will arise as this project unfolds. I also maintain great trust in the emergence of authentic leadership that will serve with grace and a reverent awareness of the intricate relationship between humanity and nature that will then truly quench the souls of those prioritizing profits over the planet. My experiences with the Southern Nevada Water Authority, and the more current unfolding of this socio-environmental war for water, have rejuvenated my resolve to evolve into being and behaving like the change I want to see in the world. May this story inspire yours.

SUMMARY

Chapter 4 is a case study depicting the environmental duplicity within the Southern Nevada Water Authority. Chapter 5 illustrates the ways in which dysfunction and homeostasis are corrected for the benefit of all generations. These include understanding how job descriptions are used to control information and how positions of authority are used to control direction for political purposes, as detailed in the foregoing case study. The correction of a dysfunctional agency culminates with the crafting of a vision, goals, and objectives, as well as understanding and implementing the seven steps in monitoring the vision's effectiveness.

NOTES

1. Great Basin Water Issues. http://www.greatbasinwater.net/issues/index.php (accessed June 12, 2011).
2. Steve Erickson. The Vegas water grab. *The Salt Lake Tribune*. http://www.sltrib.com/sltrib/opinion/52313472–82/blm-vegas-basin-grab.html.csp (accessed November 3, 2011).
3. Mojave Desert. http://en.wikipedia.org/wiki/Mojave_Desert (accessed November 4, 2011).
4. Desert USA. http://www.desertusa.com/ (accessed November 4, 2011).
5. Kyle Hansen. Las Vegas records 2.7 percent increase in visitors for 2010. February 8, 2011. http://www.lasvegassun.com/news/2011/feb/08/las-vegas-records-27-percent-increase-visitors-201/ (accessed February 8, 2011).
6. Southern Nevada Water System. http://www.snwa.com/about/history.html (accessed February 10, 2011).
7. Las Vegas Valley Water District. http://en.wikipedia.org/wiki/Las_Vegas_Valley_Water_District (accessed February 15, 2011).

8. (**1**) Michael Weissenstein. The water empress of Vegas. *High Country News*. http://www.hcn.org/issues/200/10404 (accessed March 5, 2011); (**2**) WaterWired. http://aquadoc.typepad.com/waterwired/2008/06/pat-mulroy-101-a-profile-of-the-chosen-one.html (accessed March 5, 2011); and (**3**) Patricia Mulroy general manager. http://www.snwa.com/about/board_eteam_mulroy.html (accessed January 18, 2012).
9. Cy Ryan. Driest December in 130 years prompts fire warning from state. *Las Vegas Sun*. January 6, 2012. http://www.lasvegassun.com/news/2012/jan/06/driest-december-130-years-prompts-fire-warning-sta/ (accessed January 6, 2012).
10. Green Building Council. http://www.usgbc.org/DisplayPage.aspx?CategoryID=19 (accessed March 30, 2011).
11. Molasky Corporate Center. http://www.molaskycenter.com/ (accessed March 30, 2011).
12. Richard Holmes, deputy general manager of engineering/operations. http://www.snwa.com/about/board_eteam_holmes.html.
13. Springs Preserve. http://www.springspreserve.org/about/sustainability.html (accessed June 30, 2011).
14. Freemont Street Experience. http://www.vegasexperience.com/ (accessed July 21, 2011).
15. Sustainability projects. http://www.snwa.com/env/sustain.html (accessed July 21, 2011).
16. (**1**) Groundwater. http://www.snwa.com/ws/groundwater.html; and (**2**) In-state water resources. http://www.snwa.com/ws/groundwater_instate.html (accessed July 21, 2011).
17. Response 2 SNWA; no water theft 4 developers! (Part 1). http://www.indybay.org/newsitems/2011/10/08/18692741.php (accessed October 8, 2011).
18. World Wildlife Fund. http://www.worldwildlife.org/sites/earthhour/index.html (accessed March 28, 2009).
19. Mytae Carrasco, Kyle Hansen, and Cassie Tomlin. Las Vegas unplugs for Earth Hour. *Las Vegas Sun*. March 28, 2009. http://www.lasvegassun.com/news/2009/mar/28/live-blog-las-vegas-gets-ready-go-unplugged-earth-/ (accessed March 28, 2009).
20. Sustainability Partners. http://www.sustainabilitypartners.com/ (accessed December 13, 2011).
21. Meeting minutes. Southern Nevada Regional Planning Coalition. January 15, 2009. http://www.snrpc.org/Agendas_Minutes/Old_SNRPC_Minutes/011509Minutes (accessed December 15, 2011).
22. SNWA propaganda machine cranks out its "trust us" message. http://www.greatbasinwater.net/news/news_display.php?id=526 (accessed July 7, 2011).
23. Kay sera sera. http://chanceofrain.com/2010/05/kay-sera-sera/ (accessed May 4, 2010).
24. SNWA fails to cover-up Watergrab-gate. http://noshootfoot.blogspot.com/2010/03/snwa-fails-to-cover-up-watergrab-gate.html (accessed March 2, 2010).
25. (**1**) Nevada Legislature. http://www.ofnotenevada.com/nevada-legislature/ (accessed June 20, 2011); (**2**) ProclamationSpecialSession2010Amend1. http://www.leg.state.nv.us/Session/26th2010Special/Governor/ProclamationSpecialSession2010Amend1 (accessed February 3, 2010); (**3**) Water bill may resurface in 2011 session. http://www.lvrj.com/news/water-authority-officials-not-surprised-at-legislative-inaction-85898857.html (accessed December 19, 2011); and (**4**) SNWA manager: Snake Valley pipeline "last resort." http://www.deseretnews.com/article/705374173/SNWA-manager-Snake-Valley-pipeline-last-resort.html (accessed December 19, 2011).
26. Rose Strickland. Science & law vs. power & money. http://nevada.sierraclub.org/conservation/articles/PumpingDriesUpSprings.html (accessed March 30, 2011).
27. (**1**) Time to put up or shut up. August 8, 2009. http://noshootfoot.blogspot.com/2009/08/time-to-put-up-or-shut-up.html (accessed June 10, 2011).

28. Southern Nevada Water Authority board of directors meeting. http://water.nv.gov/hearings/upcoming/springetal/exhibits/Southern%20Nevada%20Water%20Authority/SNWA_Exh_211_SNWA%20Board%20Aug%2020,%202009.pdf (accessed November 18, 2011).
29. Southern Nevada Water Authority Board meetings. Note limitations under "Public Comment." http://www.snwa.com/about/board_meetings.html (accessed October 1, 2011).
30. Fostering sustainable behavior. https://register.cbsm.com/workshops/workshops (accessed October 1, 2011).
31. Hail Mary pass. http://en.wikipedia.org/wiki/Hail_Mary_pass (accessed December 17, 2011).
32. *Las Vegas Review-Journal*. January 19, 2012. http://www.lvrj.com/news/judge-kills-water-ruling-66826217.html (accessed January 19, 2012).
33. Mary Manning. Water pipeline timeline. Las Vegas Sun. http://www.lasvegassun.com/water/pipeline/timeline/ (accessed January 19, 2012).
34. Project Snake River. http://protectsnakevalley.com/water.html (accessed January 19, 2012).
35. **(1)** Opening Day Notes. September 26, 2011.http://chanceofrain.com/2011/09/opening-day-notes/ (accessed January 3, 2012); and **(2)** Emily Green. http://chanceofrain.com/2010/02/whereas-we-screwed-up-the-first-time/ (accessed January, 4 2012)
36. **(1)** *The Las Vegas Sun*, http://www.lasvegassun.com/news/2012/mar/22/nv-vegas-water-5th-ld-writethru/ (accessed March 22, 2012); and **(2)** *The Las Vegas Sun*, http://www.lasvegassun.com/news/2012/mar/22/water-authority-gets-ok-pump-rural-water-las-vegas/ (accessed March 22, 2012).
37. *The Salt Lake Tribune*, http://www.sltrib.com/sltrib/news/53775289-78/nevada-rulings-valley-lake.html.csp (accessed March 22, 2012).
38. *Las Vegas Review Journal*, http://www.lvrj.com/news/state-clears-snwa-to-pump-water-from-rural-nevada-143886766.html?ref=766 (accessed March 22, 2012).
39. The preceding four paragraphs are based on: Rob Mrowka, Center for Biological Diversity, http://www.biologicaldiversity.org/news/press_releases/2012/nevada-groundwater-03-22-2012.html (accessed March 22, 2012).
40. The preceding two paragraphs are based on: **(1)** Emily Green, http://chanceofrain.com/2012/03/nevada-state-engineer-refills-vegas-pipeline/ (accessed March 23, 2012); and **(2)** KLAS-TV Las Vegas, http://www.8newsnow.com/story/17170664/i-team (accessed March 15, 2012).

5 Breaking the Dysfunctional Cycle of Environmental Agency Decision Making

I change myself, I change the world.

—**Gloria Anzuldua**[1]

To break the cycle of a dysfunctional agency, we must deal with all of the pieces in context with the whole, which is something we seldom do. We try instead to fix individual pieces in intellectual isolation—symptomatic thinking, which cannot work. An agency can be restructured, for example, but that does not fix the cause of the dysfunction because it does not heal the people who make up the agency—beginning at the top.

An agency can be given a new charter, but that does not fix the dysfunction, which once again is caused by the people who make up the agency. If the basic structure of an agency—such as a U.S. Forest Service, U.S. Fish and Wildlife Service, or Environmental Protection Agency—is not remodeled from the top down by *in-agency executive choice*, it will revert to its original dysfunction within 6 months to a year because the agency's culture is *within its walls*, so to speak.

Because we (in the generic) are the agency, we are the problem, the dysfunctional components. Therefore, *we are also the solution.*

The main problem is the constant struggle to retain one's dignity and integrity in an agency that has become a machine intent primarily on its own survival by maintaining the status quo through putting out "crisis fires." Such an agency is indifferent not only to its real mission of serving the people but also to those who want to carry it forward as it was originally intended.

It is up to the individual to learn how to work within the agency while maintaining personal and professional integrity. We cannot blame externals for our failure to maintain our integrity, and we must beware of using statements that seem to absolve us of our responsibility for our personal behavior because they can only lead us into the pit of powerlessness. Whatever the external obstacles, it is our personal responsibility to face them and retain our empowerment.

Any kind of professional impotence is a condition that feeds on itself. When we, as professionals, as public servants, abdicate our personal power, we assume the role of victims and develop the cynical attitude that "we can't fight the system"—"you can't fight city hall," as it were. We thus justify the prophecy that all our efforts are doomed, and that nothing we do matters or makes a difference. And therein lies the seed of the problem.

When we surrender our power by placing all the responsibility for the failures of our efforts outside ourselves, we are in jeopardy of having our work devitalize us. If, on the other hand, in spite of the obstacle, we assume responsibility for the professionalism of our work and the content of our character, it can vitalize us and some of those around us.

We simply cannot get away from ourselves as individuals. We have all brought ourselves—as the extensions of our familial backgrounds—into the organizations of which we are part, which includes our community. We must therefore recognize that, in a sense, we have become part of an "extended family" filled with "relatives" both "near" and "distant," all of whom are doing their best to cope with their ever-changing experiences of themselves and of one another.

To break any cycle, we must first be aware that we are caught up in the cycle and how it functions. Admitting, owning, and accepting the problem brings it into the light of consciousness, where it can be recognized and addressed. The first step of recovery for an alcoholic, for example, is the conscious admission, "I'm an alcoholic." The same dynamic applies to other dysfunctional coping mechanisms.

Thus, one of the keys to unlocking the inappropriate behavior of an agency is understanding our addiction to particular dysfunctional coping mechanisms on an individual basis, perhaps by participating in a series of seminars or workshops that help us to: (1) see our relationships with coworkers as possibly similar to those existing in our own families; (2) gain an insight into the workings of our dysfunctional coping mechanisms; (3) be encouraged to examine our coping mechanisms and allow them to fade away in an atmosphere of mutual revelation and exploration; (4) have opportunities to learn about ourselves and one another; (5) learn that it is okay to have and express intense feelings we may have repressed; (6) be encouraged to join with our coworkers to re-create our past so we can become liberated from the restricting influences of our early childhood, which we have brought with us into the agency; and (7) experience the support of others in discovering the universality of our struggles, which will help us to become our own persons.

Although I can do nothing directly about the fear that is so pervasive in the world today, I can do something about the ignorance, which is the purpose of this book. If one's ignorance can be lifted to some extent, then perhaps one's fear can be alleviated in like measure. Ignorance is nothing to be ashamed of; we all have our own share. But, ignorance can be lifted if we, as individuals, are willing to see ourselves as one another's keepers, which means we must individually change our thinking so that our collective social behavior can change.

Here is a story of the difference in behavior between Fear and Love that is instructive with respect to being one another's keeper:

> A man is given permission to see both heaven (Love) and hell (Fear) while he is still alive. He chooses to visit hell first. To his surprise, he finds an enormous gathering of people at a feast. They are seated at a long table covered with every imaginable delicacy. Yet the people, all lamenting loudly, are slowly starving to death.

> As the man studies this strange scene, he observes that the handles of their eating utensils are so long that they are unable to bring the food to their own mouths. The man leaves with sorrow in his heart and goes to visit heaven.
>
> In heaven he finds an almost identical scene: the same eating utensils with handles much too long for the people to reach their own mouths. But the people of heaven are laughing and rejoicing, because, rather than try to feed themselves, as those in hell had chosen to do, the people in heaven are choosing to feed one another.[2]

In the final analysis, the choice of our individual behavior is ours. We must therefore accept that by our thoughts, which we put into actions, we can keep our current, largely destructive, ways of doing things or we can create new ways based on mutual caring and social-environmental harmony. If we keep our current ways of environmentally destructive competition and conflict, we, like the people in hell, will be the authors of our own increasing physical (environmental) and psychological impoverishment and that of our children, grandchildren, and beyond. On the other hand, we can choose to create a new way of living through other-centered cooperation and coordination, like the people in heaven feeding one another, and so offer our children and grandchildren an unconditional gift of life with a sense of well-being that equates to human dignity.

HOW JOB DESCRIPTIONS LIMIT PROFESSIONAL DECISION MAKING

Another key to unlocking dysfunction in an agency is to understand the dynamics of job descriptions and how they are used to control individuals. A dysfunctional agency seeks to control its employees because commodity production to satisfy the industrial/political elite has become more important than the process of cementing human relationships. Yet, we must ask ourselves: Is human dignity not more important than commercial products? What is an agency if it is not people organized to accomplish a mission? How can we rectify the problem?

We can rectify the problem by having the courage to take responsibility for our own thoughts, the decisions we make, and the actions we take while working within our job descriptions. And, rather than being solely unconscious functionaries, we can question the validity of what we do for the good of society beyond a few special interest groups.

We joined the agency to serve people through our professional expertise. Human dignity must therefore be the foundation and primary product of an agency—especially an agency of public service. But, as it now stands, to get any kind of "real" attention in a dysfunctional agency, as in a dysfunctional family, a person must somehow threaten the homeostasis.

There seem to be two reasons for this. One is that we no longer know clearly for whom or for what we are working, or so we are told. For example, a high-ranking laboratory official in the Environmental Protection Agency told me that I worked for the Bush administration, and therefore I had to comply with the Bush administration's wishes. "No," I said, "I don't work for the [first] Bush administration. You may, but not I. I work for the public. In fact, I work for the children." The other reason for this sad state of affairs is that we are lacking—sorely lacking—leadership and are overrun with managerialship.

Leadership is of the heart and deals intimately with human values, with human dignity, because one must lead by example, as noted by Francis Bacon when he said, "He that gives good advice [from the intellect], builds with one hand; he that gives good counsel [from the heart] and example, builds with both."[3]

LEADERS AND MANAGERS MAKE FUNDAMENTALLY DIFFERENT KINDS OF DECISIONS

A leader *knows and does the right thing* and thus leads with an other-centered, ethical conviction, usually expressed as justified enthusiasm, which causes people to want to follow with action. Essentially, a leader is one who gives people value and can motivate people by sensitive negotiation so that a perceived need is raised to strong desire. A leader leads from the heart as though human dignity mattered. The irony is that such leadership is perceived to interfere with homeostatic control of any politically oriented agency geared to maximizing the production of an economically desired product. Consequently, we have few leaders today as the heads of agencies designed to be the trustees of social-environmental sustainability.

Managerialship, on the other hand, is of the intellect and pays minute attention to detail, to the letter of the law, *doing the thing right—even if it is* ***not*** *the right thing to do*. A manager relies on the external, intellectual promise of new techniques to solve problems and is concerned that all the procedural pieces are properly accounted for, hence the epitaph: "bean counter."

Good managers are thus placed at a disadvantage when put in positions of leadership because all such people can do is rise to their level of incompetence and remain there, in which case an ounce of image is worth a pound of performance. Similarly, a leader placed in the position of managerialship is equally likely to be a dismal failure because the two positions require vastly different skills.

We need excellent leaders who have excellent managers to support them in a team effort, and both are *always* in exceedingly short supply. What we have, therefore, is primarily agency-oriented, careerist managers, who often are in over their heads in leadership positions—and no leaders.

In my experience at least, a dysfunctional agency carefully, and for the most part unconsciously, screens all candidates and selects careerist managers who lack vision and will do as they are told with little interference from their feelings of right and wrong. If, however, a misjudgment occurs on the part of the corporate/political elite, and the person chosen does, in fact, have vision and the courage to follow it, the person is soon shunned (shoved into an intellectual closet of isolation, as it were) and thereby rendered impotent—as only the "machine" can do. I have watched this process many times. It is called "out-of-sight, out-of-mind dehumanization."

Another point to consider in breaking the cycle of a dysfunctional agency is that we must ask new, other-centered, future-oriented, questions, which means that public agencies must *restate—reenvision*—their missions for the future. To accomplish this, we must clearly define the professional boundaries of expertise in a functionally integrated way—*and* ***adhere to them***—if we are to have a sustainable planet for human society, present *and* future.

Although it can be and often is argued that new people entering an agency bring new thoughts and ways with them, whereas older people leaving the agency take once-pioneering thoughts and ways with them, this influx of "new blood" does not absolve us as individuals from the responsibility of so changing ourselves that we can and will honor the dignity of each and every person with whom we work, whether we agree with them or not. In the end, it is you and I who must learn to value each other and to live by one simple principle—to accept responsibility for our own feelings, thoughts, decisions, and actions. We must learn to be kind and respectful to ourselves and to one another without forfeiting our principles. And, each agency must have a clearly stated vision, as well as clearly stated goals and objectives, that honor both the intent and the heart of the highest principles embodied in the laws of the land—laws honoring the integrity of nature's biophysical principles and our place within them—if we are to replace our dysfunctional agencies with functional ones, which includes understanding the *categories of value*.

CATEGORIES OF VALUE

The Ch'an masters who carried Zen to Japan brought Confucian ethics with them. In discussing these fundamental values as a guide to personal behavior, Confucius said, "If a man will carefully cultivate these in his conduct, he may still err a little, but he won't be far from the standard of truth."[4] When we, as individuals, clearly understand and can explicitly articulate our personal values, then we can live in keeping with them.

Let us consider three categories of values: universal, cultural, and individual. Universal (or archetypal) values reveal to us the human condition and inform us of our place therein. Through universal values, we connect our individual experiences with the rest of humanity (the collective unconscious) and the cosmos. Here, the barriers of time and place, of language and culture disappear in the ever-changing dance of life. Universal values must be experienced; they cannot be comprehended. Can you, for example, know a sunset? Fathom a drop of water? Translate a smile? Define *love* or *truth*?

Universal values are the timeless constants brought to different cultures at various times throughout history. "Even as the hands of a clock are powered from the center, which remains ever still, so the universal values remain ever at the center of human life, no matter where the hands of time are pointing—past, present, or future."[5] These are the basis of the human condition toward which people aspire (such as joy, unity, love, and peace); of these, the sages have spoken in many tongues.

Cultural (or ethnic) values are those of the day and are socially agreed on. They are established to create and maintain social order in a particular time and place and can be highly volatile. Cultural values concern ethics and human notions of right and wrong, good or evil, in terms of customs and manners.

In culture, we see reflected the ideas and behaviors that a society rewards or punishes according to their perceived alignment to its values. Hence, cultural values are for an individual a mixed bag, especially in a highly complex society that has lost its sense of family, community, and mythology, like that of the United States, where there is much that may resonate with an individual and much that may not.

Every culture is a person in a sense, and like people, there is the potential for creative interaction or conflict when cultures meet. Although we are all too familiar with cultural conflicts and the destruction they have wrought, it is good to remember that a meeting of cultures also triggers tremendous explosions of creativity in such things as language, ethics, education, law, philosophy, and government.

Individual (or personal) values are constituted by the private meanings we bestow on those concepts and experiences (such as marriage vows or gender equality) that are important to us personally. These meanings are in large part a result of how we are raised by our families of origin and what of our parents' values we take with us in the form of personal temperament. These meanings may change, however, depending on our experiences in life and how much we are willing to grow psychologically as a result of our experiences. As such, individual values are reflected in such things as personal goals, humor, relationships, and commitments.[6]

Thus, how well a people's core values are encompassed in a vision depends first on how well the people understand themselves individually and as a culture, which means how well they understand their core values, and second on how well that understanding is reflected on paper, where there can be no question about what has been stated and how.

WHY A VISION IS IMPORTANT

> Teamwork is the ability to work together toward a common vision. The ability to direct individual accomplishment toward organizational objectives. It is the fuel that allows common people to attain uncommon results.
>
> **—Anonymous**[7]

Every enterprise, be it an author writing a book, an entrepreneur building a business, or a community planning its future, needs to have the organizing context of a vision as a guiding social-environmental milepost in life's journey toward which to strive. As an organizing context for social-environmental sustainability, a community must have a shared vision with these traits: (1) It focuses a wide range of human concerns; (2) it is strongly centered in the community; (3) it can use alternative scenarios to explore a possible future by depicting in words and images that which a community is striving to become; (4) its creation relies on the trust, respect, and inclusion of interpersonal relationships; (5) it is ideally suited to, and depends on, public involvement; (6) it is ideally suited to the use of creative, graphic imagery; (7) it must honor nature's biophysical principles by accepting to live within their constraints of sustainability. Although a shared vision does not replace other kinds of social-environmental planning, it is the organizational foundation within which all other sustainable planning fits—a reality that is too often neglected. Although the word *vision* is variously construed, as used here it is a strong organizing context with four separate, but overlapping, aspects: worldview, perception, imagination, and commitment.

Worldview

Our worldview is our individualized way of seeing how the world works. It is our overall perspective from which we interpret the world and our place in it. However, it can also be seen as a metaphysical window to the world, which cannot be accounted for based on empirical evidence any more than it can be proved or disproved by argument of fact. *Metaphysical* is from the Greek *meta* "beyond" and *physika* "physical," of which Albert Einstein said, "The more I study physics, the more I am drawn to metaphysics."[8]

There are, in the most general terms, two worldviews: the sacred and the material (= commodity). One need not be religious in the conventional sense to view life as sacred with a focus on the intrinsic value of all life and its biophysical processes. Such a view gives birth to feelings of duty, protection, and love while emphasizing joy, beauty, and caring, which in turn erects internal constraints to destructive human behavior against nature. *Sacred* comes from the Latin *sacer*, which has the same root as *sanus*, meaning "sane." A sacred view of life is therefore a sane view, which corresponds to the Sanskrit: *sat*, *cit*, *ananda*, or "being," "consciousness," and "bliss."

A material or commercial view of life is one invested in domination, control (= management), and profit and seeks to "gain the world" by subjugating it to the will of the industrial mentality. At the core of this worldview are several economic concerns, such as self-interest, the economy-versus-ecology dilemma, the growth/no growth tug-of-war, Rational Economic Man, and others.[9] It is vitally necessary, with respect to a commercial worldview, to protect the functional integrity of the environment in the present for the present *and* the future through constraints placed on the destructive human behavior in exploiting Earth. In other words, we must understand the interconnectedness of the political, economic, and technological problems that cause people to feel so disempowered with respect to a peaceful political process that they resort to terrorism as a political weapon to be heard and simultaneously vent their frustration.[10]

Eighteenth century British philosopher and statesman Edmund Burke, considered the founding father of modern conservatism, understood well the need for such constraints on the destructive appetites of humanity when he penned:

> Men [people] are qualified for civil liberty in exact proportion to their disposition to put moral chains upon their own appetites. … Society cannot exist unless a controlling power upon will and appetite be placed somewhere, and the less of it there is within, the more there must be without. It is ordained in the eternal constitution of things that men [people] of intemperate minds cannot be free. Their passions forge their fetters.[11]

If, therefore, we are to improve human life appreciably, we must begin *with attitudes, not facts*. An outer change always begins with an inner shift in attitude, which Albert Einstein called "a new level of thinking." "The world we have made," Einstein said, "as a result of … [the] level of thinking we have done thus far creates problems we cannot solve at the same level at which we created them."[12] As an example of the elevated thinking Einstein is intimating, consider that the installation of dams in American rivers has been widely supported in the past. In fact, "American rivers are collectively the most closely controlled hydrologic system of its size in the world."[13] But now, there is a growing shift in thinking, which is stimulating the removal of dams in American rivers to recapture their biophysical integrity and thus the natural wealth of nature's services.

Nature's services are an unconditional gift and must be accepted with the understanding that both the quality of human life and the health of our human economy depend on the services performed "free of charge" by ecosystems, services that are worth many trillions of dollars annually, such as pollination of plants by insects, production of oxygen by green plants, purification of water by soil, and so on. Economic activities that destroy habitats and impair or eliminate services performed free of charge by ecosystems will create costs to humanity over the long term that will undoubtedly exceed in great measure the perceived short-term economic profits—and our children, their children, and their children's children will pay the price. Yet, because most of these services, and the benefits they provide, are not traded in economic markets, they carry no visible price tags that would alert society to their relative value, changes in their supply, or deterioration of the underlying biophysical processes and ecological systems that generate them.[14]

These biophysical costs are usually hidden from traditional economic accounting but are nevertheless real and progressively borne by society at large—especially the children. Tragically, a short-term economic focus in current decisions concerning land use often sets in motion great costs that, again, are bequeathed by adults not only to their own children but also to all the children of the future.[15] Unfortunately, humanity, as history shows, finds the real value of something it has taken for granted only when that something is lost. Hence, the real value of common things is too often found only in hindsight.

This is but saying that *we can design our world* in such a way that culture both creates its own harmony and is in harmony with nature. To do this, however, we must understand the dimensions of diversity and consciously, purposefully protect those dimensions that govern the biophysical sustainability of our home planet for the benefit of all generations. British Prime Minister Tony Blair summed it up succinctly:

> This is the moment to seize. The kaleidoscope has been shaken. The pieces are in flux. Soon they will settle again. Before they do, let us re-order this world around us. … Today, humankind has the science and technology to destroy itself or to provide prosperity to all. Yet science can't make that choice for us, only the moral power of a world acting as a community can.[16]

Perception

Our perception is the vision with which we see the world and interpret what we see in that we create our own world by both our attitudes toward it and our perceptions of it, a point clearly made by Army Major Susan P. Kellett-Forsyth, one of the first female graduates of West Point: "It matters less what you read than where you live and where you come from, because that determines how you interpret what you read."[17] *Perception* comes from "perceive," which is from the Latin *percipere*, "to seize wholly." "It is one of the great marvels of consciousness," writes author

Laurence Boldt, "that whatever situation we clearly perceive, we improve."[18] (For a further discussion of this concept, see *The Perpetual Consequences of Fear and Violence: Rethinking the Future*.[19])

To see wholly is to see in a better way. Therefore, to perceive a problem clearly is to begin formulating a solution. To solve a problem or resolve a conflict, we need the wisdom to keep searching and the confidence to hold what we find up to the light of understanding. It is when we doubt our capacity to love and to create, which we then replace with fear and isolation, that we begin to distort our perceptions of the world. When our view of the world is based on the love and confidence of clear perception, the world becomes a better place.

Nevertheless, we most commonly move from avoidance and confusion to attention and clarity only when we perceive the necessity of doing so. "Necessity," wrote Greek author Plato, "is the mother of invention."[20] Necessity, in this sense, is the perception that a situation has become intolerable and *must be addressed* by the perceiver.

Once necessity is accepted, we begin searching for a solution to our problem by examining our old perceptions. Such action forces us out of the "proverbial box" of our prejudices and conceptual limitations so that we can sift through old ideas and concepts that have heretofore been overlooked or discarded. Necessity must be accompanied, however, by intense urgency unless we too easily and too often give up searching for a better way of being or doing.

Like beauty, necessity is in the eye of the beholder. No matter how deplorable a situation seems to an observer, the individual in the situation finds no reason to improve it until such improvement becomes a personal necessity. Until our discontent is moved to necessity and we demand a better way, we will accept that which is of lesser quality. Because we fear, and thus hate, the things that seem to trap us, we find no way out until we supplant our fear with confidence.

These notions call forward our individual and collective choices and decisions, namely, to yield to the comfortable blindness of ignorance *or* to summon our courage and make resolute our determination to search until we find a better way. It has been wisely said somewhere in time that anything will reveal its secrets if you love it enough.

Imagination

Imagination, or seeing that which can be, is the third aspect of a vision. Even as we open our physical eyes and see the world as we think it is, with all its problems and opportunities, so we can open the eye of our mind and see the positive or negative potentials as yet unrealized. To open our physical eyes fully, we must learn to trust so we can accept *what is, as it is* through the eyes of love. To open the eye of our mind, we must learn to trust that what we see in our imagination, we can bring forth in the physical world.

Whereas perception involves seeing that which already exists in the outer world, imagination involves seeing the inner world of possibilities that can be realized in the outer world. Albert Einstein was correct when he said, "Your imagination is your preview of life's coming attractions,"[21] to which William Butler Yeats added, "In dreams begin responsibility."[22] Consider, therefore, that everything humanity has ever created (or ever will create), both tangible and intangible, began as an idea in the privacy of someone's mind, be it this book, a religious order, going to the moon, or a beautiful city. Our imagination is the source of our creative power and the driving force behind our choices and decisions—the prerequisites of a shared vision toward which to build.

To solve our social-environment problems, therefore, we, as a community, as a society, must have a destination in the form of a vision toward which to journey. Can there be such a journey to a better world? I believe there can be since we have a guide in Cassie Campbell, who was in the fourth grade at Adams Elementary School in Corvallis, Oregon, when she wrote

> I wish for world peace because some children go to sleep at night and are afraid that they might get kidnapped or murdered. And that is very possible.
>
> If there was world peace, then there wouldn't be shootings at schools. The president of the United States of America would not have to work so much. Parents wouldn't worry about their children walking to the park without an adult. People would be safe at night even if they didn't lock their doors. Little kids wouldn't be scared of horror movies because they can't [come] true. People wouldn't be afraid of the dark. There wouldn't be such a thing as soldiers. You would live longer lives, possibly. People would treat each other with kindness and they would be happy. They wouldn't know any other way to act. There would be no suicides.
>
> Everybody would share and give. We all would share and give. We all could stand up straight and be proud. So, if I had one wish, it would definitely be this.[23]

To even begin planning the journey Cassie Campbell would have us set our compass by, we must resolve some major differences within our social ranks by focusing on a systems approach to our thinking that both accepts and values—as necessity—the contributions of symptomatic thinkers because they also are part of the whole. Author Michael J. Sandel frames the issue well: "Despite their disagreements, liberals and conservatives share an impoverished vision of citizenship, leaving them unable to address the anxiety and frustration abroad in the land. If American politics is to recover its civic voice, [in any sort of a civil manner], it must find a way to debate questions we have forgotten how to ask."[24]

Framing a debate sends a clear and powerful message to the public about the breadth, depth, and authenticity[25] of the issue and the sincerity of those who address it. The way a debate is framed and the choices posed set the boundaries of discourse, beyond which politicians stray at their peril because they run the risk of appearing irrelevant or radical and thus effectively lose their voice.[26]

Despite how well a debate is framed, most people cling to their personal beliefs regardless of debating points won or lost. And no amount of persuasion makes any difference. Money, on the other hand, moves mountains because, for most people, the ultimate measure of success is their monetary accumulation.

"But," you might ask, "can one person really make a difference anyway?" The answer is emphatically, yes, as exemplified by Jamie Lerner, a 63-year-old engineer, architect, city planner, and former mayor of Curitiba, capital of the state of Paraná, Brazil. Using his childlike innocence, which is at once the cradle and midwife of possibilities, he imagined Utopia and set sail.

Just 50 years ago, Brazil was primarily a rural country in which only 36% of its people lived in towns. Today, 78% live in metropolitan areas. This rapid urbanization was accompanied by formidable problems of sprawling slums, pollution, a strain on energy and public services, and traffic congestion. But not so in Curitiba, where foresight and careful planning made this city a world-famous example of environmentally sound people-oriented solutions. Curitiba has the best quality of life in Brazil.

This city of 1.6 million people is known as the "Green City" because of its beautiful parks and wooded areas. Free of pollution, with clean streets and no sign of beggars or "street children," which are so common in many Latin American capitals, the literacy rate in Curitiba is 94%, compared to a national average of 81%. How did this come about? It came about through the creative vision of Jaime Lerner.

Lerner, who, as governor of Paraná, was cheerful, easygoing, and informal—more at home working and receiving guests in shirtsleeves while surrounded by forest and waterfalls at the rustic Curitiba Institute of Research and Urban Planning than in the governor's palace. Lerner started the changes in Curitiba during his first term as mayor, from 1971 to 1975, and continued them during subsequent terms from 1979 to 1983 and 1989 to 1992. The subsequent mayor and his secretary of urban planning shared the vision with Lerner, so it continued to unfold.

Lerner's inspired transformation of Curitiba is referred to as "the largest set of joint urban changes made in a single city in the world." Unlike many people who feed their fears with a disaster mentality and thus see only one view, that of cities becoming increasingly unlivable, Lerner has hope for the urban future. "My optimism," he says, "comes from believing in human beings, in their capacity to create and to transform." What humanity does now, he believes, will define our collective future, which, in the last analysis, is based on how we choose to behave.[27] Our behavior, in turn, is based on what we learn, which makes education paramount if we are to replace the reign of Fear with that of Love.

All human values must be taught, nurtured, and lived. They can be neither legislated nor coerced. One can force compliant behavior, but not compliant values. Until we, as parents, teachers, citizens, and decision makers and leaders become psychologically mature in our own right, we will continue to teach our children—and one another—how to attempt avoiding the symptoms of our problems, rather than how to embrace their causes and thus find their potential cures.

In true community, people collectively create a context in which each member can feel sustained, nourished, stimulated, engaged, and appreciated while maintaining a sense of personal autonomy within the protection of an interdependent whole. Interdependence means focusing on the gift, the value each person brings to the common weal instead of focusing on what we perceive to be a lack in this person or that because when one person is elevated, all are elevated in like measure; when one person is diminished, all are diminished in like measure. We, as decision makers, must seek strengths in others rather than dwelling on perceived flaws and weakness that we all possess, which brings to mind a French proverb: "To know all is to forgive all." We must pull when needed and push when required if the goal of moving a community toward social-environmental sustainability is to be realized.

Genuine communities take seriously the life situations, aspirations, identities, and necessities of their members and support them as they struggle individually and collectively toward the community's shared vision of its desired future. Next to personal growth, building, maintaining, and nurturing community are perhaps the hardest work we will ever do because they require us not only to engage our hearts and souls fully but also to be vulnerable, open, and humble and to find the intestinal fortitude to be consistent in our efforts with a commitment of bulldog tenacity to see them through to fruition.

UNDERSTANDING A VISION

Crafting a clear vision provides decision makers with: (1) a better understanding of the core values of the people involved in the process and how to use their ideas, talents, and skills as a basis for planning; (2) identification of the trends, circumstances, and forces affecting an organization, such as a community or agency; and (3) articulation of a wide-angle picture in time and space to guide short-term decisions in relationship to long-term outcomes and initiatives. Contrarily, the greatest single cause of failure to achieve desires is not grasping the importance of a vision as the initial decision point for guiding the course of decisions and actions or a commitment to its implementation. This is especially true of communities and environmental agencies and their social-environmental decisions. Therefore, it is necessary to understand that a vision has three basic parts: (1) goals and objectives; (2) negotiability of constraints; and (3) monitoring to test the effectiveness of constraints and accomplishment within those constraints.

Vision, Goals, and Objectives

Defining a vision and committing it to paper goes against our training because it must be stated as a positive in the positive, something we are not used to doing. Stating a positive in the positive means stating what we want and mean directly. For example, a local community has an urban growth boundary that it wants to keep within certain limits, which can be stated in one of two ways: (1) We want our urban growth boundary to remain within a half a mile from where it is now situated (a positive stated as a positive), or (2) we do not want our urban growth boundary to be any bigger because it will look like that of our neighbor (a negative that one is attempting to state as a positive).

Further, to save our planet and human society as we know it, we must be willing to risk changing our thinking to have a wider perception of the world and its possibilities, as well as to validate one another's points of view or frames of reference. The world can be perceived with greater clarity when it is observed simultaneously from many points of view. Such conception requires open-mindedness in a collaborative process of intellectual and emotional exploration of that which is and that which might be, the potential result of which is a shared vision of a possible future.

Although a vision may begin as an idea, at some point it becomes enshrined in people's hearts as a palpable force. It then becomes impossible to turn back and accept that which was before. Few, if any, forces in human affairs are as powerful as a shared vision of the heart. Consider Mahatma Gandhi's inspired fight to free India from British rule. The secret to his success was to produce a shared vision among the people of an independent India.

In its simplest, intellectual form, a shared vision asks and answers these questions: What do we want to create? Why do we want to create it? Beyond that, it becomes the focus and energy to bring forth that which is desired. As John F. Kennedy said, "Those who anticipate the future are empowered to create it,"[28] which is similar to Gandhi's statement, "The future depends on what we do in the present."[29] It is also similar to Franklin Delano Roosevelt's statement, "The only limit to our realization of tomorrow will be our doubts of today."[30]

A statement of *vision* is a declaration that describes what a person, group, business, community, agency, or nation is striving for. A vision is like a "vanishing point," the spot on the horizon where the straight, flat road on which you are driving disappears from view over a gentle rise in the distance.

As long as you keep that vanishing point in focus as the place you want to go, you are free to take a few side trips down other roads and always know where you are in relation to where you want to go—your vision. It is therefore necessary to have at hand a dictionary and a thesaurus when crafting a vision statement because it must be as precise as possible. Throughout the vision statement, you must *say what you mean and mean what you say*, which is the most difficult part of a visioning process—especially a shared vision.

In contrast to a vision, a *goal* is a general statement of intent that remains viable until it is achieved, the need for it disappears, or the direction changes. Although a goal is a statement of direction, which may be vague and not necessarily expected to be accomplished, it serves to clarify the vision statement further. A goal might be stated as follows: My goal is to create a successful ecotourism business.

An *objective*, on the other hand, is a specific statement of intended accomplishment. It is attainable, has a reference to time, is observable and measurable, and has associated costs for achievement. Additional attributes include that it: (1) starts with an action verb; (2) specifies a single outcome or result to be accomplished; (3) specifies a date by which the accomplishment is to be completed; (4) is framed in positive terms; (5) is as specific and quantitative as possible and thus lends itself to evaluation; (6) specifies only what, where, and when and *avoids why and how*; and (7) is product oriented.

For example, my goal is to create an ecotourism business. Let us now make it into an objective: I will have a successful ecotourism business in my hometown by my 25th birthday. The stated objective is action oriented: I will have. It has a single outcome: a successful ecotourism business. It specifies where: my hometown. It specifies a date: my 25th birthday, and it is framed in positive terms: I *will have*. It lends itself to evaluation of whether the stated intent has been achieved, and it clearly states *what*, *where*, and *when*. Finally, it is product or outcome oriented—to have a successful ecotourism business in my hometown. (For a more in-depth discussion of this section, see *Vision and Leadership in Sustainable Development*.[31])

The Negotiability of Constraints

The vision of some future, desired condition, by its very nature, elicits the singular social constraint (the fixed point around which everything else turns, like the hub of a wheel) that must be met if the terms of the vision are to be fulfilled. A constraint in this sense means being restricted to a given course of action or inaction, which connotes something that restricts, limits, or regulates personal behavior.

A vision does not create constraints where there were none before because all things are constrained by their relationship to everything else; thus, nothing is ever entirely free. What a vision determines is the degree to which a socially chosen constraint is negotiable. In addition, a fulfilled vision blurs interdisciplinary lines because it rests with those who created it and those who are inspired by it. That does not necessarily extend to those whose job it is to administer the bits and pieces of everyday life, however important they might be.

To "negotiate" is to bargain for a different outcome, to cut the best deal possible. Nature does not negotiate, however, which means that some of the conditions we encounter are non-negotiable; we cannot cut a "better" deal, one more to our liking. Therefore, the challenge is to determine what is negotiable and what is not. It is important to accept, with humility and grace, that which is not negotiable and learn to account for and accept responsibility for the price of that which is negotiable because negotiability has costs as well as benefits.

When we negotiate, we trade one set of behavioral freedoms for another in that we impose a particular constraint on ourselves through a vision to alleviate or free up some other potential constraint for the common good in the future—the desired outcome of our vision. But, if "freedom is not linked to morality," contends Mikhail Gorbachev, "it is not freedom. It is permissiveness. It is just self-seeking, rather than freedom."[32] Hence, the question is: What does the creation of a shared vision do to the negotiability of our self-created, self-imposed theories, rules, and regulations?

A shared vision determines, by its defined outcome, the degree of negotiability that can be afforded to those of our self-created, self-imposed theories, rules, and regulations that are in question. In other words, a vision determines the negotiability of any particular social constraint, both in terms of social justice and in terms of nature's inviolable biophysical principles. The constraints with which we have to deal in everyday life are, in a human sense, social decisions because they are behavioral, be it how one interprets the rights of private property, how one conducts oneself in church, or where someone chooses to build a house.

Because a community's visioning process is a public one, it gives the people the right to comment on all aspects of the process (from creating the vision itself through its implementation and monitoring), which in effect places control of the process directly in the hands of the people should they choose to accept responsibility for the outcome. It is the responsibility for the outcome that demands an understanding of and exacts the accountability for how people accept the social constraints dictated by the vision. And, because people are ultimately responsible for their own behavior, they require a mechanism through which they can measure the appropriateness of their performance, which is the express purpose of monitoring.

Monitoring Tests a Vision's Effectiveness

Although the word *monitor* is used in various ways, its meaning here is to scrutinize or check systematically with a view to collecting specific kinds of data that indicate whether you are moving in the direction you intend. Monitor has the same origin as *monition*, which means a warning or caution and is derived from the Latin *monitio*, "a reminder."

With respect to social-environmental sustainability, monitoring means to keep watch over and warn in case of danger, such as straying from the desired course. On the one hand, monitoring is to highlight and so remind us of activities that we already know might be too harsh and could offend the system; on the other hand, it is to help us conserve the options embodied within the system for future generations and ourselves, but this requires the ability to ask relevant questions because monitoring is dependent on questions.

Framing a Relevant Question

Framing effective questions is critical to crafting a collective vision and for monitoring what is necessary to achieve the vision. Questions can, when used wisely, open the door of possibility. For example, there could be no journey to the moon until the question was posed: Is it possible to go to the moon? Asking the question was the first step in actually going to the moon. Consideration of a possibility is the essential first step in any journey.

Effective questions must: (1) have a specific purpose, (2) contain a single idea, (3) be clear in meaning, (4) stimulate thought, (5) require answers to address the human relationship that prompted the question effectively, and (6) explicitly relate to previous information.

A question that elicits "right" versus "wrong" answers is a pointless, hopeless exercise because it calls for moral judgments, and that is not a valid question to ask of either an ecosystem or science. If, however, the question were addressed to whether a proposed action was "good" or "bad" in terms of a community's or agency's collective vision, that is a different matter and a germane question.

For example, a good, short-term, economic decision may simultaneously be a bad, long-term ecological decision and thus a bad, long-term economic decision. To find out, however, the question must be asked.

In essence, questions lead to the array of options from which you can choose. Conversely, without a question, you are blind to the options. Learning about the options is the purpose of monitoring. In turn, knowing what to monitor and how to go about it, you must know what questions to ask because an answer is only meaningful if it is in response to the right question, one that sheds light on the "invisible present" in which we all live.

Seven Steps of Monitoring

Because the future is fraught with uncertainties, the best monitoring and the best adjustments (target corrections) based on that monitoring provide several preplanned actions, such as the following: If A happens, I will do B; if C happens, I will do D; and so on. Otherwise, we monitor only outcomes, which we use to illuminate the need for corrective actions.

For example, we have assumed that placing fish ladders in dams will sustain the migration of salmon and thereby perpetuate their successful spawning and survival of the smolts (young salmon). Over time, however, we learn that the reservoirs created by the dams (in addition to a host of other human activities) also affect the salmon's survival and migration. Therefore, monitoring a single variable, no matter how seemingly logical, may be adequate to ensure viability of the fish runs. This being the case, it is well to understand that ecosystems, and the species involved, are inevitably more complex than we are able to foresee. Monitoring provides information essential for better understanding an ecosystem's feedback loops and what corrective actions, if any, are necessary.

Monitoring has seven steps: (1) crafting a vision, goals, and objectives; (2) preliminary monitoring or inventory; (3) modeling our collective understanding; (4) preparing a plan; (5) implementing the plan; (6) monitoring effectiveness of the plan; and (7) monitoring to validate the outcome.

Step 1. Crafting a Vision, Goals, and Objectives

Crafting an adequate vision statement, with goals that clearly and concisely state desired future conditions, is the first step in monitoring. This statement makes it clear that you know what you want, where you want to go, why you want to go there, and what you think the journey will be like. The vision and its goals form the context of the journey against which you measure (monitor) all decisions, actions, and consequences.

Step 2. Preliminary Monitoring or Inventory

Preliminary monitoring is the careful observation and best understanding of the initial circumstances or present conditions—taking inventory, as it were. This step requires answers to the following questions: What are the present conditions? What is the prognosis resulting from that assessment? Though preliminary monitoring may necessitate multiple questions, the outcome could require a single action.

Step 3. Modeling Our Collective Understanding

Modeling is configuring knowledge into a conceptual framework, an explicit map of people's collective understanding. It is assumed at the outset that such a model represents their best understanding of the situation. It is critical, however, to have the humility to face the fact that the model is at best an "informed estimate" and is certain to be flawed in some aspects; in addition, the collective understanding of the current situation and the interacting variables is incomplete. Therefore, the viability of the vision, goals, and objectives (the overall model) must be continually tested and improved. As the model is improved, so is the knowledge of how the project is functioning. In turn, this provides information that can be used to make improvements in the model and thereby improve the collective knowledge of how the project is likely to function.

With respect to models, Swiss physiatrist Carl Jung made an important point when he said that, while a mean is a valid, unassailable, fundamental fact in theory, it need not occur in reality. In addition, the exceptions of either extreme, although equally factual, do not appear in the final result because they cancel each other out. If, for example, you were to weigh each pebble in a jar of pebbles and get an average weight of 5 ounces, you would still know little about the real nature of the pebbles. If you think, based on this average, that you can pick up a pebble weighing 5 ounces on the first try, you will be disappointed. Indeed, however long you search, you might never find a pebble weighing exactly 5 ounces.[33] (For an in-depth discussion of this topic, see *Evaluating Sustainable Development: Giving People a Voice in Their Destiny.*[34])

Step 4. Preparing a Plan

The next step is to write a plan of how you propose to achieve the vision and its attendant goals within some scale of time in which all the particulars, including objectives, are laid out. For your convenience, I will reiterate attributes of an objective, which is a specific statement of intended accomplishment. It is attainable, has a reference to time, is observable and measurable, and has an associated cost. An objective must: (1) start with an action verb; (2) specify a single outcome or result to be accomplished; (3) specify a date by which the accomplishment is to be completed; (4) be framed in positive terms; (5) be as specific and quantitative as possible and thus lend itself to evaluation; (6) specify only what, where, when, quantity, and duration; and (7) be product oriented.

Once you have determined your objective(s), you not only will be able to but also must answer the following questions concisely: (1) What do I want? (2) Where do I want it? (3) When do I want it? (4) How much (or how many) do I want? (5) For how long do I want it (or them)? If a component is missing, you may achieve your objective by default but not by design. Only when you can answer all of these questions concisely do you know where you want to go and the value of going there and only then can you calculate the probability of arrival. Next, you must determine the cost, make the commitment to pay it, and then commit yourself to keeping your commitment.

Step 5. Monitoring Implementation

Monitoring implementation requires an answer to this question: Did we do what we promised to do? Although this is simply documentation of what was done, it is critical because the question makes it possible to know what may have gone awry (if anything did), how and why it went awry, and how to remedy problems.

Step 6. Monitoring Effectiveness of the Plan

Monitoring effectiveness involves assessing the achievement of the objectives—not achievement of the goals or vision. A vision and attendant goals describe desired future conditions and are qualitative in nature. Objectives, on the other hand, are quantitative and thus specifically designed for quantification.

Such monitoring asks the following: (1) Is the objective adequately specific? (2) Are the results clearly quantifiable? and (3) Are they achieved within the specified timelines? Monitoring effectiveness with indicators provides information (feedback) with which to assess whether progress to achieve a desired future condition (achievement of collective vision) is on track, maintaining current conditions, or moving away from the desired future condition.

Monitoring for effectiveness involves systematic evaluation of indicators relevant to achieving the vision. A good indicator helps in the recognition of potential problems and provides insight into possible solutions. What is chosen for measurement, how that is to be accomplished, and how the outcome is interpreted will have a tremendous effect on the quality of life in the long term. (Again, for an in-depth discussion, see *Evaluating Sustainable Development: Giving People a Voice in Their Destiny.*[35])

The results of such monitoring can lead the involved community back to the beginning premise and provide answers to whether they are better off due to the project and, if so, how; if not, why not; and if not, how the situation can be improved; if so, how; if not, why not; and so on ad infinitum.

Step 7. Monitoring to Validate the Outcome

Monitoring for validation of the outcome is sometimes to be considered "research," as opposed to monitoring, because it involves testing assumptions that went into the development of objectives and the models on which they are based.

Monitoring for validation may require asking such questions as: Why did the results not meet expectations? What does this mean to the validity of the conceptual model employed? Will altering the approach produce a different outcome? If not, why not? If so, how and why? What corrections might make the model more reflective of how the system really functions?

Validation is a necessary component of any monitoring plan because it provides insights useful in making corrections in the model. Moreover, validation is useful not only in the current application but also in the design of future applications in other projects.

Because outcomes relative to shared visions are uncertain, human activities that are based on a collective vision can be thought of as tentative probings into the various aspects of nature and our reciprocal relations. Such probings are best taken and tested one step at a time. With each such probe, potentially adverse or unpredicted effects may be detected and corrected before serious, widespread, damage occurs.

SUMMARY

Chapter 5 illustrated the ways in which dysfunction and homeostasis are corrected for the benefit of all generations. These include understanding how job descriptions are used to control information and how positions of authority are used to control direction for political purposes. The correction culminates with the crafting of a vision, goals, and objectives, as well as understanding and implementing the seven steps in monitoring the vision's effectiveness.

Section III is a discussion of what it takes to be a psychologically mature decision maker who actually serves as a leader with the good of all generations in mind. Here, it is pointed out that (other than natural disasters) all social-environmental problems begin with our thinking. Put differently, we create the vast majority of the social-environmental problems that we end up facing. We do it to ourselves. Understanding both this and nature's biophysical limitations, the discussion is focused on how people must change their thinking through individual, personal growth if we, as a nation and as a global society, are to have decision makers who actually lead from a position of psychological maturity for the social-environmental sake of all generations. Chapter 6 is an enumeration detailing the essential qualities of a psychologically mature leader who makes wise, long-term, systemic decisions as a wise trustee of our home planet.

NOTES

1. Gloria Anzuldua. http://thinkexist.com/quotes/gloria_anzaldua/ (accessed February 4, 2011).
2. Chris Maser. *Ecological Diversity in Sustainable Development: The Vital and Forgotten Dimension*. Lewis, Boca Raton, FL, 1999.
3. Francis Bacon. http://quotationsbook.com/quote/1246/#axzz1CIo4rDMo (accessed January 27, 2011).
4. **(1)** D.T. Suzuki. *Zen and Japanese Culture*. Princeton University Press, Princeton, NJ, 1959; and **(2)** Lin Yutang. *Wisdom of Confucius*. Random House, New York, 1938.
5. Laurence G. Boldt. *Zen and the Art of Making a Living*. Penguin/Arkana, New York, 1993.
6. The foregoing discussion of values is based on Boldt. *Zen and the Art*.
7. Anonymous. Winning with teamwork quotations. http://englishpatterns.com/community/viewtopic.php?f=2&t=746 (accessed January 29, 2011).
8. The with and wisdom of Albert Einstein. http://freaky_freya.tripod.com/einstein.html (accessed January 30, 2011).
9. **(1)** Russ Beaton, Chris Maser, and Kevin Smith. *Setting the Stage for Sustainability: A Citizen's Handbook*. Lewis, Boca Raton, FL, 1998; and (2) Russ Beaton and Chris Maser. *Economics and Ecology: United for a Sustainable World*. CRC Press, Boca Raton, FL, 2012.
10. Fritjof Capra. True security. *Resurgence,* 211 (2002):9–11.
11. Edmund Burke quotes. http://www.goodreads.com/author/quotes/17142.Edmund_Burke (accessed January 30, 2011).
12. Quotations by Albert Einstein. http://www-history.mcs.st-and.ac.uk/Quotations/Einstein.html (accessed January 30, 2011).
13. William L. Graf (chair). *Dam Removal: Science and Decision Making*. H. John Heinz III Center for Science, Economics and the Environment, Washington, DC, 2002.
14. Gretchen C. Daily, Susan Alexander, Paul R. Ehrlich, and others. Ecosystem services: benefits supplied to human societies by natural ecosystems. *Issues in Ecology,* 2 (1997):1–16.
15. Ibid.
16. British Prime Minister Tony Blair. http://www.sweetliberty.org/issues/war/reorder.htm (accessed December 26, 2011).
17. Robert D. Kaplan. Fort Leavenworth and the eclipse of nationhood. *The Atlantic Monthly*, 278 (1996):75–90.
18. Boldt. *Zen and the Art*.
19. Chris Maser. *The Perpetual Consequences of Fear and Violence: Rethinking the Future*. Maisonneuve Press, Washington, DC, 2004.
20. Plato. http://www.quotationspage.com/quote/33942.html (accessed January 30, 2011).
21. Albert Einstein. http://www.google.com/#hl=en&sugexp=ldymls&xhr=t&q=Your+imagination+is+your+preview+of+life's+coming+attractions&cp=0&pf=p&sclient=psy&site=&source=hp&aq=f&aqi=&aql=&oq=Your+imagination+is+your+preview+of+life's+coming+attractions&pbx=1&fp=3f8c1f6954acb9e1 (accessed January 30, 2011).
22. William Butler Yeats. http://thinkexist.com/quotation/in_dreams_begin/156889.html (accessed January 30, 2011).
23. Cassie Campbell. World peace would end fears. Corvallis (OR) *Gazette-Times*, April 3, 2001.
24. Michael J. Sandel. America's search for a new public philosophy. *The Atlantic Monthly*, March (1996):57–60, 62–63, 66, 68–70, 72–74.
25. Robert B. Reich. The other surplus option. *The New York Times*, August 11, 1999.

26. For a discussion of authenticity, see Chris Maser. Authenticity in the forestry profession. *Journal of Forestry,* 89 (1991):22–24.
27. The foregoing story of Jamie Lerner is based on P.J. Loughran. Creating a slicker city. *Hemispheres*, October (2000):42, 44, 46.
28. John F. Kennedy. www.adfusa.org/.../777_file_Local_Economic_Development_Strategies_ Maual.pdf (accessed January 30, 2011).
29. Louis Fischer. *The Essential Gandhi: An Anthology of His Writings on His Life, Work, and Ideas*. Random House, New York, 1962.
30. Franklin Delano Roosevelt. http://www.quotedb.com/quotes/2691 (accessed January 30, 2011).
31. Chris Maser. *Vision and Leadership in Sustainable Development.* Lewis, Boca Raton, FL, 1998.
32. Mikhail Gorbachev. http://lists.econ.utah.edu/pipermail/a-list/2006-March/060543.html (accessed January 30, 2011).
33. Carl G. Jung. *The Undiscovered Self.* Mentor, New York, 1958.
34. Okechukwu Ukaga and Chris Maser. *Evaluating Sustainable Development: Giving People a Voice in Their Destiny*. Stylus, Sterling, VA, 2004.
35. Ibid.

Section III

What Makes a Psychologically Mature Decision Maker?

> As we evolve, we're able to transform the situation and the people around us by helping them to fulfill their purpose. Our purpose is to enlist the purpose of other people. That is really the secret of leadership.
>
> **—Sufi teacher Pir Vilayat Inayat Khan***

* Pir Vilayat Inayat Khan. http://ahamprema-bliss.blogspot.com/2009_04_01_archive.html (accessed January 24, 2011).

6 The Essence of a Responsible Decision Maker

> A hundred times every day I remind myself that my inner and outer life depend upon the labors of other men, living and dead, and that I must exert myself in order to give in the measure as I have received and am still receiving.
>
> **—Albert Einstein**[1]

A decision maker cannot keep personal values and beliefs out of the relationship with those he or she would lead; if one does, then leadership is vague and hollow. Although a decision maker must be willing to discuss the issue of values openly, to do so implies that one's own philosophy and core values will at times be revealed, which must be done with total honesty and grace. A leader has a twofold, ethical obligation to expose his or her personal values in a dignified manner and to refrain from imposing them on others.

Leadership is not meant to be a form of indoctrination whereby the leader manipulates the people at large to act or feel in the "right" or "politically correct" way. Each person's sense of truth is just that: each person's individual sense of truth based on ethnicity, upbringing, experience, and his or her resulting perception of reality. But, no one knows what is right or true for anyone else. Good leadership, therefore, requires a great deal of humility.

Unfortunately, there are many well-intentioned decision makers who are overzealous in "straightening" people out, which implies that, by virtue of their greater wisdom, they will provide the answers to the troubled populace. But, leadership is not synonymous with preaching, which is not to say that a leader should simply accept whatever the people say by remaining passive and silent.

PERSONAL VALUES AND PHILOSOPHY OF LIFE

One of a leader's duties is to challenge the values of the citizenry when he or she senses that a certain behavior is unwise or destructive and thus stealing freedom of choice from generations to come.[2] It then becomes a duty to confront the people and invite them to examine the payoffs and consequences of their choices, subsequent decisions, and resulting actions because they affect not only the people who make the decision but also their children and grandchildren in the present and all children as yet unborn.

A core issue in leadership is the degree to which and the way in which a leader's values enter into their relationship with the public through decision making. Leadership is about accepting and giving counsel to arrive at mutually acceptable and sustainable decisions; it is not about dictatorship, which is no more than blatant command-and-control decision making.

Ethically sensitive leaders are openly conscious of their own values and encourage followers to develop theirs. Leaders must challenge the values of their followers, however, and help them decide whether they are truly living by their professed values or merely espousing perceived parental and societal values without consciously evaluating them. Leaders must also be alert to the possibility of manipulating the citizenry into uncritically accepting values wholesale and thereby simply becoming a substitute for a parent, something too many people today already expect their leaders and government decision makers to be.

At this juncture, it is important to emphasize that no person can hold a neutral thought in his or her mind. Hence, no person can be truly objective. Because leaders' personal values do in fact influence their relationships with the people at large, it is crucial for them, as major decision makers, to be absolutely clear about their values and how they influence both their work and the future into which they would lead the people.

People both need to know and have a right to know where their leaders stand on values and issues in order to test their own thinking. People deserve forthright and honest involvement on the part of their leaders. Because leaders themselves do not have all of the answers, leadership must be a process whereby the people are challenged to evaluate their own values honestly and then decide for themselves which direction they want their future to take, albeit the will of the majority rules in a true democracy. But how, you might ask, in all of this does a leader deal with personal needs in the light of his or her responsibility as a decision maker for the overall good of the public—present and future?

Just as leaders cannot exclude their own values from their relationships with the people at large, neither can they hope to keep their personal needs, and thus their personalities, separate from those they lead. Leaders must recognize the supreme importance of becoming consciously aware of their personal needs; areas of unfinished emotional business, most often related to their upbringing; potential personal conflicts; dysfunctional coping mechanisms; and their sense of vulnerability when it comes time to make difficult decisions. They must realize how these personal realities might prevent their followers from freely and fully exploring certain dimensions of themselves.

Unless leaders develop this conscious self-awareness, they will obstruct the ways in which followers can change or will use followers to satisfy personal needs. Leadership then shifts from what is best for the people to what is best for the leader, as reflected in the decisions rendered. The crux of the matter, therefore, is to avoid exploiting the people for the sake of meeting personal needs.

Here, one might ask what kind of personal awareness is crucial. We all have blind spots and distortions of reality. It is therefore the responsibility of leaders, both to themselves and to the people they serve, to work continually toward expanding self-awareness with the aim of recognizing areas of personal distortion, bias, prejudice, and vulnerability that might affect the systemic wisdom of their decisions. It is particularly important that leaders become increasingly aware of the nature of unfinished emotional business that might come to the fore in their relationship with those who follow. To accomplish this, one must consciously, purposefully do what Swiss psychiatrist Carl Jung called one's "inner work."

A leader must develop sensitivity to unmet personal needs so that leadership is not used as a means to satisfy them. If a leader both recognizes and works through his or her own personality problems, there is less chance of projecting them onto the ordinary citizenry through self-serving decisions. A good leader must recognize that one's effectiveness depends on one's ability to create and maintain sound personal relationships, and that personal problems may interfere with those relationships.

As Gerald Corey says of psychotherapists, leaders have other aspects of their personalities that must be examined if they hope to be instrumental in using themselves to create healing relationships.[3] These aspects include the need for control and power; the need to be nurturing and helpful; the need to change others in the direction of their own values; the need to teach, preach, persuade, and suggest; the need to feel adequate, particularly when it becomes overly important that the people confirm one's competence as a leader; and the need to be respected and appreciated.

Such needs are neither neurotic nor necessarily destructive, but they must be kept in a healthy perspective if your needs, as a leader, are to be met by helping others find satisfaction in their own lives. Many personal needs for dignity, self-worth, and respect come from the quality of your relationships with coworkers. If deep satisfaction is not derived from work, then it is likely you are in the wrong profession, and the decisions you make are equally likely to be symptomatic and thus unwise over the long term.

For this reason, ethical leadership demands that leaders recognize the central importance of continuously probing their own depths to determine in which direction their personality is leading the constituency—through decision making that leads toward betterment or stagnation.

CHARACTERISTICS OF AN EFFECTIVE DECISION MAKER

"The true test of character," according to John Holt, "is not how much we know how to do, but how we behave when we don't know what to do."[4] Too often, people who lack within themselves the inner authority of true social-environmental decision makers—hereinafter simply termed *decision makers*—find themselves in social positions of power, which they confuse with leadership and thus abuse the limited authority of the position. Such people are smitten with power and are loath to relinquish it, even momentarily, for fear of losing it altogether. These people cannot lead because effective, sustained decision making requires a great deal of trust, a clear sense of interdependence, a clear sense of high principle, a clear vision for the general good of all generations, and the courage to stick with it despite any and all personal costs.

To wit, in December 1783, General George Washington rode to Annapolis, Maryland, where the Continental Congress was meeting. Although he learned that only seven states had sent delegates, he wanted to affirm what Congress stood for, even if the present was but a shadow of the formative days of 1776.

Around noon on the 23rd of December, General Washington strode to the statehouse, where a solemn Charles Thomson, the Irish-born Philadelphian who had served as secretary of Congress since 1774, met him at the door. Thomson escorted General Washington into the chamber, where 20 congressmen and the current president, Thomas Mifflin of Pennsylvania, awaited him amidst aisles and galleries packed with spectators.

"Sir," said Mifflin, "the United States in Congress assembled are prepared to receive your communications."

Washington rose and bowed; the congressmen in turn raised their hats as a sign of respect but without bowing. By withholding their bows, the congressmen affirmed the civil superiority over the power of the military.

Washington, with trembling hands, began his statement by conveying his happiness that the United States was now a "respectable nation," enabling him to resign "with satisfaction" the commission he had accepted "with diffidence." He overcame his diffidence, he said, because he had been confident of "the rectitude of our Cause," the support of Congress, and "the patronage of Heaven."

Although it is said that he could not read the close of his speech because he was so choked with emotion, this is what he had written: "I consider it an indispensable duty to close this last solemn act of my official life, by commending the interests of our dearest country to the protection of Almighty God, and those who have the superintendence of them, to his holy keeping."

"The whole house felt his agitations," wrote one congressman to his fiancée. "The spectators all wept and there was hardly a member of Congress who did not."

Struggling with the final words, Washington said, "Having now finished the work assigned me, I retire from the great theatre of action, and bidding an affectionate farewell to this august body under whose orders I have so long acted, I here offer my commission, and take my leave of all the employments of public life."[5]

What, you might ask, is the significance of this story? The significance is that liberty is always expensive. True liberty demands that each and every leader know not only when their duty to history has been completed but also when to step down with dignity and grace so that the pivotal idea of a true democracy as the central pillar of our nation can deepen in the centuries to come.

With this in mind, it seems clear that today's means of exercising power and authority must give way to new forms of leadership. Our concept of leadership must be recast to include the ability to foster collective, systemic decision making and collective action. If leadership is to find its highest expression in service to community, society, nation, and world as a whole, we must increasingly choose leaders who are truly motivated by the desire to serve, as evidenced through the long-term wisdom of their decisions.

It is, after all, a leader's duty to create a response from the followership, not the followership's duty to respond to the leader. While we unfortunately hear most often of the abuses of power, power can be a good—even essential—thing when used constructively and toward a positive end, which means that leadership must always be handled carefully and generously—never carelessly or selfishly.

Other Centered and Authentic

A true leader is other centered in his or her decisions and is therefore concerned primarily with facilitating someone else's ability to reach his or her potential as a human being by helping that person develop personal talents and skills and value his or her experiences. Authentic leadership thus comes from the heart and deals intimately with human values and human dignity—the heart of which is social-environmental sustainability.

Authenticity is the condition or quality of being trustworthy or genuine. Beyond the dictionary definition, authenticity is the harmony within what one thinks, says, does, and what one really feels—the motive in the deepest recesses of one's heart. The adage "deeds speak louder than words" is true as far as it goes, but what is left unsaid is that "motives speak louder than deeds." One is authentic only when one's motives, words, and deeds are in harmony with one's attitude—as evidenced through one's decision and subsequent actions.[6]

Ralph Waldo Emerson wrote, "Your attitude thunders so loudly that I can't hear what you say."[7] A person's attitude is the visible part of that person's behavior, but a person's motive is often hidden from view. When, therefore, one's visible behavior is out of harmony with one's motive, this attitude points to a hidden agenda. In turn, an authentic person is one who is willing to risk shedding stereotypical roles and be a real person in a relationship.

There is, however, a current lack of public trust in leaders—a lack of trust based on a perceived lack of authenticity. Consider the following observations of Leonard Pitts, Jr. (columnist for the *Miami Herald*), "I'm hard put," wrote Pitts in 1997, "to remember the last time a public figure expressed remorse and made me believe there was really any there." It is not a shortage of wrongdoing, he continues, but rather the apologies we have seen, which are "slickly crafted written statements often followed by carefully orchestrated acts of contrition." What we have not seen is real regret from people who are genuinely sorry for their misdeeds.

Public penitence, contends Pitts, is today a stage-managed process choreographed by media consultants who always give the same advice, "Admit the wrongdoing, express contrition, and move on—preferably all in one breath." There is no apparent need to feel sorry.

"The buck stops here," read the sign on President Harry S. Truman's desk, Pitts continues. The buck today is "passed around like the common cold." When, however, there must be some accountability, the apologies come with a footnote of "self-justification, and outright insincerity."

President Clinton, for example, in responding to allegations of irregularities in campaign financing, said, "Mistakes were made." But he never said by whom they were made, leaving us, quips Pitts, to assume the mistakes occurred under their own volition. When the president of the United States either cannot or will not stand up and forthrightly admit that he "screwed up royally," why should we, as a society, expect any better from lesser figures in the parade of world leaders? If this is the fare our leaders expect us to swallow, why even bother listening?

"They pretend to be remorseful and we pretend to believe them," concludes Pitts. We are, he says, grateful for at least the appearance of propriety, even though we know that something necessary is missing—authentic humility and penitence.[8]

The current lack of public trust in leaders may, in a narrow sense, be brought about by a leader's dishonest behavior or, in a broader sense, by a leader's perceived self-centeredness. But, it may also be caused (through no fault of the leader) by unrecognized cultural evolution.

Cultural evolution expresses itself through changing values. Culture is not genetically inherited. It can only be learned from the past, modified in the present, and passed on to future generations. The notion of culture poses two questions:

1. What happens when the evolution of culture and the resulting shift in values in one part of a society tears the social fabric with great force?
2. How do we heal the social rupture that results from this shift in cultural values?

If society has learned anything from the decade of the 1960s, it must be that one cannot unilaterally destroy "the establishment" without offering a viable alternative with which to replace it. Before an old paradigm can be cast out, there must be a new one to take its place.

How do we know when a new paradigm is at hand? We know because we are suddenly faced with a crisis that tells us that our old belief system is no longer as functional as it once was. There is a point at which we must break free of our habitual belief systems and either choose to change or stagnate. The frightening thing about change is that it means letting go of familiar people and places as we move to another stage in life. Each new paradigm is built on a shift of insight, a leap of intuition.

On the collective level, this means that cultural evolution takes place with only a modicum of hard data. Those who cling to the old ways may demand irrefutable proof that change is needed, but such proof is almost inevitably lacking. The irony is that the old way was once upon a time a "new way," which was challenged to prove that change was necessary or even desirable.

Time and human effort have proven the current paradigm to be more "in tune" with the knowledge of its time than its predecessor was but still only *partially* correct in terms of future knowledge. So it is with the new paradigm; it is more in tune with new knowledge than the current paradigm. Yet, it too will eventually be proven only partially correct and in need of change.

The trap is that any paradigm that has become comfortable has become self-limiting. New data cannot fit into the old way of thinking, which has grown rigid with tradition and hardened with age. It is therefore necessary to crack open the old periodically if new thought forms are to be allowed to enter and grow, moving the individual forward to a renewed sense of authenticity.

Here, an analogy from nature may be helpful. A crab whose body has grown enough to fill its hard shell, which is really an external skeleton, must have the shell split open so the soft-bodied crab can exit what has become a limit to its growth. The crab, once out of its old, rigid shell is vulnerable for a while, as its shell gradually hardens around its soft body. This process is repeated as long as the crab lives and grows. So it is with human culture.

An old logger brought this home to me rather forcibly some 40 years ago. I was asked to give a speech on new ways to practice forestry, ways that were deemed to be sounder ecologically. Accordingly, I spent much time explaining the necessary changes in the techniques of logging if we were to have ecologically sustainable forests. I finished the session feeling that I had done a good job and was basking in the warm afterglow when an old logger came up to me.

"Sonny," he said, his bright blue eyes snapping under his thinning silvery hair, "I'm sure you have a good point there, but I jist can't find it. Mostly," he continued with a broad grin, "I think you're jist full of shit."

Taken aback, I asked him why he thought that.

"Well," he replied, "the way I think of a forest, you just don't make no sense nohow." Having said that, he winked at me and left.

As though struck by a bolt of lightning, I realized that he was absolutely correct. Without a dramatic shift in the philosophical underpinnings of our belief systems (a cracking open of the old, rigid shell, as it were), new data simply have nowhere to go.

Moving forward may be difficult for those whose belief systems and personal identities are totally invested in the old paradigm because, in their perception, there is no reason to change. For those who subscribe to a new paradigm, moving forward is easier because there is something toward which to move—a new view that hints at a more sustainable future, a view more in tune with the ecological understanding of the moment. Yet, those who harbor new ideas are no better than those who cling to the old ways; the two views are only different because each person's knowledge originates in a different time and context. However, the most current knowledge is on the side of those who embrace the new paradigm.

Historian Arnold Toynbee asked the critical question of why 26 great civilizations fell. The answer, he concluded, was that the people would not, or believed they could not, change their way of thinking to meet the changing conditions of their world.[9]

Thus, communities can move forward, evolve, if you will, only to the extent that individuals within those communities are ready and willing to grow personally and accept new philosophies and methods of doing business demanded by a rapidly changing knowledge and responding cultural norms. Communities cannot and will not remain the same—despite those who attempt to thwart change. Consequently, those who will not accept new ideas, for whatever reason, must fall by the wayside. Nevertheless, most people seem to resist change, even that which they understand to be good in the long term and for the generations of the future if it means they must forgo their immediate desires, pleasures, and "comfort zones," which they interpret as part and parcel of their personal rights, well-being, and security.

The crux of the issue is that cultural evolution ("changing times," if you will) is really about changes in how we relate to one another, to ourselves, and to the world around us. And, it is within the context of changing relationships that people experience growth from the authenticity of a true leader, who in turn may accelerate the rate of cultural evolution for the greater good by rendering wise, systemically viable decisions.

If, however, a leader hides behind "political correctness" or within the rigid walls of an established position and refuses to accept responsibility or risk change, citizens will remain hidden from that leader. If a leader is merely a technical expert who leaves his or her sense of value and self out of the equation, the result will be sterile, hollow. It is through personal authenticity and aliveness that a leader can significantly touch the citizenry.

If leaders make life-oriented choices, radiate a zest for living, are real in their relationships with the citizenry, and let themselves be known, despite personal problems and errors, which we all have and make, they can inspire and teach the people in the best sense of the word—through example by the quality of the decisions they make. This does not imply that leaders have all the answers or are without personal problems, but it does mean that leaders must be willing to look at their own life and have the courage and willingness to make the same kinds of adjustments they want the people at large to make.

A leader who "walks his or her talk," as it were, can extend hope to the people and, by example, can inspire them to strive mightily in the face of all the uncertainties embodied in the risks of change. Hope, after all, is seeing the opportunities in an uncertain future and mapping one's course toward them.

In short, leaders are models for the people. If they model incongruent behavior, low-risk activity, deceit through vagueness and duplicity, they will not be trusted, and people of like mind will take the example as permission to act in kind. If, on the other hand, a leader models realness by engaging in appropriate self-disclosure, he or she can inspire a greater following based on trust and emulation. In the end, the degree of aliveness, commitment, and psychological maturity of a leader is the crucial variable that determines the outcome of leadership, as evidenced through the wisdom and common good *or* the self-centeredness and social-environmental folly of his or her decisions. To accomplish this, however, one must have honor.

An Honorable Person

Personal honor is a gift one gives to oneself. It is a code of ethics by which one lives and conducts oneself. One's honor is his or her personal integrity, which is maintained without obligation—legal or otherwise. At this juncture, a decision maker who would be a true leader must understand that personal integrity includes each commitment one makes to oneself, and that such personal commitments are every bit as important to keep as those made to one's constituents. It is that inner essence, the inner standard of personal honor that makes the quality of one's character.

The interpersonal manifestation of honor and trust must be based on truth. To make this point, consider the following true story, which took place in northern Australia, where some reporters tape-recorded two policemen insulting an Aboriginal person. The reporters then printed the transcript on the front page of the newspaper.

The chief of police, an intelligent and open-minded man who had for some time been trying to reshape his police force, was to be interviewed about the incident on television. Hearing this, the minister of justice told the chief to fire the two policemen and warned him to be careful of what he said publicly, or he would be fired also.

But instead, on camera, the chief said: "I must tell the truth, and the truth is those two policemen are not an exception. The rest of the police could have done the same. But, I will tell you more. The rest of the population of my region of Australia could have done the same as well. We are becoming racists."

There was an explosion in the press and on television, and the chief of police was fired the next day. The firing had not been approved by the Parliament, however, and the opposition party in the Parliament applauded him for one full minute for telling the truth in public. This action forced the government party to applaud as well, which resulted in the minister being fired and the chief being reinstated.

The upshot is that the police in that area of Australia became the best in the country because the truth had been spoken. After truth has been spoken, there is not only room for but also permission for personal growth and personal authenticity, which leads to trust—the stuff that makes communities.[10]

With respect to trust, you and I can take someone else only as far along the path toward healing as we ourselves have been willing to go. Because social-environmental sustainability in community development is based on trust in human relationships, those in authority can render no greater service than to protect the people's empowerment to choose freely. The whole purpose of choice is for local people to guide the development of their own community as sustainably as possible within the reciprocal, self-reinforcing feedback loop between their decisions and the long-term productivity of their landscape. After all, the local people and their children must reap the consequences of any decisions they make.

A leader must have an inner sense of honor, such that one's word is one's bond. At one time, and among some today, a contract between two people was sealed with a handshake, and each party's word was as sound, or more so, than a legal document.

Israeli Prime Minister Yitzhak Rabin was a shining example of a man with personal honor because he was through and through a warrior, but one who had the courage and wisdom to see that peace was the only road to the security of his country, and he took that road—even though it cost him his life. While honor is normally thought of in terms of men, it is an attribute shared equally by women, which brings me to the subject of balance between a leader's masculine and feminine aspects.

Balancing the Masculine and Feminine

Although a decision maker can lead someone from here to there in the outer, physical-spatial landscape without actually having been there before, psychologically, a person can only lead someone to a given place of consciousness in the inner landscape if he or she has in fact been there. For example, linear, left-brain thinking, which is a masculine trait, is more prevalent and comfortable in our society than is right-brained thinking, which is a feminine, relationship-oriented trait.

While tapping into the right side of the brain is more powerful than using solely the left side, it is difficult to appeal to the right side because it is so often dormant in our society, where relationship and creativity are by and large dashed on the social rocks of patriarchal conformity and quantification. Therefore, to appeal to the right side of the brain, the conceptual side, a decision maker either must have a strong, innate balance between the two modes of thinking or must consciously develop it before being able to stimulate it in others.

It is a necessity for a good leader to be well balanced between both sides of the brain, the feminine and masculine aspects of the self, because a truly effective leader must be able to make an emotional connection with those who follow. I say this because a leader cannot change people's minds but can bring personal convictions and values to the surface. Before this is possible, however, an emotional connection must be made, and that lies within the realm of the right side of the brain. Consider the Arabic proverb, which says something to the effect that a wise man knows the scale of justice can be tipped by a feather's weight and, therefore, is most careful in thought, word, and deed.

English philosopher James Allen said it beautifully:

> Law, not confusion, is the dominating principle in the universe; justice, not injustice, is the soul and substance of life; and righteousness, not corruption, is the molding and moving force in the spiritual government of the world. This being so, man has but to right himself to find that the universe is right; and during the process of putting himself right, he will find that as he alters his thoughts toward things and other people; things and other people will alter toward him.[11]

To understand the necessary equality of justice, read the same passage, as it is transcribed for women by Dorothy J. Hulst, and do your best to sense the difference:

> Law, not confusion, is the dominating principle in the universe; justice, not injustice, is the soul and substance of life; and righteousness, not corruption, is the molding and moving force in the spiritual government of the world. This being so, woman has but to right herself to find that the universe is right; and during the process of putting herself right, she will find that as she alters her thoughts toward things and other people; things and other people will alter toward her.[12]

A common human trait is to make decisions based on emotion and justify them with "fact." Although you may disagree with this statement, believing yourself to be totally rational and objective because you do thorough research before you make a decision, think back to your last major decision, your last big purchase—a car, house, an important article of clothing. Why did you select the one you did? Was it color, size, how it made you feel, or what you thought it might do for your social standing that caused you to select it? After your initial reaction, did you feel that you needed to justify the price? This same principle applies to concepts and ideas. We buy the emotion before we accept the "facts," which lends credence to the observation by American author Charlotte P. Gilman that "a concept is stronger than a fact."[13]

Therefore, leaders who appeal to the right side of their followers' brains, who touch them emotionally, make a lasting connection. Such leaders not only create experiences for their constituency or cause a long-dormant memory to be vividly recalled but also make a memorable connection that has the power to motivate.

But, before any of this can happen, before a new idea will be accepted, the leader must be likable as a person. Do we feel good about him or her? Is he or she authentic and trustworthy? This is critical because genuine authenticity is rare on the one hand and the way to the emotions and the soul on the other. Truly committed leaders are often described as dynamic, charismatic, riveting, or engaging because the authenticity of their message, their vision, is palpable, especially if it is one their constituency can also believe in.

Such a leader makes people feel good about themselves, which in turn makes them want to follow, if for no other reason than to be around the leader. Although a powerful leader can leave people on a natural high, to be genuine, a leader must also be therapeutic as a person.

A Therapeutic Person

A therapeutic person is one who has healing powers, for, as Carl Jung said, it is the personality of the doctor that has the curative effect. But why, you might ask, do you use the term *therapeutic*? I use the term because a good decision maker not only must bear the uncertainties and fears of those who follow but also must help them reach beyond their fears and accomplish more than they thought possible. In doing so, everyone involved becomes part of something greater than himself or herself.

In this sense, an effective leader treads, albeit lightly, in the realm of psychotherapy. Because a leader asks followers to risk looking within themselves, it is critical that the leader exhibit the courage to search within and hold his or her own life open to the same scrutiny asked of his or her constituents—all the while knowing that, in moments of weakness, we have all done things of which we may be ashamed.

Professor Gerald Corey has provided an excellent synopsis of a therapeutic person in terms of being a psychotherapist, and that synopsis can be adapted and be useful in this discussion.[14] As Corey says of psychotherapists, leaders must continually ask:

1. What do I personally have to offer those who are struggling to find their way?
2. Am I living my life honestly, freely, and boldly, as I am urging others to do?
3. How can I become a therapeutic person?
4. How can I be an instrument of awareness and growth for those who follow?

Here, I return to the notion of a leader as a public example whose life, as Mahatma Gandhi once said of his own life, is their message—a message based on the social-environment viability of the decisions they make.

Although the best leaders are born as such, all those who find themselves in positions of leadership can acquire extensive practical knowledge and make that knowledge available to their constituents. But until that knowledge is integrated into their own life, it is seldom, if ever, inspirational because, in the end, one always becomes an example to constituents, which then manifest the unique human qualities and experiences that have molded the would-be leader. This is a critical concept because a true leader must, above all else, inspire others to follow the noble path of service for the benefit of *all* generations.

If, therefore, leaders are to promote inner growth and change in their constituents, they must be willing to grow and change within themselves by exploring their own choices and decisions and striving to be aware of, accept, and act on their own potential for growth. A leader's willingness to live in accordance with his or her own truths, and thereby set an example in how he or she lives, yields a positive model that makes leaders effective, "therapeutic persons" for their followers.

As Corey states, and I emphasize, the following list is *not* proposed as a model for attaining perfection as a leader. Its importance lies in the struggle to attain these dimensions, even if the efforts fall short. It is the leader's willingness to remain open to the struggle of continual personal growth that is crucial, for both the ever-developing leader and those who follow.

I propose this evolving list of characteristics of a therapeutic person not as some dogmatic itemization of what is "right," but to provide a foundation for examination of your own concept of a therapeutic leader. You are free to add any characteristics that you feel may be missing. To better allow you to *feel* the list, it is presented in the first person and framed in the positive:

1. I can only lead effectively where I have personally traveled. I am always in the process of developing a style of leadership that reflects both my philosophy of life and my style of living. Although I sift through and borrow ideas from others, I remain true to myself in how I apply that which I borrow.
2. I respect myself and appreciate both what I am and what I am consciously becoming. I give out of my own authenticity, rather than seek a sense of fulfillment from others. I have the humility to ask of and receive from others without isolating myself as a means of control.
3. I am comfortable with my own sense of personal power and am secure enough to allow other people also to be comfortable with their sense of power. I have no need to diminish others or encourage them to maintain a subordinate stance. I consciously use my power to provide a healthy model for those who follow me.
4. I am in touch with myself and thus open to change and willing to take calculated risks. I am willing to plunge into the unknown, where I find within my uncertainties my previously untapped potentials.
5. I strive continuously to expand my awareness of others and myself, while realizing that limited freedom comes with limited awareness and vice versa. To enrich my own life and that of others, I focus on new experiences that expand my awareness of cause and effect, thereby improving my relationships.
6. I am both willing and able to accept ambiguity. I understand that personal growth means leaving the perceived safety of the familiar for the uncertainty of the unknown. Entering the realm of unknown territory emphasizes life's ambiguities, which can be perceived as a threat to my existence, but which I view as my hidden potential to which I am inextricably drawn. As I learn to trust my intuition, I become increasingly trustworthy to others.
7. I learn to know what I am, what I am capable of becoming, what I want from life, and what is essential. I am willing to continuously and consciously reexamine my values and strive to get in touch and stay in touch with my inner core and live from the authenticity of my own center. My standards are internalized, and I have the courage to act in a way that is consistent with my principles. I know the only way to communicate my inner truth to others is to live that truth.
8. I am conscious of my own struggles and pain from which I gain a frame of reference that allows me to identify with others, yet helps me avoid getting mired in someone else's emotional quicksand.
9. I am committed to living life to the fullest and refuse to settle for mere existence. As Winston Churchill once said about the road to success, I go from "failure to failure without losing enthusiasm."[15] At the same time, I know that success or failure is *my interpretation* of an event—not the event itself.
10. I am authentic and refuse to hide behind masks of persona, sterile roles, or facades. I will risk being genuine.

11. I am able to give and receive love from the fullness of my soul. I am thus vulnerable to those I love and have the capacity to care for others.
12. Recognizing that shame can only live in the past and fear can only live in the future, I choose to live in the present, where there is neither shame nor fear. I choose the eternal now and invite others to join me.
13. Although I make mistakes, I willingly admit them. I do not lightly dismiss my errors but choose to learn from them, rather than be encumbered with pointless shame and misery.
14. My "work" is a labor of love and a way of living. Although I find deep, intrinsic meaning in my work, I am not enslaved by the "identity" it provides. My work is coupled with other dimensions of my life, which collectively provide a sense of purpose and fulfillment.
15. Unbound by my past ways of being, I am constantly changing as I struggle to become that which I want to be. In so doing, I constantly revitalize and re-create significant relationships.
16. Being open to my emotional experiences, I can be emotionally present for others, both in their pain and in their joy.
17. Aware that I must make choices that shape my life, I am not a victim of my past decisions. If I err, I can always choose to choose again, and yet again if necessary.
18. I openly challenge unreasonable assumptions and self-destructive beliefs and attitudes rather than submit to them. Moreover, I focus on the positive rather than limit myself by dwelling on the negative.
19. In the end, all I have to give of value is my love, trust, respect, and the benefit of my experience. I am also aware that people do not care about how much I know as a leader and decision maker until they know how much I care about them. Someone once said that soldiers will follow a leader whom they know is willing to die for them if need be. This is the kind of caring that really counts.
20. Choice—whether conscious or unconscious—lies at the root of human relationships and directs their outcome. All choices are the new beginnings of a never-ending story of cause and effect. So, I must choose wisely on behalf of all concerned—present and future.[16]

Clearly, this snapshot of a leader might seem unrealistic. Who can be all these things? These are just characteristics of psychological maturity for which to strive to reach one's potential for leadership entrusted with the responsibility of making social-environmental decisions that benefit *all* generations. As the Chinese proverb points out, the journey of a thousand miles must begin with a single step.[17]

I emphasize the word *strive* because acceptance means taking the other person as the person is, without reservation, which includes your own relationship with yourself. Mere tolerance, on the other hand, is not acceptance. When people say they like this but not that about another person, their attitude reflects their biases concerning the "assets and liabilities" of that person, which is not love. Love is unconditional acceptance of all a person's virtues *and* faults—including your own.

Why struggle to achieve these aspects of leadership? A truly competent leader must be more than just technically proficient. He or she must have a genuine love of people and a considerable degree of psychological maturity, which translates into a greater-than-average degree of wisdom and *inner soul quality*.

I hope you will find this view of leadership insightful and useful in developing your own concept of leadership—one you deem essential to those people whom you would lead, so that one day you would be comfortable following them.

Detachment and Equanimity

Detachment from an outcome is unconditional acceptance of what is, which is a critical concept in true social-environmental decision making. Detachment is checking your ego at the door as you come into the room, which, at best, is incredibly difficult for most people.

When I was younger, for example, I was deeply upset by the clear-cut logging of the old-growth forests in the Pacific Northwest, where I grew up. I would argue long and loudly about the need to save them and the greed and stupidity of those whose actions would liquidate them—for whatever reason. I tried to convince anyone and everyone that the old-growth forests needed to be saved. I became so rabid in presenting my point of view as the right one that few people cared to listen unless they already agreed with me. Consequently, I became frustrated, cynical, and self-righteous, which only made matters worse. I became enraged at the "greedy bastards who were clear-cutting *my* forest," but I never thought to ask them how they felt about the old-growth forests they were liquidating, ostensibly for all time.

One day, as I was giving a passionate speech on the need to "preserve" the ancient forests of the Pacific Northwest, I suddenly felt the sword taken from my hand, and a sense of peace come over me. The tenor of my presentation changed, the challenge in my voice was gone, replaced with quiet respect, and the shift was immediately reflected in the attitude of the audience. Several people came up to me later and said they had never thought about it that way, and that what I said made sense. It was then I realized that to speak for the forests or for anything else, *I had to change*—not the people in the audience. If I wanted people to listen, it was incumbent on me to change, to say what I had to say in a way that would allow them to hear. But how? I did not know how.

A few weeks later, I saw the movie *Gandhi*. Then, I read a couple of biographies about Gandhi in which he was often quoted, and through his writings he gave me the answer. I had to detach myself from the outcome of my endeavor, a truly difficult task.

If Gandhi was correct, in detachment lay acceptance of the outcome. Expectation is the attachment, the vested interest in the outcome, because the person with the expectation thinks he or she possesses the means to achieve the right and justifiable result. If, on the other hand, I act willingly out of duty to a Higher Authority, I can act with detachment because the Higher Authority is acknowledged as possessing the wisdom to govern the outcome justly. Detachment here does *not* mean that I act without commitment. Quite to the contrary, while I am firmly committed to the principle that serves all people in the greatest good, I am detached from an outcome that would serve only the desires of my own ego.

If I am detached, I have no vested interest in the results of a given process as it might affect my personally desired outcome. This means that I can treat all sides, all points of view, and all possible outcomes with equanimity—but only to the extent that I can set aside my own ego-desires. *Equanimity* is the kernel of peace in *detachment* just as surely as *anxiety* is the kernel of agitation in *attachment*. (Equanimity is from the Latin *aequanimitās* "even or equal feelings," whereas, anxiety is from the Latin *anxius*, "to torment.")

For example, a person who has worked passionately for a cause may suddenly have the insight that passion placed before principle is a house divided against itself that cannot long stand. Because of this new understanding, the person now becomes focused on the principle as a process and becomes detached from the passion—the desired result. The reaction of their peers most often is: "How can you give up the cause? We've believed in it for so long."

For these people, attachment to the cause has become life itself, their very identity. Therefore, even as they ostensibly fight to "win," they cannot afford to win because, if they were to actually resolve the issue at the heart of their cause, they would have to find a new identity, something most people are loath to do.

If a leader is truly detached from the outcome, he or she will find equanimity to be his or her touchstone. Equanimity, the outworking of detachment, is reflected in the calm, even-tempered, and serene personality of one who is simply open to accepting what is. Such a person can lead without wasting energy through either the need for or the expectation of approval or a predetermined outcome. Such a person acts out of peace.

In turn, the peaceful action allows others to see an alternative way of perceiving something because no one is trying to convince them of anything. They are given the ideas and the space to consider them. Then, if they so choose, they can change their minds in privacy while retaining their dignity intact. Here, it is critical to understand that I cannot convince anyone that I am "right" without first convincing the person that he or she is "wrong," in which case I have already stolen the person's dignity, and all the person can do is defend him- or herself.

The leader who is detached from the outcome is part of the principle for which he or she stands and is therefore part of the resolution or transcendence of the problem. On the other hand, a leader becomes part of the problem when attached to a point of view and its *necessary* outcome. A leader's detachment and equanimity serve to make followership an exciting perspective because people feel safe in the care of one with equipoise of character.

A Good Follower

> I must follow the people.
> Am I not their leader?
>
> **—Benjamin Disraeli, 19th century British prime minister**[18]

You must learn to follow before you can learn to lead. Learning to be a good decision maker means starting at the bottom of the ladder as a simple follower and learning about each rung and its relationship to every other rung as you climb the ladder to the top. In an inspired moment, President Theodore Roosevelt said, "To me there is something fine in the American theory that a private citizen can be chosen by the people to occupy a position as great as that of the mightiest monarch … and that … after having filled this position, the man … goes back into the ranks of his fellow citizens."[19] Back into the ranks from which one rose goes the citizen—down from the pinnacle of power to the status of a private individual.

Elizabeth Sherrill gives an excellent example of a man who rose from followership to leadership and then went back to followership with grace. As roving editor for *Guideposts* magazine, she once interviewed Harry Truman at his home in Independence, Missouri. As she and her family left the Truman home, her 5-year-old son, Donn, began to cry in the back seat of the car.

"What's the matter?" she asked Donn.

"I wanted to see the president!" he wailed.

"You did. You met Mr. Truman," she replied.

"No." He shook his head. "He wasn't a *president*. He was just a man."

"Just a man," mused Sherrill. "Just a man who for a time had wielded more power than anyone else on earth. Just as Theodore Roosevelt would have said approvingly, an American president."[20]

Beginning as a follower at the bottom of the leadership ladder is critical because, as leader, you must never ask anyone to do something that you yourself are not willing to do—and one does not lead forever. It is therefore imperative that you know what it feels like to be a follower.

The best example I can think of is a movie *The Doctor*, in which William Hurt, an emotionally shut down, rather arrogant, surgeon is diagnosed with cancer and suddenly finds himself a patient at the mercy of doctors like himself. Having seen himself as a doctor through his own eyes as a patient, he makes all the interns with whom he subsequently works spend time as patients so they will understand—with a great deal of humility—what it feels like to be on the receiving end of their services. This is an example of why one must learn to follow before one can learn to truly lead.

A good leader always searches for signs of leadership talent among his or her followers and in so doing looks for certain characteristics that help identify a potential leader to encourage. These characteristics include: (1) authenticity, (2) other centeredness, (3) using adversity to advantage, (4) persistence, (5) learning from one's mistakes, (6) focusing on positive thoughts, and (7) seeing, recognizing, and seizing an opportunity.[21]

Authenticity can be characterized as "what you see is what you get." In other words, a person is simply who they are without a hidden agenda or false persona. The best leaders are other centered, which means the person is more interested in serving rather than garnering power.

Potential leaders have their share of failures and frustrations but learn how to take advantage of such experiences. What might be an obstacle to a follower represents an opportunity to a ripening leader.

Persistence, the relentless pursuit of a goal, is manifested in developing leaders because it is a prime ingredient for success. William Penn observed, "Patience and diligence, like faith, can remove mountains."[22] And all leaders must overcome mountains of rejection, dismissal, and repudiation to achieve the vision of their ideals, the personal fulfillment of the highest principles they live by.

Rather than beat themselves for their mistakes, developing leaders learn from their mistakes. As the actress Rosalind Russell pointed out, "Flops are a part of life's menu."[23] In such learning, errors become the raw material out of which future success is forged.

Consider James E. Burke, chairman of the Johnson & Johnson Company, as an example. When he first became head of the division of new products at Johnson & Johnson, he was responsible for developing a chest rub for children, which failed on the market and cost hundreds of thousands of dollars.

Burke expected to be fired over his failure, but when summoned by the chairman of the board, he received a surprising reception, "Are you the one who just cost us all that money?" asked Robert Wood Johnson. "Well, I just want to congratulate you. If you are making mistakes, that means you're taking risks, and we won't grow unless you take risks."

Jim Burke remembered the lesson, and years later, when he became chairman of the board, he reminded other junior executives of the importance of taking risks and learning from mistakes.[24]

A follower is one who focuses on positive thoughts by filtering out negative ones on the trail of leadership. Helen Keller, who became blind and deaf shortly after birth, often counseled people to "keep your face to the sunshine and you cannot see the shadow."[25] Similarly, President Dwight D. Eisenhower was fond of saying, "No pessimist ever won a battle."[26]

But, it was tennis player Billie Jean King who provided the most graphic example of positive thinking. Every time she made a bad swing with her racket, she followed it with the correct swing, even in a televised tournament. She corrected mistakes as she made them and thus embedded the thought and feeling of a positive swing in her psyche.

Seeing, recognizing, and seizing an opportunity is a behavioral trait that stands out because it is rarely encountered. People who affect society and make great leaders are those who recognize an opportunity and seize the moment. It is an important quality of leadership because, as author Dennis Waitley reminds us, "Opportunity rarely looks like an opportunity. Often opportunity arrives incognito, disguised as misfortune, defeat, and rejection."[27] People who recognize an opportunity and seize the moment are worth looking for in the crowd.

Taken in isolation, these traits do not identify a potential leader. But, in the collective, they often identify those with potential to become servant leaders. Followers who want to become leaders must develop and nurture such qualities.

In addition to these qualities, a good follower must consciously exhibit the following characteristics:

1. *Loyalty*—implies commitment not only to the leader and the vision of service but also to the principles of sound leadership. Loyalty manifests itself in one's willingness to work daily with enthusiasm on your leader's behalf. But, as German politician and cofounder of the German "Green Party," Petra Kelly reminds us, "Loyalty toward the whole of life is far more important than any ideology."[28]

2. *Understanding*—the ability to articulate and integrate into your daily life the vision and principles espoused by your leader.
3. *Candor*—the courage to speak your mind clearly, succinctly, and authentically to your leader and fellow followers, but gracefully and in private.
4. *Listening*—demands attention and care and must be coupled with observing the subtle nuances of a leader's speech and behavior if maximum clarity of understanding is to be achieved.
5. *Predictability*—being accountable for your own behavior in such a way that your leader knows who can be counted on when the need arises.
6. *Creativity*—having a beginner's mind that allows you to discover or help discover novel solutions to the problems of leadership as they arise.
7. *Effectiveness*—getting things done in a manner that helps accomplish the intent of an action.
8. *Efficiency*—getting things done in the most expedient, effective, and cost-responsible manner without compromising either the quality or principles of sound leadership.
9. *Insightfulness*—the ability to ask relevant, probing questions and foster innovative ways of seeing and thinking about ordinary things; the ability to advance new perspectives.
10. *Honesty*—allows a leader to know you can be trusted to represent the vision or principles accurately with the highest standards of integrity.
11. *Persistence*—stay with the task until it is resolved or all conceivable approaches have been exhausted.
12. *Practicality*—being grounded enough to face problems directly and develop thoughtful, positive suggestions for their resolution.
13. *Communication*—personal commitment to inform your leader of significant developments before they come as surprising news, which put your leader at a disadvantage.
14. *Helpfulness*—the constant willingness to assist in achieving the vision and upholding the principles underpinning that vision.
15. *Complementarity*—willingness to assist in such a way that your thoughts and actions complement, rather than compete with, those of your leader.
16. *Cheerfulness*—the choice and determination to maintain an even-tempered disposition come what may; hence, the ability to smile in the face of adversity and make things a little brighter for everyone.[29]

The foregoing characteristics are also those of a servant leader because a servant leader must at times be a follower.

Servant Leader

As auto manufacturer Henry Ford said, "Coming together is a beginning; keeping together is progress; working together is success."[30] In the end, it is the collective heart of the people that counts; without followers, there is no need for leaders. Chinese philosopher Lao-tzu thought a good leader was one who "when his work is done, his aim fulfilled, they will all say, 'We did this ourselves.'"[31] Such is servant leadership.

A good servant leader has a fine balance between the masculine and feminine aspects of his or her personality. Such balance is critical in authentic social-environmental decision making because servant leadership often means putting relationships ahead of immediate achievement and knowing when each is important.

A servant leader intuitively knows that service is an attitude based on credible social-environmental principles, not a function. Hence, such a leader in today's world does what is right from an ethical conviction supported by sound evidential scientific knowledge of systemic processes that further social-environmental sustainability, usually expressed as enthusiasm, which causes people to want to follow with action.

A leader is one who values people and helps them transcend their fears so they might be able to act in a manner other than they were capable of on their own. This is the essence, and the first rule, of true leadership. As such, it calls to mind a scene from the movie *Karate Kid II*, in which Miyagi, a Japanese man who is the central character, is translating the rules of karate displayed on the walls of the Miyagi family dojo in Okinawa: "Rule number one: karate is for defense only. Rule number two: first learn rule number one." One might translate this to leadership as: Rule number one: Leadership is service to others built on a foundation of inner strength of character. Rule number two: First learn rule number one.

Leadership involves authority, or control—the right or power to command, enforce laws, exact obedience, determine, or judge. Two kinds of authority are involved, that of a person and that of a position.

The authority of a person begins as an internal phenomenon of the heart. It comes from one's belief in one's higher consciousness; when listened to, it acts as a guide in life. As British author James Allen noted, "As a man thinketh in his heart, so is he."[32]

Thus, a person who has only the authority of position may have socially accepted power over other people, but only to the extent that people submit to that authority. A person who holds a position of authority yet does not live from the spiritual authority within can only manage or rule through coercion and fear, as exemplified by Mao Tse-Tung's view that, "Political power grows out of the barrel of a gun."[33] They cannot lead, however, because, as American mathematician Norbert Wiener observed, "A conscience which has been bought once will be bought twice,"[34] to which British historian and author Lord Acton might have added, "Power corrupts, and absolute power corrupts absolutely."[35]

As stated, a leader's power to inspire followers comes from a sense of authenticity because the individual has a vision that is other centered rather than self-centered. Such a vision springs from strength, those inviolable biophysical principles that govern all life with justice and equity, as opposed to the relatively weak foundation of selfish desire. Authenticity is what people respond to, and in responding, they validate their leader's authority.

Here, it might be useful to reiterate that managerialship, as opposed to leadership, is of the intellect and pays minute attention to detail, to the letter of the law, and to doing the thing "right" even if it is *not* the right thing to do. A manager relies on the external, intellectual promise of new techniques to solve problems and is concerned that all the procedural pieces are both in place and properly accounted for.

Good managers are thus placed at a disadvantage when put in positions of leadership because all such people can do is rise to their level of incompetence and remain there, in which case an ounce of image is worth a pound of performance. Similarly, a leader placed in the position of managerialship is equally inept because the two positions require vastly different skills.

Only an effective leader can guide the process of social-environmental sustainability within community development; whereas an effective manager is the one who keeps it running smoothly. By way of example, I remember driving herds of 200 head of cattle when I was younger. There are three basic positions in driving cattle, with two basic functions: point (leader) and flank and drag (managers). The person riding point is the leader, the one out front guiding the herd. The flankers (the people riding along the sides of the herd) manage the cattle by keeping them moving in the desired direction while preventing individuals from leaving the herd. Riding drag means bringing up the rear or keeping the cattle moving at a given speed while preventing individuals from dropping out of the herd, which is part of good managerialship. Together, leader and managers are responsible for moving the whole herd safely from one place to another.

Within the human community, however, a leader must be a servant of the parties involved. Servant leadership involves a unique mix of idealism and pragmatism.

The idealism comes from having chosen to serve one another and some higher purpose, appealing to a deeply held belief in the dignity of all people and the democratic principle that a leader's power flows from commitment to the well-being of the people. Leaders must often help followers bear the pain of uncomfortable circumstances, but without inflicting it, such as the necessity for compromise. Such leadership is also pragmatic. It has been observed over and over, as previously stated, that the only leader soldiers will reliably follow when risking their lives in battle is one they feel is competent and concerned with their relative safety.

A leader's first responsibility is to help followers examine their senses of reality. The last responsibility is to give credit and say "thank you." In between, one must provide and maintain momentum and ensure effectiveness.

Most people confuse effectiveness with efficiency. Effectiveness is doing the right thing, whereas efficiency is doing the thing in the most expedient manner, although at times it may not be the right thing to do.

When the difference between effectiveness and efficiency is understood, it is clear that efficiency can be delegated, but effectiveness cannot. In terms of leadership toward social-environmental sustainability, effectiveness enables others to reach toward their personal potential through participation in the process and leaves the world a little more sustainable at the same time. In so doing, a leader leaves behind a legacy of assets invested in other people.

A leader is also responsible for developing, expressing, and defending the followers' civility and values, even if they do not prevail totally in the ultimate decision. Paramount in the process of pursuing social-environmental sustainability is good manners, mutual respect for one another, and an appreciation of the way in which we serve one another. In this sense, civility focuses on identifying values, but without insisting that established formulas be followed to coerce the people to "go with the flow."

For participants to be denied an opportunity, not feel needed, and have the beauty and novelty of ideas denigrated is demoralizing. It is likewise crushing to the spirit of cooperation for a leader to ignore the dignity of the interpersonal relationships, the elegance of simplicity, and the essential responsibility of serving one another. In a day when so much energy seems to be spent on mindless conflict, being a leader affords the special privileges of complexity, ambiguity, diversity, and the challenge of including others in a meaningful way when it comes to making decisions that affect all generations.

Shared Leadership

Shared or revolving leadership comes about in two ways: first, when "subordinates" break custom and become the decision makers, and second, when someone's particular expertise is needed and that person takes over the decision making during that time. Revolving decision makers are indispensable in our lives because they take charge in varying degrees, as necessity dictates.

Such leadership relies on three things:

1. Inclusivity, which presumes that lasting solutions require participation of all affected parties
2. Mutual accountability, which presumes that sustainable solutions depend on all sides taking responsibility for answers (mutual blaming is inadequate)
3. Cultivation of democratic skills, which presumes that people must be taught the art of participation, ranging from active listening to negotiation, evaluation, and compromise

When government is viewed as distinct from civil society, it deems itself exempted from playing a role in the practice of inclusivity in interpersonal, participatory approaches to the relationships of governance.[36]

Revolving leadership is the basis of day-to-day expression in the participative democratic processes required for social-environmental sustainability. Such participation provides both opportunity and responsibility for participants to have a say in the future of their communities. This comes about through the example and accountability of personal behaviors, which influence governmental actions. In turn, citizen participation in the democratic process requires a participant's willingness to accept ownership of the outcome.

Because no one person can be expert in all things, the person in the official position of overall leadership must have the good sense, humility, and grace to support and follow the lead of a person whose expertise is momentarily in demand for specific decisions. Such a response may have to be cultivated because it is difficult for some to step aside temporarily in favor of one with superior expertise in issue-oriented or problem-oriented leadership.

In the last analysis, leadership must be shared (but neither given away nor sold) because a time will inevitably arise when it becomes necessary to call on the special decision-making competence of others. If we think about the people with whom we share our community, it becomes apparent that we must be able to count on one another if our community is to meet our needs while protecting our deepest values. Alone, we are severely limited, but together we can combine our knowledge, experiences, and values to create something bigger and better than ourselves.

According to Max DePree, chief executive officer of Herman Miller, Incorporated, a furniture manufacturing company, "The condition of our hearts, the openness of our attitudes, the quality of our competence, the fidelity of our experience—these give vitality to the work experience and meaning of life."[37] Freely and openly shared revolving leadership is one of the vehicles we can use to help ourselves and one another reach our potential both as human beings and as citizens of our respective communities.

But, you might say, I am only one person. What can I do? My actions account for so very little. Again, the answer is: You can do *something constructive* in support of making the world a little better for everyone.

Willingness to Delegate Authority

Delegating authority and responsibility is a vital, dynamic aspect of leadership. When people share the work, the satisfaction of a job well done, a victory well earned, is likewise shared. Delegation of authority challenges one's leadership to communicate, listen, plan, make decisions, and solve problems. It compels decision makers to strengthen their rapport with subordinates while enhancing productive relationships with those destined to be affected and those destined to lead. Delegating has six basic steps:

1. *Choose people who are willing to get the job done and then support them with incentives and motivation.* When searching to delegate to people, remember that motivation and dependability are more important than technical level, which can be learned on the job if necessary.

The ways in which leaders present tasks to subordinates can influence their willingness to participate in team exercises and then follow through, especially when infused with a shared vision and enthusiasm for the work ahead. It is important for the person to know how participation will meet their values and psychological needs. For example, saying, "We want you to join us," appeals to a person's need for belonging. "Your unique gift of skills and talents is critical to the completion of the project," appeals to a person's self-esteem. "If you are willing, this project can help you expand your comfort zone by stretching your abilities into areas you may never have thought you could master," appeals to personal self-mastery.

Self-actualization (fulfilling one's individual potential), as psychologist Abraham Maslow pointed out, is the highest form of human need. When people are self-actualized, they are capable of maximizing their potential, which enhances their self-concept. One's self-concept, in turn, is based on a sense of a person's core values. To motivate a person toward the level of self-actualization, it is crucial to appeal to their highest personal values.[38]

Because each person's values are unique, you must get to know another person by asking the person about him- or herself and then listening actively to what the person says. A person's answers will inevitably reveal personal values if you really listen.

2. *Match a particular person to a particular task.* Such matching can be motivating because people enjoy using their expertise, especially in a worthwhile cause, but first the leader must get to know the person to ensure a proper match by asking about his or her education, training, profession, family, interests, and hobbies.

When organizing a team or committee, it is critical to match people both to tasks and to one another. Why? Teams and committees must work as an entity to be effective and of value. Some people require little or no supervision, whereas others want nothing less than a recipe to follow. Some are technicians and like details; others are dreamers and thrive on creativity and global concepts. Some excel in the limelight, while others are more comfortable in the shadows. Some are good with numbers, others with words. Some are good speakers, others good writers. It is therefore important for a leader to structure a team or committee in such a way that the diversity of personalities, talents, and skills creates a harmonious working balance.

3. *Define the task and communicate it concisely.* Communicate the purpose of the project clearly to each member of the committee or team and spell out their responsibilities and the project schedule. Specify exactly what the final outcome is to be and, if necessary, rewrite unclear or complicated instructions in the form of an outline or diagram structured in the positive of the desired outcome.

Explain how each task dovetails with, supports, or is contingent on the completion of other tasks to integrate them into a creative whole. Then, either provide access to relevant reference materials or tell participants how to obtain such materials if they need to do so. Ask questions to ensure they understand what is expected of them. Finally, give every member your phone number and be available throughout the project to answer questions and provide additional guidance or clarification in the form of personal support when things get overwhelming, as they often do.

4. *Monitor progress.* It is important for the team leader to keep tabs on the progress of all those who are working on the project. Such monitoring helps preclude building inadvertent mistakes into the final product. A leader must always remember to be positive in the way he or she states things. To wit: "I like the way you have done part A of the task. Do you think it would work to do part B this way? Could that bring it in line with the quality of part A?"

If work is behind schedule or is flawed, the leader must remain open, positive, and willing to participate directly in the technical aspects of the task. Blame reduces effectiveness by putting participants on the defensive. The leader must call the team or committee together, discuss the problem, and look for solutions. That could include further clarification of instructions, additional reference materials, restructuring work assignments, or getting more people involved in the decision-making process. If it turns out that a particular individual is poorly suited to a task, a mentor can be assigned to work with that person or that person could be reassigned to a task more in line with personal capabilities.

If a person must resign because of illness or conflicting priorities or responsibilities, empathize with the problem, thank the person for the work completed thus far, and allow the person to bow out gracefully. There may be another day or another project on which that person might serve effectively if today the person is treated with kindness and dignity.

5. *Encourage creativity and allow for different styles of working.* A leader must remain focused on the outcome, the final result, not on the details of how the job gets done. Encouraging creativity not only vitalizes a project with interest but also highlights an observation by Albert Einstein: "In the middle of every difficulty lies opportunity,"[39] which creativity can illuminate.

With respect to styles of working, people feel a sense of ownership when allowed to work in their preferred ways. For example, some people work piecemeal and can start with any part of a project, whereas others are methodical, starting at the beginning and completing the whole project step by prescribed step. Some are dramatic and add a flare to their work; others are more of the "no drama" persuasion. However, blending those styles in a final report can present problems that will be the leader's to solve. Nevertheless, few people will probably work as you expect them to, but if you selected them for the quality of their skills and those skills are aptly applied, how work gets done is immaterial as long as it gets done on time, is done well, and the people found satisfaction in their achievement.

6. *Always reward effort.* A good and sensitive leader always—*always*—shares credit and shows appreciation and recognition for work well done. Such a leader might send a thank you note as a token of appreciation, hold a party for all those involved, or write a letter of gratitude for inclusion in the person's personnel file, should that be appropriate.

A psychologically mature leader always accords the credit to others, thereby letting all know how valuable their work is considered to be. If people are treated well, they are more likely to be willingly available the next time their help is needed. Such treatment may allow them to give of their talents and skills in a way they find deeply satisfying. Helping others to be good followers is a prelude to developing their potential as psychologically mature leaders and decision makers.[40]

Encourage Leadership in Others

Every decision maker needs to encourage others to find within themselves their own budding tendencies toward leadership because to succeed in social-environmental sustainability, a community must become, to the greatest extent possible, a community of true social-environmental decision makers. All it takes for a leader to encourage leadership in others is to keep eyes and ears open and paper and pencil ready.

One of the most important tasks of a community leader is to continually expand the opportunities for leadership and to support those with the courage to step forward. I already do that, you might think, when I can think of something worthwhile to say. As a suggestion, I recommend writing an evaluation in a similar way every time you observe someone assuming the role of leader: What do they have to offer? What is their point of view?

Each person's point of view is unique and important. The same is true of their thoughtful comments. Why? Because each person reacts differently to a given leader; in addition, professionals who take surveys know that feedback on a questionnaire reveals more about the evaluator (in this case you) than the person being evaluated. Thus, as you accept feedback, you help yourself to find things that can be used to enhance your own learning and performance.

Further, social-environmental sustainability, which depends on information, is based on the leader/follower relationship. Because a meeting of a committee can be a microcosm of the community, leaders need feedback from as many participants as possible. Once that is available, leaders have a better idea of how they are affecting the community as a whole. So, what could you, as a committee member, do?

First, nurture and counsel to encourage the leader. Begin by finding something the leader did well.

Second, be specific. Improvement occurs most quickly when successful behaviors are identified. Pointing out specific behaviors adds credibility to your comments, such as catching someone doing something right and pointing it out to them. For example, I was facilitating a visioning process in which I explained how a vision to which people are committed determines what happens to their community. During my explanation, I paused several times and asked: "Am I making sense to you?" At the close of our discussion, a gentleman remarked: "I really liked your asking us if you were making sense. It not only told me that you care about us as people and a community but also gave me permission to respond honestly without worrying too much about offending you."

One way to identify behaviors to emulate is to ask yourself the following: What one thing in this person's style of leadership works best? What one thing distracts the most?

The last step is to identify a specific behavior that could be improved. Although this may seem difficult, there are several ways to suggest improvements gently. One is to ask a question, such as, "Would it work better to make eye contact with the whole audience instead of focusing on one person?" Or, begin your comment with, "You might try … ." Another is to soften your comment by turning it into a positive: "Your ability to talk *to* people rather than at them is a real gift; it is particularly important that you use this gift during the committee's deliberations on the intended direction."

Identifying specifics is critical because it depersonalizes and neutralizes the observation. You could say, for example, "If you speak from your heart, saying what you really feel, your authenticity will come across," instead of saying, "You need to be more authentic."

Comments must always include encouragement for a specific behavior that is positive and simultaneously a suggestion for improvement as needed. Remember, the duty of a true leader is to lead by example, which means keeping your eyes and ears open and your pencil ready for the benefit of other leaders.[41]

SUMMARY

Chapter 6 was an enumeration detailing the essential qualities of a psychologically mature leader. Chapter 7 deals with the responsibilities and pressures of social-environmental decision making. These include such things as being and disclosing oneself, avoiding the trap of perfectionism, inspiring performance, nurturing creativity, establishing realistic objectives, and dealing with "burnout."

NOTES

1. Albert Einstein. The world as I see it. http://www.aip.org/history/einstein/essay.htm (accessed February 4, 2011).
2. Parallels between leadership and psychotherapy in this chapter are based on a most excellent book by Gerald Corey. *Theory and Practice of Counseling and Psychotherapy*, 3rd ed. Brooks/Cole, Monterey, CA, 1986.
3. Ibid.
4. John Holt, Jr. http://thinkexist.com/quotation/the_true_test_of_character_is_not_how_much_we/209819.html (accessed February 5, 2011).
5. The foregoing story of George Washington is based on Thomas Fleming. General Washington's journey to freedom. *Guideposts*, July (1997):24–27.
6. The preceding two paragraphs are based on Chris Maser. Authenticity in the forestry profession. *Journal of Forestry*, 89 (1991):22–24.
7. Ralph Waldo Emerson. http://www.goodreads.com/author/quotes/12080.Ralph_Waldo_Emerson (accessed February 5, 2011).
8. The foregoing commentary (five paragraphs) is based on Leonard Pitts, Jr. Few public figures show real remorse. Corvallis (OR) *Gazette-Times*, November 23, 1997.
9. Arnold Toynbee. 1958. *Civilization on Trial and the World and the West*. Meridian Books, New York.
10. The foregoing story is based on Marc Luyckx. The re-enchantment of politics. *Yes! A Journal of Positive Futures*, Winter (1997):16–17.
11. James Allen. *As a Man Thinketh*. Grosset & Dunlap, New York, 1981.
12. Dorothy J. Hulst. *As a Woman Thinketh*. DeVorss, Marina Del Rey, CA, no date. (Transcription of James Allen's *As a Man Thinketh* for women.)
13. The discussion of the balance between the masculine and feminine characteristics in a leader is drawn in part from Susan St. John. Making the emotional connection. *The Toastmaster*, 63 (1997):16–18.
14. Corey, *Theory and Practice*.
15. Winston Churchill. http://www.quotedb.com/quotes/2129 (accessed November 19, 2011).
16. Corey, *Theory and Practice*.
17. http://www.angelfire.com/mn2/patch/quoteso.html (accessed February 5, 2011).
18. Benjamin Disraeli. http://en.wikipedia.org/wiki/Benjamin_Disraeli (accessed February 6, 2011).
19. Joseph Bucklin Bishop. *Theodore Roosevelt and His Time Shown in His Own Letters*, Volume 2. Scribners, New York, 1920.
20. Elizabeth Sherrill. Meeting Mr. Truman. *Guideposts*, February 7 (1997).
21. The preceding three paragraphs are based in part on Victor M. Parachin. Mastering your universe—five simple steps for success. *The Toastmaster*, 63 (1997):8–10.
22. William Penn. http://en.wikipedia.org/wiki/William_Penn (accessed February 6, 2011).
23. Rosalind Russell. http://thinkexist.com/quotation/flops_are_a_part_of_life-s_menu_and_i-ve_never/224221.html (accessed February 6, 2011).
24. James E. Burke. http://www.referenceforbusiness.com/biography/A-E/Burke-James-1925.html (accessed February 6, 2011).
25. Helen Keller. http://www.mentalhealthministries.net/inspiration/quotes.html (accessed February 6, 2011).
26. Dwight D. Eisenhower. http://www.internetviz-newsletters.com/PSJ/e_article000991637.cfm?x=b11,0 (accessed February 6, 2011).
27. Dennis Waitley. http://johnreynoldsonline.com/blog/my-favorite-quotes-vol3 (accessed February 6, 2011).

28. Petra Kelly. http://www.encyclopedia.com/topic/Petra_Kelly.aspx (accessed February 6, 2011).
29. The characteristics of a good follower are based on Richard G. Ensman, Jr. How to be a great follower. *The Toastmaster*, 63 (1997):19.
30. Henry Ford. http://thinkexist.com/quotation/coming_together_is_a_beginning-keeping_together/146314.html (accessed February 6, 2011).
31. John Heider. *The Tao of Leadership*. Bantam Books, New York, 1968.
32. James Allen. *As a Man Thinketh*. Penguin Books, New York, 1901.
33. Mao Tse-Tung. http://everydayrisk.org/power.html (accessed February 6, 2011).
34. Norbert Wiener. http://everydayrisk.org/power.html (accessed February 6, 2011).
35. Lord Acton. http://www.brainyquote.com/quotes/authors/l/lord_acton.html (accessed February 6, 2011).
36. Frances Moore Lappé and Paul Du Bois. A place for democracy. *Yes! A Journal of Positive Futures*, Winter (1997):37–38.
37. Max DePree. *Leadership Is an Art*. Bantam Doubleday, New York, 1989.
38. Abraham Maslow. *The Farther Reaches of Human Nature*. Penguin Books, New York, 1985.
39. Albert Einstein. http://www.quotedb.com/quotes/2575 (accessed February 6, 2011).
40. The preceding discussion is based on Judith E. Pearson. Dare to delegate. *The Toastmaster*, 63 (1997):8–11.
41. The preceding discussion is based on Jean Marsh. Keep your eyes and ears open and your pen ready! *The Toastmaster*, 63 (1997):20–21.

7 Coping with the Responsibilities and Pressures of Social-Environmental Decision Making

INTRODUCTION

This chapter is based on circumstances described by Corey[1] for psychotherapists coupled with my experiences in leading workshops on sustainable community development, development of vision statements, and resolution of environmental conflicts in a number of different cultures. Situations confronted by psychotherapists and leaders have much in common. Caroline Myss affirms that: "So much of the way we respond to the external challenges is determined by how we respond to ourselves. In addition to all the relationships we have with people, we must also form a healthy and loving relationship with ourselves."[2]

I have come to believe that leaders soon realize, in addition to their study and training, that all they really have to work with is themselves—their life experiences, values, and humanity. Armed only with themselves, they face the vagaries of circumstances that all social-environmental decision makers inevitably face.

CIRCUMSTANCES FACED BY DECISION MAKERS

A circumstance is a condition or factor that accompanies an event and has some bearing on it. As such, you must consider the circumstance in determining your course of action as a social-environmental decision maker so you can bring about a desired outcome. Although no one can directly control a circumstance, each one of us can control how we *respond* to a circumstance and thereby de facto control the circumstance by controlling ourselves. Self-knowledge—and thus self-control—is among the greatest challenges of social-environmental leadership.

I therefore caution aspiring leaders not to be seduced into mimicking or consciously copying another's style. *There is no "right" way to lead, only ethical principles to follow.* Although the principles remain the same and inviolate, there are as many ways to lead as there are people to become leaders, which represents a wide variation of effective approaches to the art of leadership. In fact, one leader cannot reach all people because not all people respond equally to the personality of any given leader.

Aspiring leaders may accept the influence of another effective leader, but they must choose carefully and keep clear the boundaries (discussed further in this chapter) that protect and nurture their *own* developing style of leadership. If the aspirant allows those boundaries to blur, he or she can be no better than a carbon copy and, more likely, a poor imitation of someone else. In so becoming, the aspirant's gift is lost, while the other leader's limitations are accepted in addition to the aspirant's own.

Although I neither have nor know of a formula for leadership, I do know beyond a doubt that to lead well you must work diligently on resolving your own emotional and psychological issues because you can only take people as far along the path as you have personally traveled. With this in mind, I advocate selecting the highest, other-centered principles to follow and then adhering tenaciously to them, regardless of the boundless temptations to take the easier, more traveled, self-centered road.

Beyond that, I recommend finding that place of quiet within, where you can hear the message of your own heart, because that message (intuition) is the knowing beyond knowledge, the personal truth of your being. It is from this still place within that a true social-environmental decision maker leads into the uncertainty of time's distant horizon. And, it is from this still place within that a leader must seek the realm of dispassion if the individual is going to unfold as a leader and deal effectively with the myriad circumstances inevitably encountered, beginning with anxiety.

Dealing with Anxiety

Most budding decision makers, and even some seasoned ones, still tremble inwardly with anxiety ("stage fright" or "performance anxiety") when they anticipate standing in front of an audience. They ask themselves: What will I say? How will I say it? Will the people believe me? Will I make a fool of myself? If I do, how will I ever face them again? Will they accept me or trust me if I make a mistake?

Such anxiety can be traced to a lack of leadership experience, which can produce uncertainty about the reactions of those who will be led. After all, the approbation of the public at large is portrayed as exceedingly fickle. What, therefore, are you to expect over time? Clearly, social-environmental decision makers must continually prove themselves if they are to lead successfully. That requires decision makers to be firm—but neither arrogant nor coercive—in their leadership.[3]

Presenting yourself effectively is perhaps one of the most challenging skills of leadership and communication. If you have the wrong attitude and poor verbal skills, you will likely do more harm than good.

Paradoxically, the first step in effectively presenting yourself is the realization that any problem you are experiencing is your own. It is you who are annoyed by a situation or the behavior of someone else, which triggered some unfinished business within you. Therefore, instead of becoming impatient with the situation or blaming, judging, or attacking the other person and putting that person on the defensive while you express your feelings, try the following approach:

1. Go within and ferret out *why* the situation or person is bothering you.
2. Resolve the issue within if at all possible.
3. If a situation persists in disturbing you, find the quiet center of understanding and dispassion within and then examine the potential solutions, select one, and activate it.
4. If, however, it is a person who annoys you, consider the following:
   a. State the problem, situation, or behavior politely and calmly in a discussion.
   b. Explain how the problem affects you.
   c. Discuss a potential solution.
   d. Describe the consequences of acting on that solution.

Beyond this, your firmness will be most effective if you follow a few simple rules:

1. Speak up immediately rather than waiting hours, days, or weeks before approaching the other party with your problem. It will be less stressful and more productive if you deal with the person immediately rather than letting your fear inflate it to gigantic proportions, which it most certainly will do.
2. Be direct—but kind—in telling the other person what your problem is. Beating around the bush usually leads to an imprecise understanding by the other party and fosters unnecessary confusion over an initially simple issue.
3. Be pleasant. A smile and friendly tone of voice will accomplish more than any kind of aggression ever will.
4. Be calm and maintain your composure, which can only help to retain your credibility and to elicit empathy from the other party, perhaps by allowing the person to step momentarily into your shoes and see the issue from your point of view.

Nevertheless, because leadership is a most serious business, some anxiety is probably normal, but uncontrolled anxiety can be disabling. Moreover, if you compare yourself either with peers or with leaders of great stature, you will always lose. Remember, you are unique in the world, and your talents are therefore unique also. No one else has the gift you personally have to offer and vice versa.

The willingness to accept and deal with your anxieties is a mark of courage. Although your self-doubts seem normal, how you deal with them is what counts. If you lead from the heart and intuition *as well as* the intellect, you will find over time that self-doubt gradually disappears until, for the most part, it is no more. As self-doubt fades, your sense of authentic inner personal power will grow.

Use of Power

"From the time we are young," writes Caroline Myss, "we test ourselves and our capacity to learn what and who has power, to attract power, and to use power."[4] If we decide that we cannot attract power, says Myss, we begin living in a "power debt" and imagine ourselves living off the energies of other people. "The challenge," Myss continues, "is not to become 'power celibate' but to achieve sufficient internal strength to interact comfortably with physical power without negotiating away our spirits."[5]

"Power breeds isolation. Isolation leads to the capricious use of power. In turn, the capricious use of power breaks down the normal channels of communication between the decision-maker and the people whom he leads. This ultimately means the deterioration of power and with it the capacity to sustain unity in our society. This is the problem we face today." So wrote George Reedy, press secretary and special assistant to President Lyndon B. Johnson.[6]

Power, or the ability to control oneself and others, is possessed by psychologically mature decision makers. A vital component of such decision making is the recognition that citizens can empower themselves by sharing the power of an effective leader.

There are two struggles associated with power: attainment of and resistance to. The struggle for power embodies the danger that "absolute power corrupts absolutely." Of the converse, Milan Kundera wrote: "The struggle of man against power is the struggle of memory against forgetting."[7]

A leader must be a good role model in the sharing of power (= decision making), which means a leader must be a potent person in his or her own right and have clarity in an other-centered purpose that is vital to those who follow. Leaders who have genuine inner authority have no need to dominate the lives of others or dwarf others to demonstrate superiority. They are, instead, able to appreciate potency simultaneously in others and themselves. British statesman Benjamin Disraeli, put it well: "Next to knowing when to seize an opportunity, the most important thing to know in life is when to forgo an advantage."[8]

Clearly, the fact that power can be and is used inappropriately raises ethical concerns. For example, a leader who fears losing control because of a personal need to retain control over his or her constituency may resort to all kinds of destructive strategies (both consciously and unconsciously), as witnessed every day in local, regional, national, and global newscasts. But, leaders are also subject to such abuse by their constituencies through the ego defense mechanism of projection, whereby a person attributes both their unacceptable (fearful and thus villainous) and their most worthy (hero) traits to others. However, one sees others as deceitful, kindly, or generous only if one has those qualities within oneself.

Criticism in the Form of Projection

Criticism is really projection, which is casting outward something one perceives to be within oneself as a means of coping with personal, fear-based discomfort. It is the externalization of an inner thought or motive and its subsequent behavior, which is then attributed to someone else. Such behavior is termed *scapegoating*, which, as previously illustrated, has been officially in existence since biblical times and is still thriving.

Projection as a Means of Avoiding Personal Responsibility

Projection is the "externalization" of blame, guilt, or responsibility as a defense against anxiety and is so used as a coping mechanism. Just as an empty movie projector casts only light until a specific film is put in it, we can project onto other people only what we think about ourselves because without the thought, there is nothing to project. Like a movie screen, which reflects the projected image, people are mirrors, which reflect the projected image. Thus, we see in others what we both consciously and unconsciously see in ourselves—nothing more, nothing less. As such, judgment, that which we see in ourselves, is the projectile we cast at the mirror in which we see our own reflection, but in the form of another person or thing.

Judgment, therefore, is the projection of that which you see in yourself. It is the projectile you cast outward in the word *should*—you should do this; you should do that. "You should" is thus a common missile fired by the opposing sides in a conflict.

In reality, however, *should* is the stuff of someone else's standard of operation, someone else's concept of right and wrong, of what you should or should not be or do. Someone else's should is only yours if you choose to accept it. If, however, you choose to ignore another person's should, it has no effect.

Leaders have probably always been subject to such projections as criticism, especially from those who would rather blame out of fear than take responsibility for their own actions, as noted by President Theodore Roosevelt:

> It's not the critic who counts, not the … [person] who points out how the strong … [person] stumbled or where the doer of deeds could have done them better. The credit belongs to the … [person] who is actually in the arena; whose face is marred by dust and sweat and blood; who strives valiantly; who errs and comes up short again and again; who knows the great enthusiasms, the great devotions and spends himself [or herself] in a worthy cause; who at the best knows in the end the triumph of high achievement; and who at the worst, if he [or she] fails, at least fails while daring greatly so that his [or her] place shall never be met with those cold and timid souls who know neither defeat nor victory.[9]

At this juncture, it bears repeating that a person who serves the people as a leader must pass the tests described in the eulogy that Senator William Pitt Fessenden of Maine delivered on the death of Senator Foot of Vermont in 1866:

> When, Mr. President, a man becomes a member of this body he cannot even dream of the ordeal to which he cannot fail to be exposed;
>
> of how much courage he must possess to resist the temptations which daily beset him;
>
> of that sensitive shrinking from undeserved censure which he must learn to control;
>
> of the ever-recurring contest between a natural desire for public approbation and a sense of public duty;
>
> of the load of injustice he must be content to bear, even from those who should be his friends; the imputations of his motives; the sneers and sarcasms of ignorance and malice; all the manifold injuries which partisan or private malignity, disappointed of its objects, may shower upon his unprotected head.
>
> All this, Mr. President, if he would retain his integrity, he must learn to bear unmoved, and walk steadily onward in the path of duty, sustained only by the reflection that time may do him justice, or if not, that after all his individual hopes and aspirations, and even his name among men, should be of little account to him when weighed in the balance against the welfare of a people of whose destiny he is a constituted guardian and defender.[10]

Two years after Senator Fessenden delivered this eulogy, his vote to acquit Andrew Johnson brought about the fulfillment of his own prophecy. This is the test of leadership: to stand firmly by one's inner convictions even when they are the cause of one's being cast out of office because it was ethically necessary to tell the people what they needed to know rather than what they want to hear.

Projection Can Be Either Negative or Positive

The negative fears people harbor represent the human shadow into which we shove all the unwanted, nefarious qualities we think we possess. These are eagerly projected onto our public servants, albeit they deserve none of them.

There is another kind of projection, however, called "hero worship," which represents the positive part of ourselves that we choose to disown. Hero worship means that we project onto another person—usually celebrities—the positive qualities (the glow) we possess ourselves but are afraid to manifest.

Projections, especially those of hero worship of one kind or another, place an incredible burden of unknown responsibility on a leader because they are not only seen as larger than life but also expected to be perfect by those who project onto others their own positive qualities for safekeeping. Consider, for example, the often-impossible expectations of human perfection that we project onto clergy, teachers, celebrities, and elected officials. The illusions we create are dangerous because we neither see nor allow flaws in them.

What happens when one of our flawless illusions turns out to be an imperfect human after all? Are we filled with mercy, compassion, and understanding? No. Our hero is dashed to the ground, much as a sculptor might cast down a flawed statue, because there has been found in them a weakness that calls forth the fickle-heartedness of those who projected their positive qualities onto others rather than take personal responsibility for their own development.

A leader must be prepared to bear such acts and know that they are not personal. Leaders, to be effective, must learn to remain grounded in their own spiritual center while all about rage the storms of fear in the material world.

Criticism and Your Image

To make the following discussion as concrete as possible, I speak directly to you, the reader. How you personally handle criticism is related to the way in which you perceive yourself and how you want others to perceive you.

The way in which you perceive yourself is your *self-image*, which consists of the beliefs you hold about yourself and the values based on those beliefs. Like every one else's, your self-image is bestowed by your parents and siblings, as well as your extended family, teachers, peers, and friends. If you are fortunate enough to be treated well, you develop a positive, healthy self-esteem, which translates into a positive sense of self-worth and a corresponding self-image. Unfortunately, the converse is also true.

The way in which you would like others to perceive you is your "public image" or "persona," which you try to make as favorable as possible. You may, for example, want others to see you as authentic, knowledgeable, patient, funny, and so on. To achieve your desired public image, you make your actions reflect these characteristics to the best of your ability.

The problem arises when people criticize you because they are making judgments (projections) about your self-image, your public image, or both. Such judgments may well conflict with your self-perception. Being called a liar, for example, may produce a defensive reaction because you see yourself as honest, especially if raised with honesty as a value. If, therefore, some truth exists in the accusation, you become even more defensive in trying to reinstate your favorable public image.

Responding to Negative Criticism

Criticism perceived as negative can be responded to in six ways: (1) withdrawing, (2) denying, (3) ignoring, (4) rationalizing, (5) counterattacking, and (6) responding non-defensively.

Withdrawing is opting to accept criticism silently, perhaps by simply leaving the room without a response. The conflict may not escalate, but there may be a feeling of a loss of self-respect and self-esteem because you chose not to defend yourself.

Denying is a form of defense against criticism in that you flatly deny the accusation of your critic. Such denial is rampant in our federal, state, and local governments these days and thus offers many examples of the dynamic.

Ignoring criticism is not the same as withdrawing from it. You can only ignore criticism when you are truly not bothered by it, when you recognize it for what it is, consider the source, and let it pass without taking ownership. Although you can retain your self-respect and self-esteem, you may forgo a chance to correct your critic's misperception.

Rationalization as a defensive strategy admits the merit of the criticism quickly followed with an explanation of the action in question. Unfortunately, such a strategy gives the critic another chance to attack the new rationalization. It is best to recognize and acknowledge a mistake, if you have made one, and take corrective actions.

Counterattacking diverts the negative attention onto the one who criticized. Such action escalates conflict but usually avoids the real issue.

Responding non-defensively involves listening calmly and non-judgmentally to the attacker, who may be trying to communicate something to you, to stimulate a selected audience in opposition to you, or have some other reason. By listening calmly and nonjudgmentally, it may be possible to recognize the perceived problem and thereby begin the disarming of the protagonist. This is the most productive and rewarding way of dealing with criticism. Such an approach demands the greatest psychological maturity and self-control.

The following steps are useful in responding to criticism non-judgmentally: (1) listen, (2) acknowledge, (3) elevate the level of the discussion, (4) always tell the truth, (5) be friendly, (6) show your opponent a new way of viewing the issue, (6) ally yourself with positive symbols, (7) tactfully refute the opposition, (8) listen carefully to all questions and repeat them aloud, (9) ask questions, (10) paraphrase, (11) agree with the truth, and (12) answer directly:

Listen: Although we tend to take criticism at face value, automatically *assuming* we are somehow in error, there may be a lot of buried feelings beneath the spoken words, such as fear of a new idea or unwanted change. It is thus critical to listen with an open mind without forming any kind of mental rebuttal. Only then are clarifying questions in order. Try to ascertain what feelings are hidden by the angry words. Take notes. Do not interrupt. Once feelings have been ferreted out and addressed, it is easier to determine the real issues and resolve the conflict. As Vice President Hubert H. Humphrey said, "Freedom is hammered out on the anvil of discussion, dissent, and debate."[11]

Acknowledge: In the name of human decency, indicate to your protagonist that you recognize the criticism. Acknowledging the criticism does not mean you either agree with it or accept its content. It only means that you recognize the other party's right to have opinions and feelings that merit consideration, which allows your criticizer to feel heard and reduces the level of his or her anger.

Be careful about your tone of voice and your body language (such as your facial expressions and gestures) when acknowledging criticism because any hint of sarcasm or patronization may only make your protagonist angrier. If, however, you are truly sincere about resolving the conflict, then your words will have the ring of authenticity and signal that it is acceptable to discuss the other party's feelings.

Elevate the level of the discussion: Shift the focus from the points of conflict toward the fundamental principles underlying your perception of the issue. Once both parties agree on principles, the conflict is usually much easier to resolve. You can do this most easily if you anticipate areas in which you might be questioned or challenged.

Always tell the truth: Remember, there is only truth and untruth, as you perceive it, and the same is true for your opponent. Be totally honest, as you understand the truth, even if your position is momentarily weakened. In the spirit of truth, learn to acknowledge "I don't know" when you really do not and follow with "I'll do my best to find out for you, if you wish."

Be friendly: When the questioner is hostile, respond as if the attacker were a frightened friend asking for help. Any attempt to defame the questioner, say with sarcasm, will not only draw the audience's sympathy to the questioner immediately but also put you in a position of becoming that which you are against. The more in error you feel your opponent to be, the more incumbent it is on you to act with principle and dignity.

Show your opponent a new way of viewing the issue: Discussing the issue from your opponent's point of view can sometimes attain consensus on the fundamental principles. Then point out that there may in fact be more agreement between your opponent's point of view and yours than is obvious on the surface. In other words, strive to move away from "black and white" to "gray," from which compromise or resolution may evolve in a respectful, courteous manner. If you become argumentative, which may well seem to denigrate the opposition, whoever is listening will argue back mentally, and hostility may well return.

Ally yourself with positive symbols: Along with the presentation of underlying principles, include the emotional aspect because your opponent's fears most likely lie therein. Controversial issues involve symbols, and the side most effectively associated with positive symbols (such as law, human dignity, freedom of choice, and so on) is the more likely to prevail—and it just might be your opponent's point of view.

Tactfully refute the opposition: If you are in a public meeting, you will need to counter the criticism that may already have convinced the audience that the other person is correct, but this must be done in a non-threatening manner. Avoid any statement that might be misconstrued as a personal attack on anyone.

Listen carefully to all questions and repeat them aloud: Begin by making sure you understand the question correctly, and that the audience not only understands the question (as much as possible) but also knows to which question you are responding.

Ask questions: It is virtually impossible to uncover another person's true feelings without asking questions. Most people will initially tell you what sounds good, but it is rarely the real reason. To ferret this out, you must ask questions, and your interest must be sincere. If, however, you are facing a hostile audience, be sure to both ask and receive questions from all parts of the audience and do not allow one questioner to monopolize the available time.

Paraphrase: Repeat in your own words what you think the other person has said as a means of coming as close as possible to understanding what the other person really meant. This is critical because it is common for one party to say something and the other party to hear or understand something completely different. Paraphrasing ensures, as much as humanly possible, that this kind of misunderstanding is avoided by giving the other person the chance both to clarify his or her own thoughts and feelings and to correct your interpretation of what you heard.

Agree with the truth: If the criticism has merit, say so. While denying the "facts" is futile, agreeing with the facts does not mean that the other person has interpreted them correctly. Remember a fact—other than an irrefutable, measurable observation—is still the *interpretation of an event* as seen through the eyes of the beholder, but not the event itself. By the time this final step is reached, it is likely that the real issue will be on the table, and you will be able to discuss possible solutions with the other party. Here are some tips that may help you:

1. Acknowledge the anger and frustration by saying, "It is clear that you are really upset by this. I'm open to talking about it if you are."
2. Remain calm because a true, non-threatening response means that you have equipoise of character while you are being criticized.
3. Criticism is not personal; so do not take it as such. It is usually based on fear of some kind.

4. Smile, lean forward, nod, maintain eye contact, or otherwise acknowledge that you are listening. In this way, the speaker will know he or she is being heard, which is paramount if she or he is to move beyond the initial attack.
5. Show the other person respect even if he or she is yelling or cursing.

Answer directly: Always give simple answers to simple questions. If the question demands a lengthy reply, ask permission to answer the question now or agree to discuss it later with anyone interested.[12]

BEING AND DISCLOSING YOURSELF

Clearly, the very notion of criticism raises questions about how to respond and how much of oneself to disclose. One of the greatest gifts a public decision maker can bestow is to be authentic. This means complementing your intellect (academic learning) with your intuition.

It takes a great deal of courage simply to be you, and not everyone has such courage. Some people become so encrusted in a professional or political role that one cannot tell where the role ends and the real person begins.

The other extreme is one who labors diligently to demonstrate his or her humanity and in so doing overreacts and blurs any distinction between a leader and the public. If, as decision maker and leader, you inappropriately disclose too much about yourself, you are stealing attention from your constituency in an act of self-fulfillment. Given that, when is self-disclosure appropriate?

Disclosing persistent feelings directly related to the present transaction can be useful, even dutiful. For example, it would be both appropriate and dutiful to express your feeling that the townspeople do not have a clear sense of what they want a particular city park to be like and help them clarify their vision of the desired outcome. Regardless of circumstances, it is wise to ask why you are revealing yourself, what purpose it will serve, and to what degree it is appropriate, remembering that you are *not* omnipotent.

PERFECTIONISM AND THE FEAR OF MAKING DECISIONS

A decision maker must learn early that he or she cannot afford to burden himself or herself with the personification of perfection. Perfection and being perfect are different. Perfect, as it is meant in the Judeo-Christian tradition, means to be whole (not flawless, but emotionally healed, despite the flaws), whereas perfection is to be flawless. It is, therefore, the duty of a good leader to struggle toward wholeness, knowing it may never be attained.

A leader driven by the need for perfection is so afraid of criticism that he or she is loath to admit errors, which then become demons in the form of secrets that must be hidden and guarded, all the while increasingly ensnaring the person in the web of denial. Only when you openly and forthrightly admit your mistakes can you learn from them and in the process create and maintain interpersonal relationships built on trust. That trust allows and accounts for human frailty and human error.

When the need for perfection rears its ugly head, there is an alternative to accepting its tyranny. Building a little conscious imperfection into your endeavors frees them from the tyrant of perfection, as illustrated by a Persian story: Persian rug weavers of old, although capable of weaving a perfect rug, always inserted a single flaw because to create a perfect rug would be blasphemous since "only Allah is perfect." In this way, they honored their Higher Power and kept their "right size." That confirmed their humanity and protected them against the neurosis of perfectionism.[13] This practice of consciously accepting an innocuous personal flaw is good for a leader to follow because death of tyrannical perfectionism allows you to be more honest with your followers.

Honesty with Followers

Many decision makers fear facing their limitations because they think it will cost them the respect of their followers. "I really don't know where to go with this situation," or "I can't see any solution to this problem." Honest leaders have the best chance of winning respect from those whom they serve by frankly admitting personal limitations, which *everyone* making social-environmental decisions will and *must face.*

People overwhelmingly favor direct honesty as opposed to attempts to fake competence. Nevertheless, leaders—especially those young in experience—often burden themselves with their perceived public expectation. Namely, they should be all-knowing, skillful, and flawless in their decisions despite their lack of experience. But, even if they were *all-knowing* and possessed vast experience, it would not be enough in every situation. Ignorance would still reign supreme at times, probably more often than they would like.

Although ignorance is thought of as the lack of knowledge, there is more to it than that. Our sense of the world and our place in it is couched in terms of what we are sure we know and what we think we know. Our universities and laboratories are filled with searching minds, and our libraries are bulging with the fruits of our exploding knowledge, yet where is there an accounting of our ignorance?

Ignorance is not okay in our fast-moving world. We are chastised from the time we are infants until we die for not knowing an answer that someone else thinks we *should* know. If we do not know the correct answer, we can be labeled as stupid, which is not the same as being ignorant. Being stupid is usually thought of as being mentally slow to grasp an idea, but being ignorant really means not knowing the acceptable answer to a particular question.

When a leader *can* answer a citizen's question with a purposeful, "I don't know," that leader is freely admitting his or her limitations and thus affirming that he or she neither is nor must be in charge of the universe. Secreted in ignorance is the incredible freedom to accept the frailty of what it means to be human, to be simply what and who you are.

Society's preoccupation with building a shining tower of knowledge blinds us to the ever-present dull luster of ignorance that underlies the foundation of the tower from which all questions must arise and over which the tower of knowledge must stand. Each new brick in the tower of knowledge is molded and fired in a kiln of questions that simultaneously originate in and illuminate our ignorance. For example, do you know where the headwaters of the mighty Indus, Ganges, Brahmaputra, Irrawaddy, Yangtze, and Mekong Rivers are?

The Tibetan Plateau is the birthplace of these great, glacial-fed, Asian rivers. In fact, Tibet is sometimes referred to as the Third Pole because it stores the most ice and water after the Arctic and Antarctica.[14]

1. The Indus River is one of the chief rivers of southern Asia. From its source in Tibet, the Indus originally flowed some 1,900 miles through Indian Kashmir and Pakistan to the Arabian Sea, which is an arm of the Indian Ocean—but today it no long reaches the sea. The river's drainage basin encompasses 332,000 square miles, most of which is in Pakistan.[15]
2. The Ganges River, the most sacred river to India's Hindus, flows 1,569 miles from the western Himalayas south and east through the Gangetic Plain of North India into Bangladesh, where it empties into the Bay of Bengal. The Ganges drainage basin has the densest human population in the world, with over 400 million people or about 1,000 people per square mile.[16]
3. From its origin in southwestern Tibet, the Brahmaputra River flows 1,800 miles across the Tibet Autonomous Region of China, where it breaks through the Himalayas in great gorges to flow through the Indian states of Arunachal Pradesh and Assam before joining the Ganges in Bangladesh as it empties into the Bay of Bengal. Today, however, the Brahmaputra is under a grave threat from dams.[17]
4. The Irrawaddy River originates from the confluence of the N'mai and Mali Rivers, both of which have their source in the Himalayan glaciers of northern Myanmar (formerly known as Burma). The Irrawaddy River flows relatively straight from north to south through Myanmar before creating the Irrawaddy Delta as it empties into the Andaman Sea, which is part of the Indian Ocean. Its drainage basin is about 98,487 square miles and covers a large part of Myanmar.[18]
5. The Yangtze River flows for 3,988 miles from the glaciers in Qinghai on the Tibetan Plateau eastward through southwestern, central, and eastern China before emptying into the East China Sea at Shanghai. The Yangtze not only is the longest river in Asia and the third longest in the world but also drains one-fifth of China's land area, and its river basin is home to one-third of the Chinese people.[19]
6. From the Tibetan Plateau, the Mekong River flows through China's Yunnan Province, Burma, Laos, Thailand, Cambodia, and Vietnam, where it forms the Mekong Delta as it empties into the South China Sea. It is the seventh-longest river in Asia and the tenth longest in the world. The Mekong's length is estimated to be 3,050 miles, and its drainage basin is 307,000 square miles.[20]

Nevertheless, because the glaciers are melting faster due to warming global temperatures, both floods and water shortages will increase in the near future. If water sources dry up or become contaminated, there will be fateful consequences for over a billion people.

> Because water in this region does not have a price tag yet, we take this most precious resource and its fount for granted. … Our short-sightedness [symptomatic thinking] blinds us to the relation between our activities and their longer-term consequences. The great push for economic development in the last 50 years has been possible due to rapid use of Earth's fossil-fuel resources. However, the hidden costs have been accumulating and are borne mainly by those least able to protect themselves. Sooner or later, all of us will have to pay the price.[21]

Yet, ignorance, which often is seen as negative, is but a point along the continuum of consciousness. The same is true for knowledge and the intuitive knowing beyond knowledge.

The quest for knowledge is never ending. The quest does not imply, however, that a thoroughly schooled person is an educated person or that an educated person is a wise person. Leaders, in common with all humanity, are often blinded by their ignorance of their ignorance. The only thing worse than not knowing is not knowing that you don't know. The pursuit of knowledge does *not* guarantee wisdom. Hence, a leader is prone to becoming the personification of "the blind leading the blind" when overemphasis on competition makes looking good more important than being good. The fear of appearing incompetent is one of the greatest enemies of a leader who sincerely wants to be wise enough to learn.

Although our ignorance is undeniably immense, it is from that vastness that our sense of wonder grows. But, when we do not recognize our ignorance, we do not know enough even to question, let alone investigate, our ignorance.

A leader cannot, however, teach anyone anything. All a leader can do for someone else is to facilitate learning by helping that person discover the wonder of his or her own ignorance. By asking an appropriate question in an appropriate way, a wise leader may be able to help a person become aware of his or her ignorance in a given area without stealing either the person's dignity or sense of wonder.

The Greek philosopher Socrates observed that a teacher is but a "midwife" because once a person realizes his or her ignorance and begins in earnest to search for understanding, that person comes to see that such understanding can only be drawn out from within.[22] Understanding, after all, is the unique perspective of each and every person, and that includes understanding silence.

Understanding Silence

Most people are profoundly uncomfortable with silence and feel compelled to speak. It is not uncommon for a decision maker to become so threatened by silence that he or she does something counterproductive to break the silence and thus relieve his or her anxiety, especially when addressing a small group of people or in a one-on-one conversation about social-environmental sustainability. Silence, when allowed to flow unimpeded through indeterminate seconds and minutes, however, can draw people out and cause them to engage both uncomfortable circumstances and one another.

Silence, in addition to its potential for drawing people out, can have many outcomes, including: (1) agreement; (2) quietly considering things just discussed or evaluating some insight just acquired; (3) waiting for another person to break the silence; (4) boredom, distraction, preoccupation, or just nothing to say at the moment; (5) hostility; (6) hidden or unexpressed disagreement; (7) communication without words, where words are inadequate or silence is refreshing; or (8) communication on a superficial level with some fear or hesitancy to reveal real concerns.

A leader would do well to explore the meaning of silence when it occurs. He or she can be the first to acknowledge the silence, tell the others how he or she feels about it, and then pursue the meaning of the silence rather than pretending that it does not exist by making useless small talk to regain a lost sense of comfort. When someone breaks the silence, however, it is mandatory to listen—really listen—because we all need to feel the validation of being heard.

Understanding the Need to Be Heard

Listening is the other half of communication. Communication is a gift of ideas but only if one person gives the gift through speaking and the gift is accepted by another through listening. Spoken words that fall on "deaf ears" are like drops of rain that evaporate before reaching Earth. "Intolerance of another's ideas belies one's faith in one's cause," as Gandhi observed.[23]

The watchword of listening is *empathy*, which means imaginative identification with, as opposed to judgment of, a person's thoughts, feelings, life situation, and so on. The more a leader can empathize with a person, the more that person feels heard, the greater the bond of trust, and the better the leader understands the situation. This means, however, actively, consciously listening with a quiet, open mind, without forming a rebuttal while the other person is speaking. Such listening is an act of love.

For example, the intent of a television program that I was part of some years ago was to discuss the issue of ancient forests in the Pacific Northwest. An elderly lady on the program tried in vain to be heard, but the moderator consistently ignored her. Even after the program was off the air, she tried again to tell the moderator how she was feeling, but he continued to ignore her. In the end, just to be heard, perhaps only by herself, she spoke out loud to no one. She spoke into space. She may as well have been alone in the world.

Not listening can be a conscious act of passive violence, a purposeful way of invalidating the feelings—the very existence—of another person. Each of us needs to be heard and validated as a human being because sharing is the bond of relationship that makes us "real" to ourselves, nurtures trust, and gives us meaning in the greater context of our respective communities, society, and the universe at large. We simply must be in a relationship with one another to find meaning.

Therefore, only when a person has first been validated through listening as an act of love can that person really hear what a leader is saying. Only then can a leader share another's truth. Only then can a leader's gift of ideas touch receptive ears.

A person's gift of leadership cannot be received, however, if there is no one to receive it, if there is no one to hear. Therefore, if a leader listens—truly listens with his or her heart as well as ears—and validates the other person's feelings, even if the leader does not agree, he or she can begin to resolve differences before they become disputes. Moreover, the leader's role will be much easier, especially if clear boundaries are established.

Establishing Your Boundaries

Boundaries are those lines drawn with silent language that allow communication while protecting the integrity of everyone's personal space. The language created by boundaries transcends individual space to include familial space, cultural space, and even political/national space. Understanding personal boundaries among individuals of the same culture is difficult enough, but expanding that concept into a fluid working ability among different cultures is a most difficult task. Working through a translator in an unfamiliar language can further compound the difficulty.

A simple way of looking at boundaries is the adage "good fences make good neighbors." Take the case of cliff swallows, for example, which attach their mud nests to such surfaces as the faces of cliffs, the sides of buildings, and underside of bridges. These enclosed, globular nests share common walls, which not only strengthen the nests but also prevent the inhabitants from peeking into each other's abodes. If, however, a hole appears in the common wall and the swallows can see each other, they bicker and squabble until the hole is repaired and tranquility is restored.

A more complicated way of dealing with psychological boundaries is to compare them to the home ranges and territories of animals. A home range is that area of an animal's habitat in which it ranges freely throughout the course of its normal activity and in which it is free to mingle with others of its own kind. A territory, in contrast, is that part of an animal's home range that it defends, for whatever reason, against others of its own kind. This defensive behavior is most dramatic and noticeable during an animal's breeding season.

How does this concept apply to us? Suppose it is Saturday morning, and you leave your home to take care of a few errands. You simply go about your business engrossed in your activity without paying much attention to what is going on around you or to the people you pass unless you happen to meet someone you know. When you have finished your errands, you start home.

The closer you get to your neighborhood, the more alert you become to changes around you, such as people moving in two blocks away. This "protective feeling" intensifies as you approach your home and notice a car with an out-of-state license plate in your neighbor's driveway. You get out of your car and notice, perhaps with irritation, that the neighbor's dog had visited your lawn again. If your neighbor's dog had anointed someone else's grass, you would have paid scant attention.

The same general pattern extends to your home. Inside your home, how well you know someone and how comfortable you feel in his or her presence determines the freedom with which that person may interact with you and your family and move about your house. You are the most particular about your ultimate private space—your physical being.

Leaders, particularly those with little experience, are often puzzled about how to deal with an overly demanding citizen. Because caring leaders typically feel they *should* be helpful, they often burden themselves with the unrealistic standard of giving unselfishly, regardless of how great the demands.

These demands can manifest themselves in a variety of ways. Suppose a person calls you frequently at home and expects to talk at length, demands to see you more often and for longer than is necessary or can be accommodated, wants to socialize, wants you to assume his or her responsibilities, expects you to "pull strings" for him or her, demands that you make him or her "special" in your life, or demands that you make decisions for him or her. There are people like that. So, leaders must draw clear boundaries in relationships with subordinates.

Leaders must review relationships with overly demanding subordinates. Were inappropriate demands made? How were such situations handled? Did you say "no" when appropriate or desirable? Did you value yourself enough to protect yourself by not acceding to unreasonable requests or demands? Do you allow yourself to be manipulated for whatever reason(s)?

Demanding subordinates can feed the ego of an inexperienced leader—to the leader's functional detriment. Unaware leaders can be deluded into feeling an exaggerated sense of importance by thinking that they must at all times be available to those who are in need, seemingly asking: What would the people do without me?

There are two imperatives in dealing with demanding subordinates: first, being aware of the nature of the demands and subsequent reactions; second, demonstrating the courage to confront the situation. It is the leader's task to set behavioral boundaries through establishing the rules of conduct—especially when it comes to social-environmental decision making. The rules of conduct are the infrastructure of any society that makes effective leadership possible. Building and maintaining trust is dependent on establishing, understanding, and respecting boundaries. In turn, interpersonal boundaries are necessary to clear communications between leaders and subordinates.

Dealing with the Uncommitted

The uncommitted represent the flip side of the overly demanding. The uncommitted person generally lacks motivation and has little investment in the democratic process. This lack of motivation is commonly evidenced by a stated indifference or an unwillingness to assume any personal responsibility in the ongoing process.

Inexperienced decision makers may be drawn into unproductive games with uncommitted followers. These followers receive inordinate amounts of attention in an effort to convince them of the value of involvement, the validity of the decision maker's point of view, or both. Leaders must therefore draw clear boundaries with uncommitted participants or risk wasting time and resources.

Accepting Slow or Delayed Results

No leader should expect instant results, although they may occasionally arise. From my experience, a leader must remember: (1) whatever situation people find themselves in, they did not get there overnight, and the situation will not be changed overnight; and (2) the best way to engage people's resistance is to ask them to change from a known situation (almost regardless of how bad it seems) to an unknown future, even if it can be demonstrated to be significantly better. In other words, people prefer the devil they know to the devil they don't.

People in a group, even a political group, often act like patients in psychotherapy in that, when confronted by change, they seem to get worse before they get better. After they decide to work toward change, they must exhibit greater self-honesty in dropping defenses and façades in an act of personal transformation.

Having decided to work toward change, people may experience an increase in anxiety and its accompanying disorganization, which in turn can lead to depression or a panic attack. When confronting change, people often say, "Maybe the old way wasn't so bad after all. At least I knew what to expect. I feel so vulnerable now not knowing what's ahead. I may have been better off before. Maybe there's still time to back out."

A leader, especially if inexperienced, must learn patience when dealing with outcomes because they are rarely predictable. You must also learn to trust yourself enough to allow the outcome to manifest itself without trying to control it. To give an unconditional gift as a leader, you must learn to be dispassionate.

A leader may be sensitive and feel acutely the effects of the words and actions that frightened people hurl when confronted with the necessity of change, particularly that thrust on them by circumstances. The leader must learn to accept the often-disparaging comments wisely and dispassionately by displaying inner strength and poise. These comments are based on fear and are not personal—unless you make them so. The leader's thoughts must be turned outward toward the well-being of those affected instead of brooding over feelings of self-doubt, helplessness, impotence, uncertainty, and even ambivalence about leading, which unkind words and actions can stimulate.

Some leaders waste too much time on such introspection. This weakness must be overcome. Although everyone makes mistakes, a wise person learns from them rather than brooding over them. It is what a leader contributes in love, compassion, and truth by example that is all-important.[24] Leading by example means dealing honestly with your limitations.

Learning Your Limits

It is still necessary to remember that even an experienced decision maker can become glum and begin to doubt his or her value as a leader when forced to admit that there are people, and sometimes entire audiences, who he or she cannot touch in a significant way. A person must have the self-honesty to admit that he or she cannot work successfully with everyone. For example, I have found with audiences that a general rule might be stated as such: Some of the audience will think you are great, some will think you are wasting their time, and some will seem relatively indifferent.

For a leader, there is a delicate balance between recognizing realistic limits and challenging those "limits." To find out how real your perceived handicap in leadership is, your perceived limitations must be tested by opening up to the person or group subjected to your leadership. Daring to risk feeling temporarily uncomfortable can, sometimes, create grounds for identifying with the person or group. Thus, before deciding that you lack the necessary life experiences, philosophical compatibility, or personal qualities to work successfully with a particular group of people, it might be good to find mentors in the group and develop the habit of asking for, receiving, and using their ideas and counsel.

Cultivate appreciation for information proffered by people who can help you improve your leadership abilities. Develop a habit of systematically asking for, receiving, and using ideas, impressions, and counsel from those around you.

Listen carefully when you ask for counsel. Learn to ask the right questions for a maximum return and, when possible, let people know how you have used what they told you. They will be more inclined to help you again if you need it.

With colleagues, it is wise to establish a flow of helpful feedback by asking questions or restating information to make sure you understand the counsel they give, which also informs the colleague that you are truly considering what they say. "Did I understand you when you said … ?" or "Let me get this clear … ," or perhaps "If I understood you correctly, I thought I heard you say … ." Again, let them know when and how you use the wisdom they have so generously shared.

Choose counselors carefully. If you ask for counsel from a position of strength and desire to improve, most people will prove willing to help.

When you are fortunate enough to receive good information, accept it at face value in a defenseless demeanor. If it feels overly critical, accept it gracefully nonetheless. Assess it fairly from a point of personal detachment. Remember, you asked for it, and it is given as a helping gesture.

It is also important to show genuine behavioral change. People feel gratified when they see their counsel is heeded and are more likely to continue to act like your mirror as you improve your leadership.

Practice self-evaluation. Ask yourself some questions after important meetings. Debrief yourself by asking: Did I open the session well? Did I say the right things and ask the right questions? Did I speak from my heart? Did I get the desired outcome for the good of the whole?

You might debrief yourself at the end of each day or at least at the end of days with important events: What did I do today? What did I need to do that I left undone? How can I do better next time? Debrief similarly at the end of each week, month, and year.

Although difficult times are inevitable, any good mentor will tell you they produce personal growth and maturity. Appraise your efforts when you encounter difficulty. Determine what you learned from the experience because every difficulty has hidden within a lesson you need to learn. Good judgment, after all, can evolve from bad experiences. You might also ask friends and colleagues how they handled similar problems or circumstances.

When confronting a problem, imagine likely scenarios you could use to resolve it. A flowchart of the options open to you could be useful in this process. As your skills of simulation deepen, you will gradually become adept at predicting solutions to problems before they occur.

It may be helpful in developing principles or conclusions about your leadership to record "case histories" in a journal kept for that purpose. Over time, that "book of lessons learned" will become a powerful source of information and a record of personal growth and accomplishments.

However much you may think you require help in overcoming your own perceived limitations of leadership, you undoubtedly have talents and skill that you can offer to someone else. Act as a mentor to less-experienced peers, and you will find that you are also helping yourself by sharpening your own skills of human relationships. In turn, your efforts will motivate you to learn and excel in your own ability to lead while helping to avoid self-deception.[25]

Avoiding Self-Deception

Self-deception is not ordinarily a conscious process. It is more commonly both subtle and unconscious. A leader's motivation for self-deception is the need to make the relationship with followers worthwhile and productive. A person's need to see change in the form of forward movement may blur reality and cause him or her to be less discerning than is wise because his or her role as a leader carries with it an investment of succeeding as a decision maker.

What would happen if the person had the sense of being a flop as a leader? His or her need to feel instrumental in assisting others to lead fuller lives is partly based on a need to feel he or she is making a significant difference in the world, a need that can lead to self-deception.

At such times, a leader may look for evidence of progress, however slim, and rationalize away any elements of failure. In so doing, the leader assumes credit for the apparent progress when it may be largely due to another variable, something unrelated to leadership. When it is understood that success or failure is not the event but rather a person's interpretation of the event, it becomes possible to explore the phenomenon of self-deception consciously and thus lessen the chance of its occurrence, which in turn lessens the likelihood of becoming over invested in those served.

Overinvestment in Followers

Inexperienced decision makers are prone to worrying too much about the welfare of those they serve. By identifying so closely with followers or causes, a decision maker loses his or her own identity and takes on that of the people. Uncontrolled empathy with fellows can distort a leader's view and become counterproductive.

The most leadership-oriented thing a person can do is to be present—really present in the *eternal now*—with the people, feeling with them and experiencing their struggles with them, but *without* assuming their responsibility for living their own lives responsibly. A decision maker who becomes lost in the struggle and confusion of their community cannot hope to lead the people toward positive change. Thus, a leader who assumes the responsibility of his or her followers does them a great disservice by stealing their struggle. To do so steals their experience of their struggle and their ability to grow and master themselves through the gift of insight that inner struggle offers.

Another mistake leaders, even seasoned leaders, make is attempting to derive personal fulfillment for their own needs—or unfinished psychological business—by becoming inappropriately enmeshed with individuals in their own interest groups. A few illustrations of such enmeshment include the following:

1. The personal need to be liked, appreciated, and approved of
2. The fear of challenging followers lest they think poorly of the leader and abandon the cause
3. Sexual feelings and sexually inappropriate behavior on the part of the leader toward an individual in their working group
4. Extreme reactions to certain individuals who evoke old feelings in the leader, such as individuals who are perceived as judgmental, domineering, controlling, and so on
5. The need to alleviate an individual's pain or struggle because his or her work experience is awakening old wounds or unrecognized inner conflicts in the leader
6. Compulsively giving advice, with the leader always in the superior position of "teacher" and thus continually dictating how the individual should think, act, choose, and live

Because it is not acceptable for you, as leader, to use your fellow citizens to resolve your own feelings of inadequacy and the emotions they evoke, it is all the more important that you work diligently on your own unresolved psychological problems in an appropriate, professional setting. Although recognizing how your personal needs can intrude into your leadership is a first step, you need to willingly, consciously explore your inner self *or* be in danger of losing yourself in the people you serve and thus use them, individually or collectively, to accommodate your unfilled needs. In this circumstance, a sense of humor is helpful.

The Value of Humor

Social-environmental decision making, although a serious responsibility, need not be dull. A cause can be enriched, even enhanced, when a leader and citizens can laugh together, but most particularly when you, as decision maker, can laugh at yourself in front of others.

Genuine humor can help build unity, produce a kindred spirit, and reveal the leader as human. There are times, of course, when laughter is used to cover up discomfort and to escape from facing a threat, such as the need to change. It is the leader's task to distinguish between humor used as a crutch, which distracts, and that which unites people and enriches their lives, such as dealing with difficult situations or persons.

COPING WITH SOMEONE YOU DISLIKE

Have you ever encountered a person whose behavior was personally offensive, a person you wished would simply go away so you would not have the urge to chew the inside of your cheeks raw every time the person spoke? A common failing of decision making is to split your energy among blaming, berating, and bemoaning someone else's behavior, which often causes you to lose sight of the vision toward which you are supposed to be leading. All is not lost, however; there are things you can do to "save your sanity," as it were.

Identify Those Specific Behaviors that Trigger Your Own Emotions

Identifying your behavioral triggers is important because we each respond uniquely to those around us and to the world at large. Each time we react strongly, either positively or negatively, to another person, it says more about us than it does about the other person. It is important, in this sense, to accept personal responsibility for your own thoughts and feelings because what drives you to distraction may not bother anyone else or may actually be enjoyable to others.

Talk to the Individual Privately and Use the "Sandwich Technique"

Once you have identified the specific behaviors you find annoying, share them with the person involved, but sandwich your comments between positive statements. Everyone has at least some positive attributes, although at times (depending on your state of mind) it may seem like you have to dig pretty deep to find them.

Once you have ferreted out the person's positive attributes, visit with the person, but be *firm* rather than passive, passive-aggressive (indirect passively expressed aggression), or aggressive. Use "I" statements and address the specific behaviors that annoy you, as opposed to generalized "you" statements. For example, say, "I felt uncomfortable when you told that sexist joke," rather than, "You're a sexist."[26]

Timing Is Critical

Choose a time when you are calm and can stick to the issue at hand in your response to the person's specific unsuitable behaviors. Plan to meet in a relaxed, neutral setting, perhaps for a cup of coffee. Make sure there is enough time for a friendly visit and be prepared not only to express yourself directly and kindly but also to listen empathically.

Remember Your Role as Decision Maker

You, as decision maker, must be willing to hear and bear the fears of your fellow citizens if a mutually supportive environment is to be forthcoming. There are neither exceptions nor addenda to leadership that excuse you from dealing with someone you dislike.

A significant challenge to leadership is making inclusive behavior come alive when faced with someone who is self-centered, critical, and generally obnoxious in ways that strip others of their self-confidence. Well-documented circumstances can be a powerful tool when coupled with well-communicated feedback. When such feedback is handled with quiet dignity, it has the potential to benefit both the group as a whole and the person being confronted because of his or her behavior.

A Leader Does Not Quit When Things Get Tough

Although it is not always necessary, a brief "time-out" may help, provided someone, such as a trusted subordinate, is capable of taking over for a while. There are times when the most able decision maker is just too close to a situation to see it clearly—especially social-environmental issues. At such a time, you must take a deep breath, stand back, refocus, and then resume leadership.

If Nothing Else Works, Find a Facilitator

Use a facilitator if nothing else works. It is critical that the facilitator be transformative and free of personal investment in the outcome. (See *Resolving Environmental Conflicts* for a thorough discussion of transformative facilitation.[27])

The ideal outcome of dealing with someone you do not like is that he or she becomes your teacher. Although you may not be relieved of your annoyance in the way you might wish, you can always change yourself for the better in response to the problems posed by the person. An unmistakable sign of a true leader is the ability to change yourself in such a way that your protagonist becomes your teacher and—perhaps—your friend, because, in the end, as Goethe said, "Kindness is the golden chain by which society is bound together."[28] When that kindness takes over, you can, for a moment, see yourself through the eyes of another person.

IMAGINE YOURSELF AS DIFFERENT PEOPLE

Being a good social-environmental decision maker requires being aware of many different things happening simultaneously. That includes the various ways different people respond to any given circumstance, especially when fear of loss or change is involved. Consider, for example, what even the U.S. Central Intelligence Agency saw clearly as the handwriting on the wall:

> The rising tide of the global economy will create many economic winners, but it will not lift all boats. … [It will] spawn conflicts at home and abroad, ensuring an even wider gap between regional winners and losers than exists today. … [The] evolution [of globalization] will be rocky, marked by chronic financial volatility and a widening economic divide. … Regions, countries, and groups feeling left behind will face deepening economic stagnation, political instability, and cultural alienation. They will foster political, ethnic, ideological, and religious extremism, along with the violence that often accompanies it.[29]

Although conflict often arises where fear resides, if you can imagine yourself in the shoes of another you can tap into what the other person may be feeling and why. To do this, you must ask: How would I feel in that person's situation? Why would I feel that way?

Some people assume this imaginary role naturally and automatically; others can learn to do so. If you have the ability to step into another's shoes, you can often defuse a potential conflict before it starts or before it gets out of hand and requires special resolution. You can also anticipate where events are going and thus help people to have compassion and understanding for and patience with others. In addition, by putting yourself in another's shoes, you can often help people to pool their talents in a complementary (instead of competitive) way that actually inspires people to perform better than they thought possible.

INSPIRING PERFORMANCE

Excellent performance can produce outstanding results, which may lead to strong participation. But, remember that listeners, as individuals, may not understand their own feelings, let alone the feelings of others. It is thus important for the decision maker to help the group understand that it is human nature to achieve, to build, and to want success.

Success is greatly aided by enthusiasm. The person with enthusiasm generally prevails to the end, whereas the person who lacks enthusiasm is easily discouraged. Generating enthusiasm can overcome negative attitudes. For example, a meeting that is well organized and flows easily provides a meaningful, if not enjoyable, time for all and generates enthusiasm in visitors who join in future activities. Although enthusiasm can be contagious, a poorly organized and focused meeting can be depressing, which is the antithesis of engendering enthusiasm.

Behind all of this must be the hard work of a few people dedicated to the proposition of a shared vision of social-environmental sustainability at all levels of society. Success of such endeavors requires insistence that all concerned share in the work. Clear guidelines of acceptable conduct and firm enforcement of those standards will go a long way to ensure equal and meaningful experiences as people strive toward the creation of a shared vision for a sustainable community within the context of a sustainable landscape. That can be realized only in small steps and in the accumulating results of individual projects coupled with the necessary changes in attitudes and behaviors.

To foster the participation and responsibility necessary to create a shared vision for a successful sustainable community at all levels of society, it is critical to select carefully those who will strive toward achievement of that vision and cultivate their talents. In the same way, potential leaders must be selected carefully and their talents cultivated through training and experience as they take on more and bigger tasks and responsibilities throughout their careers.

When problems arise, as they inevitably will, it is important to work with leaders as they grow in competence to increase their commitment to wise social-environmental decision making. People usually perform well from a combination of the following factors:

1. *Good/appropriate training*—Those individuals and groups that are the most successful in fulfilling their function have, individually and in the aggregate, up-to-date training, which implies a constant updating of expertise.
2. *Adequate equipment*—Workers are only as good as their tools/equipment allow them to be.
3. *Time*—Time management is critical to ensure that adequate time is available to do the job right the first time. Adequate timing is most advantageous in the long term as it is clear that "not enough time" leads to breakdown, inadequate compliance with laws and regulations, dangerous situations, and sloppy work. The consequences of not having time to do it right leads to having to do it over at higher costs in time and resources.

4. *Motivation*—Only those who are highly motivated perform at or near their maximum potential. Work and its products decline in quality and quantity as motivation declines.

The most immediate task of a leader is to determine which of these factors is limiting successful operations. Asking the following questions may help:

1. Does the person doing the work know what is supposed to be done and when? (If not, you will need to inform, preferably show, them what to do and how.)
2. What is the specific difference between the present level of performance and the desired level of performance? (The person must know what he or she is required to do to perform satisfactorily and be motivated to do it.)
3. Does the person know what constitutes exemplary performance? (If not, you must explain what is expected and what is lacking.)
4. Does the person have the necessary skills to perform the job satisfactorily? (If not, it will be necessary to provide appropriate training and experience.)
5. Is the expected standard of performance realistic? (Unrealistic expectations result in both a poor job and a frustrated worker. It is essential that expectations be realistic, with an opportunity for exceeding those expectations.)
6. What effect does the worker's level of performance have on the others doing their assigned job? (The person needs to understand that he or she is part of a team, and the performance of the entire team depends on each member doing his or her job well.)
7. Does the worker have the necessary resources—tools, know-how, and budget—to do the job? (If not, provide them or anticipate failure.)
8. Are there obstacles affecting the person's performance beyond their control? (If so, remove them.)
9. Does a person's positive performance yield commensurate rewards? (If the person is performing well but receives an undesirable reward, such as twice as much work to do in the same amount of time, then his or her incentive to perform will likely decrease.)
10. Could the person do the job satisfactorily if he or she wanted to? (If not, identify what is missing and provide it—or simply realize that the person is not qualified for the job and get someone who is. It should be made clear, if that is necessary, that the assignment was a failure of management and not that of the person.)

Team members must be motivated if their performance is to be exemplary. Praise work well done—by both individuals and the teams of which they are part. Look for accomplishments and attributes to praise. You might begin by praising work currently being accomplished and encourage striving for excellence through sensitivity to contributions, however small, to encourage team members. An example might be the following: "Jim is doing a fine job as committee chair for participation by local government, despite his busy schedule. [Applause.] Once he gets his agenda better established, it should be easier to meet productively with local government officials."

NURTURING CREATIVITY

There is another criterion of performance that relies on inspiration. It is called creativity. Although a person's ability to lead can be enhanced by creative fellow citizens, it requires a sensitive and personally secure decision maker to nurture another's creativity because the outcome of such creativity must be shared openly. An effective leader understands that creativity requires the freedom of imagination as well as time to relax and read, have discussions with colleagues, and experiment. After all, creativity often comes in the quietude of an unguarded moment. You cannot, therefore, be creative on command or under the pressure of timelines.

When children are allowed to be children and develop their imaginations, they tend to become "productive thinkers" as adults who ask: "How many ways can I look at this problem? How many ways can I solve it?" Such thinkers have eight things in common:

1. They look at problems from all angles.
2. They make their thoughts visible.
3. They produce results.
4. They make novel combinations of ideas, images, and thoughts.
5. They force relationships, connecting unconnected things, and thereby see what others miss.
6. They think in opposites; they can tolerate ambivalence between two seemingly incompatible subjects.
7. They think metaphorically.
8. They prepare themselves for chance; chance or luck is where a prepared mind and opportunity intersect.

When a productive thinker, whose imagination is vibrant and ready, finds something interesting, he or she drops everything in progress and goes with it. Too often, people with talent fail because they are fixated on their preconceived plan as though it were set in concrete and so become stuck within its rigid limitations. But, productive thinkers see the way to achieve a desired outcome as an endless flow of possibilities, each of which holds its own promise and novelty despite where it might lead; each, therefore, is worthy of investigation. The endless possibilities with their promise and novelty are the crux of what education needs to become if our democracy is to survive in a way that serves people as equally as possible, which is one of the basic things the children want.[30]

However, creativity among your current followers can also be encouraged by appropriate recognition, such as praise before peers or such formal recognition as awards. Recognition can also take the form of an explanation of who came up with the idea and how, when, and where it will be implemented. That demonstrates the leader's sincerity in listening to an idea, grasping its essence, giving credit where credit is due, and putting the innovation to work.

Effective decision makers understand that creativity requires risk and accept that failure may ensue. Leaders must be willing to accept that creativity produces failure at least twice as often as success. If, however, you are right more than half the time, you are ahead of the game. Insecure leaders, on the other hand, cannot handle such open-ended risk and usually kill creativity.

Some leaders are mental parasites. They behave as if their asking for ideas implies that they should get credit for the ideas that emerge. If the idea is greeted with enthusiasm, they might mumble that they do not deserve all the credit, but are at the same time obviously reluctant to share it. On the other hand, if the idea bombs, they are quick to give "credit where it is due." There is an old adage: It's remarkable what can happen when you don't care who gets the credit.

MAKING DO WITH WHAT YOU HAVE

Here, a lesson from Zen cooking is in order—making do with what you have. Zen masters refer to a life lived fully and completely, with nothing held in reserve, as the "supreme meal." A person who lives such a life is called a "Zen cook" because that person knows how to plan, cook, appreciate, serve, and offer the supreme meal of life.

Cooking, like leadership, is about transformation. "When we cook," says Buddhist teacher Bernard Glassman, "we work directly with the elemental forces of fire and heat, water, metal, and clay."[31] When cooking or baking food, something almost magical happens as the heat transforms the ingredients, through chemical interactions, into a culinary delight. Although this kind of transformation, as does leadership, requires faith, a Zen cook follows the middle road by not only having faith that the food is coming along but also checking now and then to see how it is doing.

An accomplished Zen cook—or leader—is something of an alchemist in that he or she can transform poisons into virtues, not by adding a secret ingredient, but by leaving something out: attachment to the self. Anger, for example, is considered a poison when motivated by self-centered interest and the need to control. But, through detachment of the self, the same irrational emotion becomes a clearly focused, fiery energy of determination, which transforms a negative force into one that is positive. Take the self-centeredness out of greed, for example, and it is transformed into a desire to help. Drop self-indulgence from ignorance, and it is transformed into a state of sacred unknowing in which new things are allowed to arise.

How, you might ask, does a Zen cook, or a leader, find the necessary ingredients with which to prepare the meal? You simply open your eyes, look around, and do the best you can with the materials at hand in each and every moment. Consider the following story:

> A father sees a map of the world in a magazine, cuts it out, cuts it into pieces, and then gives the pieces to his son to put back together. To the father's astonishment, the boy hands him the assembled map within 10 minutes.
>
> "How did you put the world together so fast?" asks the father.
>
> "On the other side of the world is the picture of a person," replied the boy. "I put the person together, and the world came together."

Our thoughts, emotions, actions, insights, and relationships are all ingredients for our meal, but we must be open to them. Instead of openness of mind, we usually create our own boundaries, our own tiny view beyond which we refuse to look. With practice, however, we can each expand our view until everything becomes a potential ingredient for our meal.

Here, a story from Africa might be instructive:

> A little boy wanted to give his teacher a gift, but he was very poor. So, he walked 2 miles to the beach and picked up a handful of beautiful sand. He then walked two miles back. The next day, he gave the sand to his teacher.
>
> She thanked him and then said, "But you walked so far."
>
> "The journey," he replied, "is part of the gift."

As we learn to see ourselves as part of "at-one-ment" with the world, we become attuned to the unity of all life, and the whole of the world becomes available. At that point, the Zen cook (or leader in our case) knows that every aspect of life offers itself as an ingredient for the supreme meal.

Our natural tendency in cooking is not to use an ingredient and in leadership not to avail ourselves of a person we think will ruin our meal. In so doing, we all too often deny their existence or discard them through dismissal as of no value for our meal. But as Dogen, the 13th century founder of the Japanese Soto Zen tradition, instructs us, we must take the very ingredients we think will ruin our meal and figure out how to use them to improve it because they will be there whether we like it or not.[32]

"No matter who we are," observes Glassman, "we tend to reject someone or something."[33] Some Zen students found it incongruous to cook a gourmet meal and then learn to set a table properly for rich people because to the students it did not constitute traditional *samu*, or work practice, like chopping wood, carrying water, or weeding a monastery garden. Many of the students asked, "How can you serve the rich? What kind of a thing is that for a Zen center to do?"

Rejection takes many forms. A Zen student who rejects a person because the person is rich has the same problem as a rich person who rejects a Zen student because the student is poor. If you can learn to work with that which you would reject, it turns out that you are working with yourself, with those shadow parts of yourself, both good and bad, that you reject. In other words, if you can learn to work comfortably with a rich person whom you have rejected, then you can begin to accept and work with the richness rather than the poverty in yourself,[34] which merits a brief discussion of the shadow.

Many years ago, Dr. Carl Jung, the Swiss psychoanalyst, defined the dark side of our psyche as the "shadow." The shadow is a handy dumping ground for all the characteristics of our personality we choose to disown. It is that part of us we fail to see or know, that which has not adequately entered into our everyday, waking consciousness.

The shadow may be thought of as the despised quarter of our being and is paired with being wrong, bad, or evil. But, also held within the shadow of our psyche is the pure gold of our being, the noble, creative aspects of our personality of which we are afraid.

The root of this whole shadow-making process within us begins as we enter into the realm of culture and cultural ideals. We divide our lives and separate things into good and evil, acceptable and unacceptable, because culture insists—literally demands—that we behave in a particular manner.

The bad news is that these refused, unaccepted, bury-in-the-deep-dark-basement parts of ourselves do not go away, for all our characteristics must appear somewhere in our personal inventories. Nothing may be left out. That which we try to omit is simply swept into and collects in the corners of our personalities. When hidden long enough, they take on a life of their own, emerging as sudden "unexplainable" outbursts of anger, often with an energy nearly as great as that wrapped up in our egos—those parts of ourselves we know about and consciously accept as *who we are*.

When the energy of the shadow builds up too much, it can erupt as a black mood, anger, rage, harsh words spewed out of our mouths, or some indiscretion slipping past us. It can also manifest as depression, accidents, or even psychosomatic illness. Conversely, ignoring the golden qualities within ourselves can be every bit as damaging as hiding our dark sides. It may even be necessary to suffer a severe shock or illness before a person learns to let out the magnificent, inner gold.[35]

The good news is the behavioral mirror in front of us, and our work is to own every aspect of ourselves, including those disowned parts we do not want and so project outward onto someone or something else. The last would be like the aforementioned scapegoat driven from a community with all the people's shadow elements on its back.[36]

Instead of heaping all our unwanted psychological baggage onto a scapegoat, our task is to restore ourselves to wholeness by putting these fractured, alienated parts together again. If we can learn to love our inner enemies, then we can also begin, in like measure, to redeem and love our so-called outer enemies.

Many personal and collective benefits result when we each see ourselves in totality. For example, we would fall in and out of love a lot less frequently because we would not initially project the golden parts of ourselves onto another and then, as love grows thin, replace the gold by projecting instead those parts of ourselves that are annoying, distasteful, and even downright intolerable.

If we saw ourselves in totality, relationships in general would be truer and more enduring. Hero worship, in its varying degrees, would cease because we would each accept our finest qualities and be responsible for them rather than living them through someone else—hero worship. We would also stop making others the bad guys and look first within ourselves to assess what about ourselves we refuse to see, know, and accept.

The shadow is the cause of war on all levels, both within and without. In addition, the process of owning our inner heaven and hell is the highest form of creativity. A great deal of energy is released and begins to flow as we reach a place of inner wholeness. This place, where light and dark touch, is where miracles arise, where personal and collective peace is possible.

If you do not think you have a shadow, ask yourself: How many times a day do I employ compulsive substitutes (such as food, caffeine, alcohol, cigarettes, television, and so on) for emotions? How many times do I run away from facing my emotions through the use of distractions?

You cannot run away from something that is negative; it will only grow larger and stronger each time you try. You must steadfastly confront your fears by moving through them toward something that is positive.

Dogen taught that every meal has to include a harmony of the six flavors—bitter, sour, sweet, spicy, salty, and plain—just as every leader must deal with the various kinds of personalities: those who are constantly argumentative, those who depend on others for their sense of "okayness," those who are fatalistic, and those who are afraid to act and so are constantly appraising and reappraising all the data. None of them is better or more important than the other. As each ingredient has a different flavor and a different reason for being part of the meal, so each personality type has a different strength and a different reason for being part of the team. They are all important to the Zen of cooking and of leadership.[37] In addition, a leader who is truly interested in helping fellow workers use their creative ability to fashion what they need from that which is at hand must be careful to establish realistic objectives.

ESTABLISHING REALISTIC OBJECTIVES

Realistic objectives are essential to a viable relationship between a decision maker and those persons affected by the outcome. If the objectives are too ambitious to be accomplished, the participants soon become discouraged and may flee what they consider a sinking ship. On the other hand, if the objectives are realistic and most are achieved, people will likely forgive the few that may go unmet. A decision maker is therefore wise in doing the necessary homework to ensure that the objectives selected are not only attainable as stated but also further the cause while fostering a sense of urgency, such as adapting to a changing climate.

Adaptation is a process of deliberate change in response to social stress brought about by observable environmental conditions. The dominant adaptation to environmental change is primarily symptomatic and thus actor centered: Individuals respond to specific circumstances in an effort to ameliorate perceived personal and social vulnerabilities. But, the rare decision maker who takes a more dynamic, systems-oriented approach honors the resilience of ecological systems by guiding citizens to live within the long-term biophysical constraints of social-environmental sustainability.[38] If we adopt such an approach, we adapt immediately to current conditions, in part by repairing (as much as possible in the short term) the damage we have done to the ecosystems we rely on for a good quality of life so they can maintain their ability to serve our requirements. Such adaptation is a critical concept because we cannot move away from a negative idea (what we do not want), which we continually try to do. We can only move toward a positive idea (what we want). We must, therefore, think about, plan for, and consciously move toward social-environmental sustainability—one person, one community at a time.

And how, you might ask, are we to do this? In answering this question, consider that whether we as a species ultimately end up with an environment compatible with our existence or with one hostile to our existence depends on the *level of consciousness we bring to the questions we ask*. And, before we can get fundamentally new answers, we must be willing to risk asking fundamentally new questions. This requirement presents an increasingly difficult task with respect to the sustainability of our environment because there is an ongoing and underlying disengagement from nature-based recreation by the increasingly technologically oriented public, and because of this disengagement people are becoming increasingly separated from the biophysical systems that support them.[39]

Learning how to frame a good and effective question is paramount, both for the crafting of a collective vision for the future and for the process of monitoring the actions necessary to achieve the vision. A question is a powerful tool when used wisely because questions open the doors of possibility. To illustrate, I use the previous example of going to the moon, which was not possible until someone asked: Is it possible to go to the moon? At that moment, going to the moon became possible, albeit no one knew how. To be effective, each question needs to embody six characteristics: It must have a specific purpose, contain a single idea, be clear in meaning, stimulate thought, require a definite answer, and explicitly relate to previous information.

In a discussion about going to the moon, one might usefully ask, Do you know what the moon is? The specific purpose of this question is to find out whether the person knows what the moon is. Knowledge of the moon is the single idea contained in the question. The meaning of the question is clear: Do you, or do you not, know what the moon is? The question stimulates thought about what the moon is and may spark an idea of how one relates to it; if not, that can be addressed in a second question. The question, as asked, requires a definite answer, and the question relates to previous information.

Asking a question that focuses on right versus wrong, as said previously, is a hopeless exercise because it calls for human, moral judgment, and that kind of question is not a valid one to ask of either an ecosystem or science in terms of social-environmental sustainability. A good question would be whether a proposed action is good or bad for taking care of the landscape your community depends on for a good quality of life for the citizens. To find out, you must inquire whether a good short-term economic decision is also a good long-term ecological decision and so a good long-term economic decision. Such questions are important because a good short-term economic decision can simultaneously be a bad long-term ecological decision and thus a bad long-term economic decision, one that generations of the future will have to pay for. The point is that one must ask before an answer can be forthcoming.

The old questions and the old answers have led us into the mess we are in today and are leading us toward the even-greater mess we will be in tomorrow. We must therefore look long and hard at where we are headed with respect to the quality of the questions we have heretofore asked. Only when we are willing to risk asking really new questions can we find really new answers.

With respect to new questions, those that will raise the level of social consciousness about the future of humanity as a realistic objective, there is an extremely insightful article—one that can go a long way in securing a dignified future for all generations. The article, authored by 38 individuals, is titled, "The Identification of 100 Ecological Questions of High Policy Relevance in the UK."[40] These questions are far more than merely ecological, however, because they address the very heart of social-environmental sustainability.

That said, community decision makers can derive realistic objectives for policies furthering their community's evolution toward social-environmental sustainability by adhering to these five general guidelines:

1. Evidence-based policy requires researchers to provide the answers to biophysical questions that are of interest to policy makers. To identify such questions, it is necessary to invite representatives from a broad cross section of community interests, bring them together with scientists from academic institutions, and ask them to generate a list of questions from their community. If the expense is too great for a single community, those sharing a bioregion can meet together for a 2- to 3-day workshop. (A bioregion is an area constituting a natural ecological community with characteristic flora, fauna, and environmental conditions contained within the drainage basin of a river, where all the water flows into the basin on its way to the sea.)
2. There will be a plethora of questions arising from the workshop. In the end, however, a short list of questions must be decided on that address the preferences of a single community or the collective of communities from the bioregion.
3. The areas covered must focus on major issues of environmental concern, such as agriculture, fisheries (marine or otherwise), climate change, nature's biophysical services, livestock grazing on public lands, forestry, habitat fragmentation versus habitat connectivity, urban growth boundaries, open spaces, riparian areas and floodplains, water catchments, and so on.

   Here, we can use agriculture as an example: Agriculture is rightly recognized as a source of considerable emissions of atmospheric carbon dioxide, with concomitant opportunities for mitigation. Whereas future agricultural productivity will affect the emissions of carbon dioxide from both the conversion of native landscapes to crops and from the intensified use of petrochemical fertilizers, investment in agricultural research is rarely mentioned as a strategy for asking and answering the questions of how to mitigate agriculture's contribution to atmospheric greenhouse gases.[41]

   Limiting the concentrations of atmospheric carbon dioxide to low levels requires strategies for managing anthropogenic carbon emissions from terrestrial systems, as well as from private and industrial sources. A comprehensive approach, one that lowers the cost of meeting environmental goals, carries with it profound implications for agriculture: As agricultural land becomes increasingly scarce due to its conversion to major highways, housing developments, and shopping malls, as well as shifts in land use due to climate change, the prices of food crops and livestock will probably continue to rise.

   That said, most of the world's carbon emissions from private and industrial uses carry no value, explicit or implicit, in today's accounting. Although considerable research has focused on alternative mechanisms for pricing fossil fuel and industrial carbon, both explicitly through taxes or cap-and-trade regimes and implicitly through regulatory frameworks, less attention has been forthcoming in developing methods of associating carbon values with terrestrial systems. This lack is not only because such valuation is less straightforward than those dealing with carbon emissions from fossil fuel but also because the cost of implementing policies to mitigate the emission of carbon dioxide from uses of terrestrial systems, such as intensive agriculture, would probably be higher than in the energy stream from fossil fuels.

Nevertheless, developing methods for conveying carbon values to land use decision makers could substantially improve the environmental effectiveness of systems employed to limit the emissions of global carbon dioxide. Improved land use and improved agricultural practices could reduce the upward-trending costs of mitigating the emission of carbon dioxide on crop prices. However, the allocation of scarce land resources—and the competition it engenders—will remain a major challenge for the 21st century.[42]

4. The questions must be general in scope rather than narrow because broad issues, not specific ones, drive policy, whereas scientists are frequently best equipped to answer specific questions. Therefore, it will be necessary to extract narrower questions from the general questions relating to a specific policy before researchers can proceed.
5. Regarding synthesis and applications, greater communication between policy makers and scientists is required to ensure that applied ecologists are dealing with issues in a way that can feed into policy. It is particularly important that applied ecologists emphasize the generic value of their work whenever possible.[43]

Once the questions are compiled, they can be winnowed by topic to select the most pertinent or urgent ones, prioritized, and established as objectives to be answered in the furtherance of a community's strides toward social-environmental sustainability.

THE NEED FOR URGENCY

Being able to create and maintain a sense of urgency to accomplish a task indicates an effective leader. Consider, for example, that most people steadfastly refuse to deal with making out a will, let alone prearranging their own funeral (even for the benefit of their children), until absolutely necessary. And then, they balk because of the sense of finality, which brings with it a foreboding.

Part of the reason for the palpable finality and foreboding is that, until now—this instant—dealing with one's will and funeral arrangements were abstractions somewhere in the future. Suddenly, however, they are not only in the present—in the here and now—but also concrete in one's experience, which is frightening to many people and thus avoided.

Therefore, if a leader wants to accomplish something for the benefit of the people, such as crafting and implementing a vision for a sustainable future, one of the daunting tasks is to create and maintain a sense of urgency about that which is important for the public good. Without a sense of urgency, most people languish simply because it is easier, whereas working toward an excellent but unknown (abstract) future is much more difficult and demanding.

For example, as His Holiness the 17th Gyalwang Karmapa Ogyen Trinley Dorje observes:

> The gap between the rich and poor is greater than it has ever been; proponents of economic growth seem to have forgotten the poor to benefit the wealthy. Furthermore, biological diversity is greatest and environmental degradation highest in many poorer parts of the world. Can the economic development model protect or replenish our precious natural resources? Can we borrow natural resources and return them for future generations? If not, we are robbing Earth of its riches and calling our actions economic development.[44]

GIVE COUNSEL, NOT ADVICE

One must lead by example, as English philosopher, statesman, and scientist Francis Bacon notes: "He that gives good advice, builds with one hand; he that gives good counsel and example, builds with both."[45] To this, English novelist Mary Anne Evans, better known by her pen name George Eliot, might add: "Blessed is the man who, having nothing to say, abstains from giving wordy evidence of the fact."[46]

A mistaken notion of those who are naïve about the nature of leadership is to equate giving advice with giving good counsel. To provide counsel must not be confused with dispensing information or advice. The role of a decision maker as counselor is to help another person (or group of people) discover, usually through questions, their own solutions and direction and to recognize their own freedom to act. A decision maker's task is to ask questions in such a way that a person can most easily make wise choices and find sufficient courage to act for the benefit of the people at large and willingly accept the consequences of their decisions.

It is not a leader's task to direct a person's thinking by giving advice and thereby limit the person's ability to act freely from his or her own conscience. After all, every person is responsible for the outcome of any decision he or she makes.

Unfortunately, some leaders, who are inexperienced in counseling yet who are called on for counsel, fall into the trap of believing they are leading only when being prescriptive and imparting advice to their subordinates—or other leaders. It is therefore critical to understand the power of the questions we ask.

THE QUESTIONS WE ASK

Each question is a key that opens a door to a room filled with mirrors, each a facet of the answer. Only one answer, however, is reflected in all the mirrors in the room. If we want a new answer, we must ask a new question—open a new room with a new key.

But, instead of seeking something new, we keep asking the same old questions—opening the same old door and looking at the same old reflections in the same old mirrors. We may polish the old mirrors and hope thereby to derive new and different meanings from the old answers to the same old questions. Or, we might think we can pick a lock and steal a mirror from a new and different room with the hope of stumbling onto new and workable answers to the same old questions.

The old questions and the old answers have led us into the social-environmental crises we face today and are leading us toward the even-greater crises enveloped in the current dysfunctional social-environmental feedback loops we keep reinforcing—the results of which we will be facing tomorrow and the day after, and the day after. … We must, therefore, look long and hard at where we are headed with respect to the quality of the world we help shape as a legacy to all generations. Only when we are willing to risk asking really new questions can we find really new answers and thus make better, more viable social-environmental decisions. For example, how must we alter our lifestyles so that our economic system can flourish within the constraints of nature's inviolable biophysical principles and thus pass a socially environmentally sustainable world to future generations?

The answer to a problem is only as good as the question and the means used to derive the answer. There is, however, no single reality, but rather a multiplicity of realities, the representation of which depends on one's position in the process of negotiating an acceptable social-environmental view of reality.

If, therefore, we are going to ask intelligent questions about the future of Earth and our place in the scheme of things, we must be free of scientific opinions based on "acceptable" interpretations of politically oriented, scientific knowledge. In addition, we would be wise to consider the gift of Zen and approach life with a beginner's mind—a mind simply open to the wonders and mysteries of the universe.

A beginner sees what *the answers might be* and knows *not what they should be.* If, on the other hand, a person deems himself or herself an expert, he or she is "sure" of what the answers should be and can no longer see what they might be.

The beginner is thus free to explore and to discover a multiplicity of realities, while the expert grows rigid in a self-created prison of a single pet reality, which often turns into an obsession to be protected at any cost—an all-too-common fate of leaders. So it is that a beginner often understands a question better than an expert does. Therefore, a wise and effective leader keeps a beginner's mind open to a multiplicity of realities, each of which has secreted within its possibilities a heretofore-untapped opportunity.

We must keep a beginner's mind if we are going to ask intelligent questions and be open to multiple hypotheses, realities, and explanations, and we must be willing to accept a challenge to our ideas in the spirit of learning, rather than as an invitation to combat. The greatest triumphs of both science and leadership are not, after all, triumphs of facts but rather triumphs of new ways of seeing, thinking, perceiving, and asking questions.

Such triumphs of vision and thought come not only through knowing which questions to ask but also through a willingness to risk what most people think of as failure. According to Princeton University President Harold Shapiro, "An excessive zeal to avoid all risks is, in the end, an acceptance of mediocrity and an abdication of leadership."[47] We must beware, however, of giving in and "raising the flag of failure" too soon because if we do not immediately achieve our stated objective, society is quick to judge something as a failure.

As stated, success or failure is a personal view that lies not in an event itself but rather in the interpretation of the event. When, for example, Thomas Edison's 10,000 experiments with a storage battery failed to produce results—and society would surely have deemed that a failure—he said, "I have not failed. I've just found 10,000 ways that won't work."[48] This same line of reasoning is implicit in Winston Churchill's famous commencement speech: "Never, never give up! Never, never give up!"[49]

Heretofore, we have been more concerned with getting "politically correct" answers than asking scientifically and culturally wise questions. Politically right answers validate our preconceived economic/political desires. Scientifically and culturally wise questions would lead us toward a future in which environmental options are left open so that generations to come may define their own ideas of a "good-quality environment" from an array of possibilities.

A good question, one that may be valid for a century or more, is a bridge of continuity among generations. We may develop a different answer every decade, but the answer does the only thing an answer can do: It brings a greater understanding of the question. An answer cannot exist without a question, so the answer depends not on the information we derive from the illusion of having satisfied the question but on wisdom of the question we asked in the first place.

In the final analysis, the questions we ask guide the evolution of humanity and its society, and it is the questions we ask, not the answers we derive, that determine the options we bequeath to the future. Answers are fleeting, here today and gone tomorrow, but questions may be valid for a century or more. Questions are flexible and open ended, whereas answers are rigid, illusory cul-de-sacs. The future, therefore, is a question to be defined by questions asked by those who dare to lead by the authenticity of personal example, each in his or her own style. To do so, however, one must first stay viable as a leader who exercises the ability to make wise, systemic decisions for the benefit of all generations.

MAINTAINING VIABILITY AS A LEADER

In both psychotherapy and leadership, the single most important instrument we have is ourselves, as we are and what we become as we grow. The most powerful technique at our disposal is our ability to model aliveness, authenticity, compassion, and dispassion. And, because leadership is about helping others to fulfill their potentialities, it is our responsibility to take care of ourselves so we are able to continue functioning as a wise decision maker in the service of others from our own aliveness. To accomplish this, we must consciously work to deal with those factors that threaten to drain our life's essence and render us impotent. One of these lethal factors is "burnout."

Burnout leaves people feeling tired, drained, and without enthusiasm. Where they were once idealistic and intent on making their mark by changing the world for the better, now they no longer care. Their ideals fade. They feel like slaves rather than wise decision makers and leaders. What they have to offer seems neither wanted nor accepted, which leaves them feeling unappreciated, unrecognized, and unimportant. So, they go through the motions of their accepted, public obligations with the aliveness and zest of a robot or zombie.

Burnout steals one's sense of being grounded in the everyday world, where they can see some measure of concreteness in the results of their efforts. People suffering burnout begin to feel oppressed and some fight the very "system" they intended to change. But, the more they actively resist, the more the system resists change, which stifles any sense of personal initiative.

Because a person at this point in the cycle of burnout has lost sight of his or her passion and vision toward which to build, burnout begins to feed on itself. When this happens, the person feels increasingly isolated and in despair, the latter increasing the sense of isolation in a self-reinforcing feedback loop that prevents seeking help and support from others.

Because burnout can rob a leader of vitality, it is critical to examine some of its causes, remedies, and ways of prevention. Here again, I acknowledge Corey[50] for his excellent and concise listing of some of the causes, remedies, and preventions of burnout, which I have adapted and added to; these are discussed next.

Causes of Burnout

Recognizing the causes of burnout is a step toward its remedy. Among these are the following:

1. Caring for and being invested in a cause more than others are involved, which produces a need to pull the other people along as opposed to simply guiding them.
2. Giving a great deal of personal time, ideas, and energy but receiving little or no positive and meaningful responses in return.
3. Always trying to stop things from happening.
4. Being under constant pressure to produce, perform, and meet timelines.
5. Working with a difficult population, such as those who resist any kind of change unless it is somehow self-serving.
6. Experiencing disillusionment as a leader when personal ideals of service come up against greed, apathy, or corruption.
7. Being unable to deal with the calumny and criticism that inevitably come with leadership.
8. Having unresolved personal conflicts beyond the responsibility of leadership, such as marital problems, chronic health problems, financial problems, and so on.

Remedies for Burnout

Learning ways in which to care for oneself beyond recognizing burnout as a problem is critical for leaders as wise decision makers. My own experience, in concurrence with that of Corey,[51] leads me to think that accepting *personal responsibility* for my thoughts, motives, and actions is a critical factor in overcoming burnout.

To reverse the process of burnout (projecting blame for my sense of failure onto whatever sounds good outside myself), it is necessary to turn the searchlight inward and ask such questions as: Do I really want to serve as an unconditional way of extending love, *or* am I really trading my service for approbation as a way to find personal fulfillment? Turning inward means that I must honestly examine the root causes of my negative feelings and accept that they *come from within*—not without. Therefore, *the cure lies within.*

Until I understand and accept that both the cause and the cure are within, I am not only surrendering personal power to forces outside myself but also acting out the role of a helpless victim, the behavioral gateway to cynicism. Although there are obstacles to leadership that constrain actions, there are ways of learning to keep my own counsel and to act out of personal integrity and meaning.

It is pointless to blame "the system" for the dysfunction internal to myself. Moreover, problems lying within the system can be identified and addressed one at a time with diligence, courage, persistence, patience, and dispassion.

Preventing Burnout

Learning to look within ourselves to determine what choices we have, what choices we are making, and why we are making them can go a long way in preventing burnout. There are also other ways to prevent burnout, most of which include assuming the responsibility of consciously, actively nourishing ourselves:

1. Making sure I have a clear, exciting, believable, attainable vision toward which to lead
2. Taking responsibility for my own motives, thoughts, decisions, and actions and leaving others the same responsibility
3. Finding interests outside my role as leader and decision maker
4. Taking the initiative for change
5. Attending to my own health through adequate sleep, relaxation, regular exercise, proper diet, and time for reflection, fun, or meditation
6. Cultivating nourishing people as acquaintances with whom to share love, support, ideas, humor, interests, and so on
7. Learning how to ask for what I want and allowing it to manifest
8. Learning how to grow inside so that I am clearly focused, other centered, and well grounded while being dispassionate and detached from the outcomes of leadership, which renders external recognition unimportant in the validation of my accomplishments
9. Learning to play and honor the child within by giving expression to that child freely
10. Learning to be in touch with my emotions honestly, including recognizing pain and letting tears flow when necessary
11. Evaluating whether the project I am working on is still worth my time and commitment or whether it is necessary, in terms of personal growth, to move on to other challenges
12. Taking the initiative to form a support group with colleagues to share feelings and find better ways of approaching difficult situations
13. Finding a wise mentor with whom to debrief and from whom to glean counsel

This list is not meant to be exhaustive but rather to provide some insight into maintaining physical and mental health while accepting the trials and enjoying the fruits of leadership by keeping an exclusively personal scorecard about your own successes or failures. If you remain true to your ideals and retain a healthy zest for living, you will continue to grow and change and in so doing find life to be exciting and ever new.

To remain personally and professionally alive, you must periodically evaluate the direction in which you find yourself traveling and determine if you are, in fact, living in the way you really want—true to your beliefs. If, perchance, you are not, then you must decide what changes are necessary and find the courage to risk acting, rather than simply waiting for circumstances to force your hand.

By being well grounded and in tune with yourself, the experience of life can be one of harmony, by which you have the unmistakable feeling of personal power, self-control, and self-mastery. Self-mastery, in turn, allows you to integrate the experiences of life, including your feelings, with your experiences as a leader in such a way that you have empathy and compassion for those whom you serve without being captured by them. Such a personal synthesis provides a solid, insightful foundation for true servant leadership—something sorely needed in today's governments the world over. Having said this, let us examine some ways in which a government (any government) can foster exemplary leadership from within.

SOCIAL-ENVIRONMENTAL DECISION MAKING WITHIN ORGANIZATIONS

Organizations, including governments and corporations, cannot empower people, but they can give them the opportunity, proper tools, and capacity to work actively in achieving desired outcomes. For instance, incentives for good trusteeship of a local community's natural resources created by the community itself can replace tax breaks used to recruit multinational corporations with little or no stake in either the ecological sustainability of the community's natural resources or the high-quality jobs needed to sustain the community's cultural well-being.

In addition, organizations for economic development can provide expertise and financial backing, rather than merely ladling out grants, so that local community partnerships can learn to create their own vision and the plan of how to achieve their goals. Such support would include assistance in strategic planning for businesses, including marketing, to help communities retain those small businesses deemed critical to their social sustainability.

Local leaders can also employ measures of performance for agencies and their staffs to reward those actions that help the growth and sustainability of community partnerships. In this spirit, a governor, mayor, or city council can require the use of cross-agency or cross-departmental decision making and can combine funding for programs to increase the efficient *and effective* access of local communities to resources at the agency's or department's disposal. Greater devolution of decision-making authority to the local level can make participation in community partnerships more effective, which creates the opportunity for community self-empowerment in moving toward social-environmental sustainability for the common good.

A prerequisite for sustainable development in a local community, society, and nation is that it must be inclusive, interrelating all relevant disciplines and professions from all walks of life. Setting a good example is a critical function of any local organization or government involved in implementing the principles and practices of sustainable community development. Leading by example—breaking down bureaucratic barriers of turf through interdisciplinary crossing of departmental lines, recycling and buying recycled goods, providing day care, encouraging car pooling, and offering flexible working hours—increases not only the capacity of an organization to govern its own people and simultaneously improve their quality of life with the community but also the effectiveness *and* efficiency of governance.

It is thus important for organizations to identify departmental and community links concerning mutually interrelated issues and to bring all people affected or their representatives to the table in an effort to resolve shared problems collectively, which includes dealing with human diversity. Understanding and accepting diversity allows us to acknowledge that each of us has a need to be needed, to contribute in some way. It also enables us to begin admitting that we do not and cannot know or do everything, and that we must rely on the strengths of others with complete trust.

Just as simplifying an ecosystem or complicating a mechanical system increases its vulnerability to disturbance, so will segregating diverse elements within a community lead to its social, ethical, economic, and environmental decay. Those conditions may then spread throughout society, one community at a time. Conversely, overall acceptance of diversity of thought, culture, expertise, and economic status allows all persons to contribute to the development process in a special way, making their unique gifts a part of the holistic effort necessary to create a sustainable local community. Accepting diversity helps us to understand the need each person has for love, equality, identity, and opportunity. Recognizing diversity gives us all a chance to provide meaning, fulfillment, purpose, and a gift of our talents to our community and future generations.

Assuming people accept diversity, what is it they most want from the development process? People want the most effective, productive, and rewarding way of working together to achieve a common end. They want the process and the relationships forged therein to meet their personal needs for love, belonging, meaningful contribution, and an opportunity to make a commitment to a special place—their community—the opportunity for personal growth, and the ability to exert reasonable control over their own destinies.

Control over personal destinies, thus the destiny of a community, can be increased if federal agencies focus on the community's goals for achieving social-environmental sustainability. Having said this, however, it must be recognized that federal agencies (*not* a local community) have jurisdiction over *public lands* and thus are legally responsible and accountable for the biophysical sustainability of those lands as a national legacy for *all* generations. But, that does not mean a "one-size-fits-all" present-day solution (although perhaps efficient from a decision-making point of view) is either necessarily wise or effective.

Consider, for example, a recent (November 13, 2011) one-size-fits-all proposal in Congress that would allow the Border Patrol to circumvent dozens of environmental laws within the 100-mile zone of the borders between both the United States and Mexico and the United States and Canada. Supporters contend this authority is needed to cut through a bureaucratic gridlock, where border agents have difficulty dealing with environmental laws and rules governing roadless areas. Whereas such an exemption might be necessary to deal with drug smugglers and other people entering the United States illegally from Mexico, an approach designed for the southern border does not make sense along the wilderness landscapes paralleling the U.S. border with Canada, which not only are far different environmentally from the deserts of Texas and Arizona, or even the landscapes of Southern California, but also are devoid of the same human problems.[52]

It is also possible to create organization-to-community trusteeship contracts—including the government, local and otherwise—that recognize it is the results that count, not necessarily how many rules were followed. Such action can devolve authority closer to the citizens of a community and simultaneously allow employees at all levels of government to empower themselves to achieve excellent results, which may well improve citizen participation in the process of crafting a collective vision of social-environmental sustainability as the foundation for sustainable community development and thus a better world for all generations.[53]

SUMMARY

Chapter 7 dealt with the responsibilities and pressures of leadership. These included such things as circumstances faced by leaders as decision makers, being and disclosing oneself, avoiding the trap of perfectionism, inspiring performance, nurturing creativity, establishing realistic objectives, and how to remain viable as a leader. Chapter 8 recognizes the fact that, in a true democracy, children deserve a voice in determining the kind of future we adults leave for them. After all, they must live with the social-environmental consequences of our decisions and actions. In that, they have no choice—unless we consciously, purposefully give it to them, which means changing our adult thinking.

NOTES

1. Gerald Corey. *Theory and Practice of Counseling and Psychotherapy*, 3rd ed. Brooks/Cole, Monterey, CA, 1986.
2. Caroline Myss. *Anatomy of the Spirit: The Seven Stages of Power and Healing*. Three Rivers Press, New York, 1998.
3. Ibid.
4. Ibid.
5. Ibid.
6. Laurence J. Peter. *Peter's Quotations: Ideas for Our Times*. HarperCollins, New York, 1993.
7. Milan Kundera. The struggle of memory against forgetting. http://www.guardian.co.uk/commentisfree/2008/oct/15/milankundera-humanrights (accessed February 7, 2011).
8. Benjamin Disraeli. http://www.giga-usa.com/quotes/topics/knowledge_t003.htm (accessed February 7, 2011).
9. Theodore Roosevelt. http://www.theodoreroosevelt.org/life/quotes.htm (accessed February 7, 2011).
10. John F. Kennedy. *Profiles in Courage*. Harper & Row, New York, 1961.
11. Hubert H. Humphrey. http://thinkexist.com/quotes/hubert_h._humphrey/ (accessed April 3, 2011).

12. The forgoing discussion is based on Jean Marsh. Keep your eyes & ears open and your pen ready! *The Toastmaster*, 63 (1997):20–21.
13. Robert A. Johnson. *Owning Your Own Shadow*. Harper, San Francisco, CA, 1991.
14. [His Highness] 17th Gyalwang Karmapa Ogyen Trinley Dorje. Walking the path of environmental Buddhism through compassion and emptiness. *Conservation Biology*, 25 (2011):1094–1097.
15. **(1)** The Indus River. http://geography.howstuffworks.com/asia/the-indus-river.htm (accessed January 6, 2012); and **(2)** Dorje, Walking the path.
16. Ganges. http://en.wikipedia.org/wiki/Ganges (accessed January 6, 2012).
17. Brahmaputra River: **(1)** http://en.wikipedia.org/wiki/Brahmaputra_River; **(2)** http://www.britannica.com/EBchecked/topic/77154/Brahmaputra-River (accessed January 6, 2012); and **(3)** Dorje, Walking the path.
18. Irrawaddy River. **(1)** http://en.wikipedia.org/wiki/Irrawaddy_River (accessed January 6, 2012); **(2)** N'Mai River. http://en.wikipedia.org/wiki/N%27Mai_River (accessed January 6, 2012); and **(3)** Andaman Sea. http://en.wikipedia.org/wiki/Andaman_Sea (accessed January 6, 2012).
19. Yangtze River. http://en.wikipedia.org/wiki/Yangtze_River (accessed January 6, 2012).
20. Mekong: **(1)** http://en.wikipedia.org/wiki/Mekong (accessed January 6, 2012); and **(2)** http://www.britannica.com/EBchecked/topic/112224/China-Sea (accessed January 6, 2012).
21. Dorje, Walking the path.
22. The preceding discussion of ignorance is based on Chris Maser. *Of Ditches and Ponds: A Journey Through the Metaphors of Childhood and Maturity*. Woven Strings, Amarillo, TX, 2006.
23. Louis Fischer. *The Essential Gandhi: An Anthology of His Writings on His Life, Work, and Ideas*. Random House, New York, 1962.
24. The previous two paragraphs are based on (**1**) Fred Pryor. Do you have an appetite for input? *The Toastmaster*, 63 (1977):11; and (**2**) Richard G. Ensman, Jr. Become your own mentor. *The Toastmaster*, 63 (1997):14–15.
25. The foregoing discussion is based on Luann Lee Brown. Coping with a disliked member. *The Toastmaster*, 63 (1997):6–7.
26. Ibid.
27. Chris Maser and Carol Pollio. *Resolving Environmental Conflicts*. 2nd ed. CRC Press, Boca Raton, FL, 2011.
28. Johann Wolfgang von Goethe. http://quotationsbook.com/quotes/author/2895/#axzz1DOJ8DwMb (accessed February 8, 2011).
29. Jerry Mander, Debi Barker, and David Korten. 2001. Does globalization help the poor? *IFG Bulletin* 1 (2001):2–5
30. The preceding discussion of productive thinking is based on Michael Michalko. The art of genius. *Utne Reader*, July–August (1998):73–76.
31. Bernard Glassman. The sacred act of cooking. *Delicious*! November (1997):32–34.
32. Eihei Dogen. *Moon in Dewdrop: Writings of Zen Master Dogen*. Kazuaki Tanahashi (Ed.). North Point Press, New York, 1995.
33. The preceding two paragraphs are based on Glassman. The sacred act.
34. Ibid.
35. The discussion of the shadow is based largely on Johnson, *Owning Your Own Shadow*.
36. *The Holy Bible*. Authorized King James Version. Leviticus 16:21–22.5.
37. Glassman, The sacred act of cooking. *Delicious!* November (1997):32–34.
38. Donald R. Nelson, W. Neil Adger, and Katrina Brown. Adaptation to environmental change: contributions of a resilience framework. *Annual Review of Environment and Resources,* 32 (2007): 395–419.

39. Oliver R. W. Pergams and Patricia A. Zaradic. Evidence for a fundamental and pervasive shift away from nature-based recreation. *Proceedings of the National Academy of Sciences of the United States of America,* 105 (2008): 2295–2300.
40. William J. Sutherland, Susan Armstrong-Brown, Paul R. Armsworth, and others. The identification of 100 ecological questions of high policy relevance in the UK. *Journal of Applied Ecology,* 43 (2006):617–627.
41. Jennifer A. Burney, Steven J. Davis, and David B. Lobell. Greenhouse gas mitigation by agricultural intensification. *Proceedings of the National Academy of Sciences of the United States of America,* 107 (2010):12052–12057.
42. The preceding three paragraphs are based on Marshall Wise, Katherine Calvin, Allison Thomson, and others. Implications of limiting CO_2 concentrations for land use and energy. *Science* 324 (2009):1183–1186.
43. Sutherland et al., The identification of 100.
44. Dorje, Walking the path.
45. Francis Bacon. http://www.goodquotes.com/quote/francis-bacon/he-that-gives-good-advice-builds-with (accessed February 8, 2011).
46. George Eliot. http://www.quotedb.com/quotes/1419 (accessed February 8, 2011).
47. Harold Shapiro, in David D. Ho. *MIT News*, June 5, 1998. http://web.mit.edu/newsoffice/1998/hospeech.html (accessed February 8, 2011).
48. Thomas A. Edison. http://www.brainyquote.com/quotes/authors/.../thomas_a_edison.html (accessed February 8, 2011).
49. Winston Churchill. http://personal.inet.fi/business/squashkeskus/wm2003/pdf/6.issue.pdf (accessed February 8, 2011).
50. Corey, *Theory and Practice*.
51. Ibid.
52. Matt Gouras. Environmental law waiver faces northern skeptics. http://abcnews.go.com/US/wireStory/environmental-law-waiver-faces-northern-skeptics-14943025#.Tr_qtBw0i4A (accessed November 13, 2011).
53. Abraham Maslow. *The Farther Reaches of Human Nature*. Penguin Books, New York, 1985.

8 Giving Children a Voice in Their Future through Shared Decision Making

The ultimate test of human conscience may be the willingness to sacrifice something today for future generations whose words of thanks will not be heard.

—Gaylord Nelson, former governor of Wisconsin, founder of Earth Day[1]

There is a proverb among South Sea Islanders that states: It takes a village to raise a child. This maxim holds a truth that extends well beyond the village. Namely, for a village to be socially and environmentally sustainable and healthy, it requires a unifying factor to integrate the myriad interactive components into a functional whole. And, that unifying factor *is* the villager's focus on the children.

If, for example, we ask ourselves what it would take to design a community that would make children happy, we would find ourselves engaging the whole of the environment. This is a self-evident truth when one considers that every community, as an interactive system, requires a unifying center around which it turns. Moreover, for life, human or otherwise, to have any measure of good quality, the basic components of the global commons must be given highest priority: from clear air, to pure water, healthy oceans, fertile soils, sustainable forests, and healthy food, to parents who are psychologically mature enough to be loving, to asking their children what kind of future they want their parents—as trustees—to protect for them as a legacy from one generation to the next. What, you might ask, do I mean by the "global commons"?

The "commons" is that part of the world and universe that is every person's birthright. There are two kinds of commons. Some are gifts of nature, such as clean air, pure water, fertile soil, a rainbow, northern lights, a beautiful sunset, or a tree growing in the middle of a village. Others are the collective product of human creativity, such as the town well from which everyone draws water; a baseball field in a city park; the Statue of Liberty in New York harbor; the Lincoln Memorial in Washington, D.C.; or Mount Rushmore near Keystone, South Dakota, where the faces of George Washington, Thomas Jefferson, Theodore Roosevelt, and Abraham Lincoln are carved in 60-foot relief.

The commons is the "hidden economy, everywhere present but rarely noticed," writes author Jonathan Rowe.[2] It provides the basic ecological and social support systems of life and well-being. It is the vast realm of our shared heritage, which we typically use free of toll or price. Air, water, and soil; sunlight and warmth; wildflowers and butterflies; wind and stars; mountains and oceans; languages and cultures; knowledge and wisdom; peace and quiet; sharing love and community; joy and sorrow; and the genetic building blocks of life—these are all aspects of the commons.

A commons has an intrinsic quality of just being there, without formal rules of conduct. People are free to breathe the air, drink the water, and share life's experiences without a contract, without paying a royalty, without needing to ask permission. As such, a commons engages people in the wholeness of themselves and in community. It fosters the most genuine of human emotions and stimulates interpersonal relationships in order to share the experience, which enhances its enjoyment and archives its memory.

We humans have jointly inherited the commons, which is more basic to our lives and well-being than either the market or the state. We are "temporary possessors and life renters," writes British economist and philosopher Edmund Burke, and we "should not think it amongst [our] rights to cut off the entail, or commit waste on the inheritance, by destroying at [our] pleasure the whole original fabric of [our] society."[3]

Echoing the sentiment of Edmund Burke, Carolyn Raffensperger has penned what she calls the "The Law of the Commons of the Natural World":

1. The commons shall be passed on to future generations unimpaired.
2. All commoners have equal access to the commons and use by commoners will be allocated without discrimination.
3. Government's key responsibility is to serve as a trustee of the commons.
4. The commons do not belong to the state but belong to commoners, the public.
5. Some commons are the common heritage of all humans and other living beings. Common heritage establishes the right of commoners to those places and goods in perpetuity. This right may not be alienated, denied, repudiated, or given away. The Common Heritage law is a limit on one government's sovereignty to claim economic jurisdiction and to exclude some commoners from their share.
6. The precautionary principle—first do no harm—is the most useful tool for protecting the commons for this and future generations.
7. Eminent domain is the legal right and process for taking control of private lands for public use and shall be used exclusively for that purpose, which includes operations by private entities, such as pipelines, power lines, and so on.
8. Infrastructure necessary for humans and other beings to be fully biological and social creatures will reside within the domain of the commons. The positive benefits (externalities) of the commons shall accrue to all commoners.
9. The commons are the foundation of the economy. Therefore the market, commerce and private property shall not externalize damage or costs onto the commons.
10. Damage to or loss of the commons shall be compensated to all commoners.[4]

Both Burke's and Raffensperger's concepts are critical because children have an innocence that, in my experience, can seem close to intuitive wisdom—if not supplanted by adults—with respect to sharing the commons, like we once had as children, but which we have forgotten in becoming adults. In essence, we have lost our way in today's hurry, worry, and competitive materialistic world of clashing ideologies and power-hungry ideologues waging war against one another to "control" circumstances—a physical impossibility. It is thus increasingly clear to me that a society in which the corporate/political elite wield the dominant influence will do what they can to maintain the status quo and thus their grip on the power of control. Without a firm commitment to social-environmental sustainability, however, no society has a viable context within which to greet the children it brings into the world—much less nurture them.

To nurture children, we must have an atmosphere of social-environmental harmony based on social-environmental equality that translates into social-environmental justice that translates into social-environmental sustainability. Here, the linchpin is social-environmental justice, which, by its very nature, asserts that we owe something to each and every person who shares the planet with us, both those present and those yet unborn. In this context, all we have to give of real value to one another—ever—are love, trust, respect, and wisdom gleaned from our experiences, each of which is embodied in the ramifications of every decision that gives birth to an action that creates an option that we pass forward for better or ill.

Each person—whether child or adult—has a gift to give, and each gift is unique to that person and critical to the social-environmental whole. All gifts are equal *and* different. What is true for individual humans is true for cultures and societies because each is equal in its service to Earth. Each life, each culture, each society is equally important to the evolutionary success of our planet, whether we understand it or not. Each also has its own excellence and cannot be compared to any other. All differences among people, cultures, and societies are just that—differences. The hierarchies or judgmental levels of value are human constructs that have little or nothing to do with reality. Every life, culture, and society is a practice in evolution, and each is equal before the impartiality of nature. Although cultural values may change with time, basic human requirements do not appear to do so—such as obedience to nature's biophysical principles, as well as the basic human need for love, trust, respect, and equality.

And, it is exactly because options embody all we have to give the children of today, tomorrow, and beyond that social-environmental justice, in the form of basic human values, must necessarily form the context of human equality. In this sense, a decision in the present produces a future circumstance, and if the decision—wherein children have no voice—bodes ill in terms of their future options, that decision is analogous to taxation without representation, which countermands everything our democracy is reputed to stand for.

Human equality demands one of the scarcest resources we have—our willingness to *listen*—really listen—to one another as psychologically mature adults and equals. It is especially important, however, that psychologically mature adults listen to children, who are allowed to be children and taken seriously when they speak—despite being children. Not listening, as stated, is an act of violence, especially if children are involved, because it is a purposeful, albeit often unconscious, way of invalidating the feelings—the very existence—of another person. Everyone needs to be heard and validated because sharing is the bond of relationship that makes us "real" to ourselves and gives us meaning in the greater context of the universe. We simply cannot find meaning out of relationship with one another. Therefore, only when I have first validated another person through listening, as an act of love, can that person, adult or child, really *hear* what I am saying. Only then can I share another's truth. Only then can my gift of ideas enter receptive ears—and vice versa.

As I look around the world, I realize, ever more clearly, that all we have in the world of real value as human beings is one another, and *all we have to give one another is one another.* We are each our own gift to one another and to the world, but we cannot give our gifts if there is no one to receive them, if there is no one to hear. Therefore, if we listen—*really listen*—to one another and validate one another's feelings, even if we do not agree with them, we can begin to resolve differences before they become disputes and thereby share what we each have to give—the gift of personal knowledge. The gift of *knowledge* has certain requirements, however, not all of which are followed.

THE GREAT AMERICAN IRONY: CHILDREN HAVE NO VOICE IN THEIR FUTURE

> The direction in which education starts a man will determine his future life.
>
> **—Plato, *The Republic*[5]**

The great American irony is that our children do not have the First Amendment right of "free speech" because we adults neither ask them what they want us, the trustees of their future, to leave for them nor listen when they try to tell us. Moreover, we handicap them further by teaching them in the negative. Even when we take what they say seriously, they cannot tell us what they want—only what they do not want. If, therefore, social-environmental sustainability is to prevail, we adults—as trustees and decision makers of the children's future—must learn to think in the positive and *teach them to think in the positive* to make wise decisions for all generations.

Instead of educating young people in terms of positive human values (civility; peaceful cooperation; social-environmental harmony, equality, justice, and sustainability; and how to think positively and creatively), children are increasingly educated in the negative (to be individualistic and strongly competitive, even combative; to be acquisitive and ever-more materialistic; and to move away from what they do not want—a psychological impossibility). By the fourth grade, most children are no longer taught how to think but rather what to think. All the while, adults create the social-environmental problems that we expect them to solve because we are increasingly incapable of doing so.

To help illustrate this point in my 2004 book *The Perpetual Consequences of Fear and Violence: Rethinking the Future*,[6] I turned to Connie Anderson's fourth-grade class at Harding Elementary School in Corvallis, Oregon, for the counsel of her 25 students.

First, I Asked the Children What They Wanted

I visited Connie's class and inquired of the children—as I have inquired of children for years—if anyone had ever asked them what they wanted their world to be like when they grew up. Although the children represented several nationalities, races, and religious backgrounds, their unified answer was a resounding "No!" Note: I have *never* had a child answer in the affirmative to this question.

Next, I asked them what they wanted us, the adults, to leave for them. Although the discussion was instantly lively, mostly because of a few outgoing boys, the children could not answer my question because the only way they knew how to think was in the negative—to try and move away from what they did not want. For example, Grace said, "I want no wars in the world." Trevor responded, "Not being violent." Michael mixed the positive and the negative: "Peace, but not everybody being 'phony.'"

Therefore, the children and I then spent time considering why it is important to focus and think in the positive, which proved relatively easy for most of the children as long as they had positive reinforcement. Their overall response left me with a wonderful feeling of hope because so many of them could switch permanently to positive thinking if they were taught that way at home, in school, and through their reading materials.

Here, at least, is one thing that could be done for the sake of our children and the future of human society—*teach children to think in the positive*. Thinking in the positive is such a big idea that it would overarch the entire scholastic curriculum from kindergarten through PhD programs and beyond. Moreover, it would fundamentally change how we treat one another, our society, and ourselves and thereby change the way we treat the world as a whole. What could human society and the world be like if children—the future leaders—were taught to think in the positive?

Prior to my second visit, the children were asked to write down the things they wanted adults to protect for them and to rank their desires in order of importance. In addition, their statements were to be written in the positive, which all but two or three of the children did. During my second visit, the children and I spent time discussing what they had written.

Virtually all the children wanted peace in the world; that was the first priority for roughly half of the class. Peace was closely followed by clean air, clean water, more trees, everyone having enough food and a home, kindness to animals, and wanting extinct animals to live again—although they were not so sure about a return of the dinosaurs. Elise hoped that "everyone is free from slavery, everyone is trustworthy." Lauren wanted everyone to recycle, and Andrea wanted more teachers so the "homeless kids" could learn. Michael wanted more money spent on schools and education, whereas several children wanted "people to be nice to other people" and "all people to be healthy." A few boys were primarily interested in technology, such as "watches that are computers and you can ask for anything you want to eat and get it and it's [also] a phone."

One thing that surprised me was the number of boys *and* girls who wanted what they called "hovercraft." Having no idea of what they were talking about, I asked for clarification and was promptly told what "to hover" means. Acknowledging my adult ignorance, I was informed that hovercraft were like automobiles that neither touched the ground nor polluted, which I took to mean that we could remove some paved roads and convert the land back to wise agricultural use, fertile grasslands, and vibrant forests and thus greatly expand the soil's ability to capture and store water.

Because the discussions were largely dominated by a few bright boys, I asked Connie to have each child write an essay about what they wanted the world to be like when they grew up and had children of their own. Although there clearly were bright children in Connie's class, some with wisdom and farsightedness that was truly astounding, I have found that many children, especially girls, have currents that run deep and are best expressed in private on paper, hence the essays. I have also found, as mentioned, that by the time children enter the fourth grade, they are not so much being taught how to apply their imaginations through asking questions that are relevant to them and thus clear thinking, but rather are increasingly being taught *what* to think in a way that fits into adult agendas.

To help get this point across, a copy of each child's essay was given to his or her parents for safekeeping until the day its author had his or her first child for the new parent to see if he or she had helped create the kind of world described in the essay—the one the parent had wanted as a child for his or her child.

Unfortunately, it was not feasible to have every essay published in my book, so a few were selected to illustrate that the children did cover the major subjects of social-environmental sustainability—relationships, both domestic and international; personal responsibility; people's relationship to Earth; peace; safety; enough food for people and animals; animal well-being; more fish; recycling; cleaning Earth; clear air and water; clean sources of energy for clean factories and manufacturing; human health and longevity; protecting endangered species; and more trees planted in a greater variety of "types." I did not, on purpose, correct the few misspellings because that would have stolen their childhood.

Vice President Hubert Humphrey saw each child as "an adventure into a better life—an opportunity to change the old pattern and make it new."[7] The latter part of his statement echoes an observation my friend and colleague Jack Ward Thomas made about teaching children in the positive:

> First, adults would be exposed to their "better natures" and forced to confront the "forgotten voices" buried deep within them. Second, "children" would feel and understand early on that they can and should speak up, and when they do, what they say and feel makes a difference. When that occurs, their worth and thoughts are honored as important. That contributes to their self-esteem and bolsters their interests in their environment. They will be better prepared, as adults, to deal with those most important of issues—their world and their children's world.[8]

And This Is What They Said

On relationship and personal responsibility, Maryam R. wrote:

> If I got to pick what I wanted the future to be like, I would want people to be nice to other people. I want people to be nice and grateful to each other. I would also want people to be cooperative. Cooperative means to work well together. I would really like it if that would happen.

The other thing I would want is for people to help clean up the world. The reason I would want that, is because the world is getting more and more polluted lately. Almost every year people start polluting the world more. Imagine only one person cleaning up all this garbage in the world, that would take about 100 years! My only wish is for people to start cleaning up their own garbage and to stop polluting.

Catalina M. combined peace and relationship in her essay:

I want peace when I have children of my own. I want people to calm down and to be friends with other people from other countries. Peace is having parties together and going to each others houses, kids playing together at school peacefully. That is what I want when I have children of my own.

If I was in charge of the world I would do this to make it come about. I would make people friends by asking them to try to be friends. I would bring one of my friends to my other friend's house. And introduce them to each other and then they can be friends. I could ask all my friends to come to my house and then they can be friends. I could also make people friends at school by asking everybody if they want to play with everybody.

Ellen L. thought peace and animals went together:

When I have children of my own I would like there to be peace everywhere. Peace is a happy quiet feeling, to be a kind, helping, caring person. And when the paper comes out in the morning it would talk about how people in different parts of this world get along. People in this world will get along so we can all live peacefully together. We should all become peaceful and caring.

People should enjoy animals. Some animals help the environment, like slugs and snails. Animals help the air, environment, and people. Fuzzy, cuddly, warm animals like dogs, and cats, make good pets. Fish make good pets to. People and children should enjoy the animals that are still alive. We should care for and love animals.

Michael G. and Elise E. focused on peace:

If I got to chose one thing that I would like to see when I have children of my own it would be peace. I want my children to grow up with peace around them. If the world was all peaceful you would see everybody being kind to each other. You can sort of feel peace. That feeling is wonderful. But to have peace everybody would have to try.

First, everybody will have to be convinced that a peaceful world is a good world. Then, we (meaning the people who want peace) could tell them (meaning the people who don't want peace) that it will help the world a lot and everybody will want to be nice to each other. It will be hard but if everybody works hard the world just might be a peaceful place. I know that peace will make the world a better place. And that is the world I want.

Michael G.

If there was peace when I'm older and have kids they probably won't even know what war is. My kids will be happy because they wouldn't worry about stuff happening. A world with peace would be like being in heaven because people will be kind. I hope even if I'm dead there would be peace. My children will be happy they wouldn't have to worry when they go on a plane because people won't crash it into buildings and die. It would be nice it there's peace forever. Peace would help nature live, but it would be okay if they cut down some trees for like my children and other people to use for there house, but they should put more trees in that area.

Peace would make people stay home with there family. You will smell a pretty smell because nature is alive. People will share other toys because peace helped them. Teenagers will be kind instead of egg some peoples houses. If there was peace, Artists will be able to draw a beautiful picture because thing like flowers and buildings won't be recked. Kids will be happy playing on the street playing football or soccer all kinds of sports.

Elise E.

Elise C. thought safety and peace were part of each other:

I think that there should be peace throughout the world. There should be happy families because everyone is safe. I really want the world to be safe. All people should be happy. Safe is when you are protected from dangerous things. Like if someone is in a fire thats not safe. Happy is when you are cheerful and jolly. Like if you're having a birthday party you have an excited feeling that means you're happy.

I think that how we can have world peace to happen is we find all the good people gathered together. Meaning all the people that are in the world that are not perfect but they only do something bad by accident. Find all the terrorists and bad people and talk them into being good. If they disagree we will show them that they are more good than there are bad people. They will probably want to change their minds, if not, will compromise with them. There would be peace because everyone is following the Lifelong Guidelines: Truthful, Trustworthy, Listen actively, say no putdowns and do their personal best.

Mitra G. and Taylor R. were mainly concerned that people have enough food:

I would like to have everybody have enough food to eat. So everybody would live longer than they are living today. We would also see more fat on other people. Everybody would know what grapes tast like. You would see people out on the streets being able to eat more food.

We could do this by first having a fundraiser. Then after the fundraiser we would have a tow year food gathering. After that we would use the money to buy more food. Then after that we would have a fundraiser every other year. We could also teach people to grow their own food. If we did this worldwide everybody would have enogh food to eat.

Mitra G.

What I want the world to look like when I have children. I want everyone to have food because if you don't have food you can't live and you probably don't want to die. I want the earth to feel safe. I want it to smell clean and look clean. I want it to sound quiet and that the food tastes good.

How can this come about is people working harder like to make food, cook food, try to help save endangered species, keep planting trees, and try to keep clean air. If you want to know how to know how to help then clean up garbage like at the beach, at a school in you neighborhood or at a park. You can ask your friends if they will help and you should recycle. If everyone recycled then the world would be much cleaner and easyer to move around.

Taylor R.

Kipp G. was concerned about animal well-being:

When I have children of my own I want them to enjoy animals, study animals, and care of animals. I think we should leave animals alone. And if we do leave them alone, I think they might do something for us. Like maybe they might lead us to a berry that we don't know of and maybe it's an engredience to a medicine that could really help people. I think if we work together and we go to the president and say to him "we want the animals we have now and we want to stop hunters from destroying them," I think we could get his attention and he might do something about it, like stoping the hunters, or putting more animals on the endangered species list.

Both Gyun-Chang N. and Lauren K. wanted clean air and water:

I want the world to have more clean water and more clean air. There is lots of bad air so I want more clean air so we can breathe better air. I want more clean water because them we can have more clean water for the animals and people. If we have more clean water there will be more fish. If there is more clean air people can breathe easier.

It there is more clean air people will be healthier in the future because they will be breathing healther air. I wish there were hovercrafts that go as fast as cars. If there was hovercrafts that go as fast as car there will be less more gas. If there are more clean water there will be twice as much than right now. If there is more clean water people won't have to drink dirty water.

Gyun-Chang N.

When I have children I want the world to have clean air and clean water. The air and water would be poision free. In the cities, people would have clean air in their face, no masks. The air taste good and smell good and the water would be clear and taste sweet. The plants would get a lot of clear water in the cities.

You could get this to happen by saying that you have to recycle almost everything. People could take all of the cans and trash out of the rivers, streams, ponds and lakes. Another way would be to filter oil and pollution out of all polluted bodies of water. A way to get the air clean would be to have all factory workers to figure out a way to run the factory using some other fuel, like solar power or a battery. People could try to transport air from the farm lands to the cities, then take the air somewhere that is in space. If people did this, we would have a very clean earth.

Lauren K.

Finally, Sebastian L. wanted more trees in the world:

When I grow up I want the world to have a lot of trees. I want the world to smell like Pine and Beech and like all the other many types of trees. When you look around all that you will see is a whole bunch of trees. When you listen carfully you will hear little chattering squirrels and birds and you will even hear the splashing. You will touch all different types of bark and sap. I really think that we can get more trees if we try.

> I know that we can get more trees planted but only if we try. It can come about if we start cleaning and start helping mother nature. We can recycle. I know that for every tree we cut down five more are planted in it's place but they are all the same type! one of my classmates said that if we keep planting the same type of tree and a disease comes around all the trees will die. I think that it's a great idea to plant more trees because some are becoming extinct. We can save them if we try.[9]

Here, adults must ask: Is what the children want unreasonable, unethical, or impossible to do or anything more than a choice of how we think and behave as evidenced by the decisions we make? If it is possible to fulfill the children's requests, which echo our own, too-often hidden or suppressed desires, why is it so difficult for us adults to become other centered and positive in our thinking? Are we so jaded in our view of life that we are no longer capable of understanding that the wisdom exhibited by children is nothing but our own suppressed understanding, or is it simply that we choose not to make the effort required to be kind and thoughtful?

> Some see the world as it is and ask 'why?' I see the world as it could be and ask 'why not?'—George Bernard Shaw (1856–1950).[10]

Then, I Asked Their Teacher What She Had Learned

Having heard from the children, I asked Connie to share what she had learned from the exercise:

> I thought it was fine when Chris requested permission to come to my classroom and ask my fourth grade students the following question as a closing part of this book: "What would you want the world to be like when you're grown up and have children of your own?" When he asked if I would have them write an essay on this subject, I said "Yes" because it would be a different kind of writing assignment, something I'd never thought to have them write about.
>
> After Chris came and talked with them, I had my students write a list of things they would like to see in the world when they are older. When we shared these lists with one another in class, many ideas were stated in the negative. With a little coaching, the children rewrote sentences such as, "There shouldn't be any wars," to say, "There should be peace all over the world." After making a list of positively stated items, I felt my students were ready to write their essays on what they wanted the world to be like when they had children of their own.
>
> I was, however, surprised by a number of the essays. Instead of describing a perfect world, many of the students had written persuasive essays on why there should be *no* war, *no* polluted rivers, and/or *no* dirty air. This turned out to be a much harder essay for them to write than I had originally thought. I found, as I talked with them, that they could not see the world as a perfect place. They already had rigid mind-sets as to why things cannot change. By 9 and 10 years of age, we adults have already tainted our children's outlook on life to the point that most cannot view the world through a child's eye and see hope.
>
> Since the completion of this exercise, I have been questioning what we are doing to our children. Have we done them a disservice by introducing them to adult problems at such a young age? After all, we adults have created the problems, yet can't seem to

> fix them. Why, therefore, should we pass these problems forward to the children and expect them to fix what we cannot?
>
> Maybe we should concentrate on raising our children in a positive atmosphere for as long as possible. Then, when they are older and problems arise, they will have beginner's minds and can see what *might* be done instead of having to wade through the predetermined, negative mind blocks of their predecessors—us—who created the problems in the first place. Or better yet, why not ask ourselves how we must behave to avoid creating such problems as polluted water and air, and then actually be responsible for our own behavior?
>
> Might this work? I have no idea. What we have been doing for years and years has not been working; so why not try something really new, such as being positive and teaching our children in the positive?[11]

Of course, this means that the books our children read, both in school and out, must be written in the positive if our children are to learn how to think and live positively. But, writing books in the positive requires that the writer must learn to think and behave in the positive. To reiterate, Gaylord Nelson, a senator from Wisconsin when I met him, said: "The ultimate test of human conscience may be the willingness to sacrifice something today for future generations whose words of thanks will not be heard."[12] And what is the greatest sacrifice we adults can make? The answer is simple—to forgo the notion of our adult superiority over children and give them an active voice in decisions that will affect their future. After all, every decision we make—for good or ill—becomes a consequence to which we commit all generations without recourse.

CHANGING OUR ADULT THINKING

For you who think it impossible to change societal behavior because "I'm but one person, what can I do?" the answer is consistently: "I can do *something*." Ours is not to question the size or value of our individual contributions as adult decision makers. Our task is simply to give from the essence of who we are. Each gift is unique and valuable. None is more or less important than another, but rather, each complements the other. And each adds a necessary piece to our growing knowledge as a whole.

Given the innocence of childhood, how can we teach our children so they remember what it means to love—and to dream of what can be? If you were to ask biologist Rachel Carson, a kind and exceedingly gentle woman, this is what she would say:

> A child's world is fresh and new and beautiful, full of wonder and excitement. It is our misfortune that for most of us that clear-eyed vision, that true instinct for what is beautiful and awe-inspiring is dimmed and even lost before we reach adulthood. If I had influence with the good fairy who is supposed to preside over the christening of all children, I should ask that her gift to each child in the world be a sense of wonder so indestructible that it would last through life, as an unfailing antidote against the boredom and disenchantments of later years, the sterile preoccupations with things that are artificial, and the alienation from the sources of our strengths.[13]

The circumstances we adults pass forward, based as they are on the choices we make, coalesce to create the circumstances of the future to which our children must respond, a future that could be very bright if we followed the Universal Declaration of Human Rights, given here in abbreviated form:

> The General Assembly [of the United Nations] proclaims this Universal Declaration of Human Rights as a common standard of achievement for all peoples and all nations, to the end that every individual and every organ of society, keeping this Declaration constantly in mind, shall strive by *teaching and education* [emphasis mine] to promote respect for these rights and freedoms and by progressive measures, national and international, to secure their universal and effective recognition and observance, both among the peoples of Member States themselves and among peoples of territories under their jurisdiction.
>
> Article 1: Right to Equality
> Article 2: Right to Freedom from Discrimination
> Article 3: Right to Life, Liberty, and Personal Security
> Article 4: Right to Freedom from Slavery
> Article 5: Right to Freedom from Torture and Degrading Treatment
> Article 6: Right to Recognition as a Person before the Law
> Article 7: Right to Equality before the Law
> Article 8: Right to Remedy by Competent Tribunal
> Article 9: Right to Freedom from Arbitrary Arrest and Exile
> Article 10: Right to Fair Public Hearing
> Article 11: Right to be Considered Innocent until Proven Guilty
> Article 12: Right to Freedom from Interference with Privacy, Family, Home, and Correspondence
> Article 13: Right to Free Movement in and out of the Country
> Article 14: Right to Asylum in other Countries from Persecution
> Article 15: Right to a Nationality and the Freedom to Change it
> Article 16: Right to Marriage and Family
> Article 17: Right to Own Property
> Article 18: Right to Freedom of Belief and Religion
> Article 19: Right to Freedom of Opinion and Information
> Article 20: Right of Peaceful Assembly and Association
> Article 21: Right to Participate in Government and in Free Elections
> Article 22: Right to Social Security
> Article 23: Right to Desirable Work and to Join Trade Unions
> Article 24: Right to Rest and Leisure
> Article 25: Right to Adequate Living Standard
> Article 26: Right to Education
> Article 27: Right to Participate in the Cultural Life of Community
> Article 28: Right to a Social Order that Articulates this Document
> Article 29: Right to fulfill Community Duties Essential to Free and Full Development
> Article 30: Right to Freedom from State or Personal Interference in the above Rights[14]

This universal declaration is clearly based on a foundation of cultural wisdom and justice—qualities of higher human consciousness. In addition, we adults must recognize, accept, and act on the premise that a society, which loves its children, would give them a voice in their future by asking them what kind of world they want *and* protect the children themselves from abuse, as well as protect their birthright—a healthy environment as a global commons. We are, after all, the trustees of the children's welfare and their future, which makes them the beneficiaries of our thinking, choices, decisions, and actions.

Before we can give children a voice, we must treat them with kindness and respect, which allows them the safety to speak and know they will be heard and taken seriously. To this end, adults must make a concerted effort to follow the 1989 United Nations Convention on the Rights of the Child, which is presented here in abbreviated form:

Article 1: A child is a human being below the age of eighteen years unless, under laws applicable to the child, maturity is attained earlier
Article 2: Right to Freedom from Discrimination
Article 3: Right to Protection of Best Interests
Article 4: States Implement and Protect a Child's Rights to the Maximum Extent Possible
Article 5: Respect for Parental Guidance and Developing Capacities
Article 6: Inherent Right to Life and Development
Article 7: Right to a Birth Name and a Nationality
Article 8: Right to Preservation of Identity
Article 9: Right to Freedom from Unnecessary Separation from Parents
Article 10: Obligation for Reunification with Family and/or Visitation Rights when in the Best Interest of the Child
Article 11: Right to Freedom from Illegal Transfer Abroad
Article 12: Right to be heard in any Judicial and/or Administrative Proceedings that directly or Indirectly Affect the Child
Article 13: Right to Freedom of Opinion and Expression
Article 14: Right to Freedom of Thought, Conscience, and Choice of Religion
Article 15: Right to Freedom of Association and Peaceful Assembly
Article 16: Right to Privacy
Article 17: Right to Freedom of Information
Article 18: Right to Parental Recognition of Their Responsibilities, with the Aid of the State when Necessary
Article 19: Right to Protection from Abuse and Neglect
Article 20: Right to Protection when a Child is without Family
Article 21: Right of a Child During Adoption to have His or Her Best Interest be the Paramount Consideration
Article 22: Right of a Child seeking Refugee Status or who is considered a Refugee to Receive Protection and Humanitarian Aid as Set Forth Within these Articles
Article 23: Right of a Mentally or Physically Disabled Child to a Full and Decent Life
Article 24: Right to Health and Health Services
Article 25: Right of a Child under State Protection to have His or Her Situation Periodically Reviewed
Article 26: Right to Social Security
Article 27: Right to an Adequate Standard of Living
Article 28 & 29: Right to an Adequate Education
Article 30: Right to a Cultural Identity
Article 31: Right to Rest, Leisure, Recreation, and Cultural Activities
Article 32: Right to Protection from Economic Exploitation and Dangerous Labor
Article 33: Right to Protection from Drug Abuse

Article 34: Right to Protection from Sexual Exploitation
Article 35 & 36: Right to Freedom from Slavery and Other Forms of Exploitation
Article 37: Right to Freedom from Torture and Deprivation of Liberty
Article 38: Right to Protection from Armed Conflicts
Article 39: Right to Rehabilitative Care
Article 40: Right to Protection Within the Juvenile Justice System
Article 41: Respect for Standards Higher then Those Set Forth in this Convention of Child Rights
Article 42–54: Right to the Implementation of the Above Articles[15]

Whether we, as individuals, are concerned about future generations depends on our childhood experiences and the values we learned. Moreover, how we learned to cope with circumstances as children influences how we treat one another as adults. Society is thus as peaceful or combative as we are as individuals.

The more we are drawn toward peace and an optimistic view of the future, the more functional (psychologically *mature* and healthy) we are. The more we are drawn toward debilitating, destructive conflict, cynicism, and pessimism, the more dysfunctional (psychologically *immature* and unhealthy) we are.

To change anything in society, therefore, it is first necessary to look inward to confront, understand, and change myself. Self-evaluation and change put the contest of growth where it really belongs—within my own heart. As such, my inner struggles are the greatest learning experiences I will ever have. In addition, the greater the understanding of my own behavioral dynamics and unresolved fears and pain, the easier it is to understand these dynamics in others and thus introduce compassion and wisdom into how I treat other people and the world around me.

Thus, molded into the family template in an unknown and unknowable universe, the most consistently pressing existential questions since the dawn of humanity have probably been: "Who am I?" and "What value do I have in the immensity of the ever-changing, unknown, and unknowable universe?"

Each person must answer the first question for himself or herself. With respect to the second question, however, I find a more universal answer—namely, my sense of value is derived from caring for and nurturing something or someone else. What could be more important than nurturing children? That not withstanding, most of us grew up in dysfunctional families, thus the perceived necessity of creating the Universal Declaration of Human Rights and the Convention on the Rights of the Child. Now, the question becomes: How do we break the mold of familial dysfunction that made the Universal Declaration of Human Rights and the Convention on the Rights of the Child social necessities in the first place?

The last question is compounded by facing growing acts of terrorism and a spiral into social-environmental bankruptcy, which further compounds terrorism through decades and centuries of generations retaliating against one another—with no reprieve in sight. There is, however, a remedy for this situation—let the young, innocent children of the elementary grades become our *spiritual* teachers even as we become their *material* teachers. With this in mind, the following are a few suggestions whereby adults can improve the practice of familial relationships—familial landscapes, if you will—that create the human web of life:

1. Develop a set of positive, personal values that define who you are for yourself and stick with them.
2. Listen to and obey your own intuitive, conscious self because it is the guidance system of your behavior.
3. Learn to *accept* other people rather than merely tolerate them because acceptance allows others to be who, what, and how they are without residual rancor on your part. Tolerance is the repression of emotions that can lead to angry outbursts or acts of violence. Each of us is responsible for what we do, regardless of how we feel at the moment.
4. Control your attitude, or it will control you.
5. Whereas the circumstances of life influence who we are, *we* are responsible for who we become.
6. Do work that is important to you not only because it will likely consume more than a third of your adult life but also because work that serves others can create a sense of fulfillment.
7. Compare yourself to the best *you* can do, *not* to the best others can do.
8. It is not what happens to you that matters but rather what you do about it that counts.
9. Develop both the right and left sides of your brain because that will lead to an improved ability to think, see things in greater perspective, and help you to lead a balanced life.
10. Give freely without any thought of compensation. This means giving for the sake of improving the person you are, not for recognition. Giving anonymously is an easy way to practice giving freely.
11. Many people have helped you in great and small ways to create a fulfilling life. Therefore, say "Thank you" and mean it. Remember, people tend to be quick to criticize and slow to praise. So, make it a point to acknowledge the efforts of others and reward every act of kindness with an act of kindness, even if it is only a heartfelt "Thank you."
12. Although you cannot make someone love you, you can make yourself lovable.
13. Be humble. There is always someone who is wiser than you. Therefore, if and when the urge for recognition comes over you, remember that you are where you are because of all the others who have gone before and helped you along the way. Whatever praise comes to you, therefore, is partly theirs.
14. Enjoy conscious simplicity. The more content you are with simple things, the simpler and easier it is to be content and feel the joy of life well lived.
15. Take time to reflect on your actions. Action taken without reflection may cause you to do something in an instant that you will regret for a lifetime. It is a lot easier to act or react in kind than to think and *respond* wisely. In other words, it can take years to build trust and only seconds to destroy it.
16. Think big thoughts. Develop big ideas. Whatever you can imagine may be helpful to someone else or to the health of Earth. After all, we become what we think.
17. Finally, all we have of real value to give one another, including our children—*ever*—are choices and some things from which to choose. With every choice we pass forward, we give our children an unconditional gift of our

love, our trust, our respect, and the benefit of our experience. Conversely, with every option we foreclose, we withhold our love, our trust, our respect, and the benefit of our experience.

Added to this list are suggestions by Professor David W. Orr for things adults can do for the sake of children:

1. Protect the purity of the air, water, soil, and food from pollution by artificial chemicals whose combined effects are not *and, perhaps, cannot* be known.
2. Connect children with the soils, forests, grasslands, waterways, plants, and animals that are so vital to their overall well-being.
3. Do everything humanly possible to protect the children from the risks of anthropogenically related climate change.
4. View renewable natural resources as value-added products to be passed on to future generations.
5. Leave children a legacy of beauty and biophysical richness.
6. Protect prime agricultural lands.
7. Create and maintain societal priorities of establishing and maintaining educational opportunities for all citizens.
8. Protect neighborhoods and communities from ever-expanding transportation systems.
9. Protect good-quality open space in which children can safely play, explore, and discover both nature and themselves.[16]

It is to our collective benefit to remember what it is like to be a child with a child's infinite imagination of possibilities, hopes, and dreams. The young belong in body, mind, and spirit to the present and the future, whereas we adults too often cling to the past, the recollection of a time we laboriously drag with us into our perceived present. This "out-datedness" became apparent to me while I was still employed with the Bureau of Land Management, where I saw people in Washington, D.C., repeatedly make unwise decisions because they were based on circumstances they remembered as newly emancipated, idealistic professionals in the field a decade or two earlier. We tend to become encrusted in our narrowly perceived "realities" of the present (realities that have often been formulated in an earlier time and different place), which means our sense of responsibility, as it migrates through time, is largely to protect the economic comfort of the status quo for the sake of our own generation—with too little consideration for future generations.

THE ULTIMATE DEMOCRATIC REVOLUTION IN DECISION MAKING: GIVING CHILDREN A VOICE IN THEIR FUTURE

We must—for the children's sake, if no other—discard our view of Earth as a battlefield of subjective competition, where human superiority reigns over nature, where my superiority dominates yours, and where the superiority of adults subjugates children. We will all be better off when we consider Earth in terms of complementary efforts in which all gifts are equal—including the innocence and imagination of children. Each is uniquely important to the health and well-being of the whole, living system—planet Earth. I say this because life demands inner struggle and tenacity, albeit tempered by outer cooperation, which continually fits and refits each living thing to its function. Complementary efforts, such as those among adults, as well as those between adults and children, imply equality. And, equality provides the stage on which hope, dignity, and social-environmental sustainability can reign supreme in the human experience.

Accomplishing this philosophical revolution requires the psychological maturity and innovation to engage children in the democratic process, such as addressing the school board.

Address the School Board

Adults can begin to engage children in the democratic process by having members of every school board go into grade-school classrooms and ask the children what they want from their educational experience and why. Have them write essays (in the positive) to answer those questions and take the essays to the board meetings, where a surprising amount of wisdom can be gleaned from them. Then, honor the children by applying that wisdom to their educational curricula.

This is a critical exercise because we adults view education solely from our points of view—increasingly science, math, business, and global competition. Meanwhile, the softer, social aspects of the liberal arts slip farther and farther away, which renders our society evermore competitive, technological—*and* impersonal. It is imperative for a truly democratic society to be well educated and well balanced in the breadth of its knowledge if it is to survive the trials of the times and simultaneously be trusted in the realm of its international diplomacy. Trust, in turn, is earned through honesty, compassion, cooperation, and the willingness to share openly the best our culture has to offer of its humanity.

Interviewing New Teachers

I have heard of a progressive school in Texas with an enlightened way of interviewing prospective teachers. Three panels interview each applicant: one of teachers and school administrators, one of parents, and one of the students who will have the teacher who is hired. I think this is a superb idea because every teacher influences his or her students throughout their lives and can be instrumental in the life path a student chooses to follow. I have had a few such outstanding teachers in my life (both women and men), and I am profoundly grateful for each of them.

Creating a Community Vision

I always ensure that children are present at a meeting when I help a community come to grips with its vision of a desired future condition. I have set up panels of grandparents, parents, and children so that three generations are represented. I have the grandparents tell the audience what the community was like when they were children and what they wanted from it as they grew up. Next, the parents share their feelings, and finally the children tell their grandparents, parents, and the audience how they view the community and what they want from it.

In one case, a grade-school boy turned to his parents and told them to stop growing marijuana—the community's main cash crop. He told his parents that, while he was being taught to tell the truth, he had to lie to protect them from going to jail. The audience, constituted of community members, was very quiet.

I then had the audience form small groups, each with at least one child. The groups were to describe their desired future condition for the community. The adults were so impressed by the contribution of the children, they asked them to present the groups' results to the whole audience, which the children did with a great deal of enthusiasm.

Next, representatives from the various groups were to cooperate in drafting a vision statement that encapsulated the shared vision of a desired, future condition. Children from each group were delegated to help draft the vision statement for the whole community—a vision statement that is still in effect more than a decade since it was drafted.

In another instance, Dr. Dean Button, director of program development, School of Humanities and Social Sciences, Rensselaer Polytechnic Institute, Troy, New York, wrote about a visioning process I facilitated:

> Maser recalled, during our interview, that the group was struggling to verbalize their vision of where they wanted to go: "I said, 'Okay, now I want from each of you a one- or two-sentence statement of what you think this all means, what you want your community's future to be like.' They wrote. Some of them had to write a paragraph, but a vision statement has to be one or two sentences. … The challenge is that most adults come from their heads. So, we went around the room and they read their statements, but each one was rejected—until they came to a girl who was a sophomore or junior in high school. (I had told them in the beginning that I wanted some children to participate. I wanted young folks.) When the young woman read her statement—which came straight from here [refers to heart], unadulterated—they all said, 'Yes! That's it!'"
>
> This moment was singled out by several of the participants as one of the most memorable of the entire process. Months later, it was recalled with a mixture of incredulity and pride. Even some of the "old-timers," who expressed impatience with the emphasis paid to "process," had a noticeable shift in energy and enthusiasm when making the recollection.
>
> An official with the Forest Service recalled: "They went through all kinds of gyrations at the meetings—identifying their mission and value systems, those kinds of things. Some of them worked and some of it was probably wasted energy—but the interesting thing is a lot of the stuff in here came from a couple of high-school kids. I was at that meeting, and what came out of those kids was really like a light went on to that committee! Gee! That's what we're really here for! That was of value to me. I thought, man! That's great! When the local kids see what they want out of their landscape—and they live here—not the adults—it's the kids. …"[17]

Resolving Conflicts

I have also found that having children participate in the resolution of environmental conflicts has a sobering affect on the adult disputants, who often tell kids, in effect: Do as I say—not as I do. The presence of children not only helps the adults to act with greater psychological maturity than they might otherwise but also gives the children a voice in the outcome. After all, whatever the future brings, it is the children who must be able to navigate its circumstances—the legacy of the disputants' decisions.

Every conflict must end in a shared vision of the future if it is to be resolved. Moreover, crafting a vision statement prior to the eruption of a conflict can act to achieve resolution and thereby preempt conflict—for both adults and children.

Healing the Land

When I help a community produce a vision and subsequent plan for repairing some aspect of its degraded landscape, I have the person in charge contact the community's school teachers and get permission for volunteers to visit with the children. Once in the classroom, a volunteer explains to the children what is going on, discusses with them what they want the land in question to look like, and finds out why that is important. The children, in turn, write essays (stated in the positive) that inform community leaders of what the children want the land to look like when the task is completed and why that is important.

As with the other visioning processes, the children help decide what to heal, how, and why. Through the essays and personal participation in the visioning process, as well as repairing the landscape, the children have both a voice in the future of their community and a vested interest in the outcome. In this way, whatever the landscape becomes, it is also partly their gift to the community—present and future, which may someday include their own children.

If children are given a voice in addressing the school board, interviewing new teachers, helping to create a vision for their community, participating in the resolution of conflicts, and helping to heal the land, that is participatory democracy. Such inclusive participation moves a society toward social-environmental sustainability and helps to fulfill the South Sea Islander's vision: It takes a village to raise a child. And, it inspires developing leaders to continue the process across generations.

Children's Advisory Councils

While children in the United States technically have First Amendment rights—the right of free speech—that is of little avail if no one listens. Thus, forming a "children's advisory council" will give children a voice in how they want *their* resources to be cared for.

I have worked with a lot of grade-school children to help me understand what kind of future they want us, today's adults, to leave for them. I had them draw pictures and explain in writing what kind of landscapes they want and why. I had them write essays on the kind of world they want when they are adults and have their first child. I have had them make lists of the attributes they want in their future. I have had discussions to find out how they see the world today. As the beneficiaries of our thoughts, decisions, and actions, the children want:

1. people to be kind to one another
2. people to be responsible for their own behavior and how it affects other people, animals, and the environment
3. peace in the world
4. people to enjoy and care for animals
5. everyone to feel safe
6. everyone to have enough food
7. clean air and water
8. a lot of trees
9. forests to be healthy homes for animals
10. forests to produce wood for homes
11. forests in which to hike, camp, and play
12. a place to really see the stars
13. cars that get a lot of miles but do not pollute
14. all children to have an education, including "homeless children"

Is there anything unreasonable in that list? Is there anything impossible to achieve if we adults were to act in a way that was psychologically mature? After all, growing old is inevitable, but *"growing up" is optional.* Is there anything in the list that everyone would not like to be reality? Is there anything that has not been sought through the ages? The German poet Johann von Goethe said it well: "All truly wise thoughts have been thought already thousands of times; but to make them truly ours, we must think them over again honestly, 'til they take root in our personal experience."[18]

Because I find the children's desires reasonable, responsible, and just plain common sense, representing, as they inevitably do, the inclusivity with which we are all endowed before it is stolen from us by adult conditioning, I recommend the formation of a "children's advisory council" (hereafter called the "children's council"). Children would be told that the world needs their dreams—and needs them to hold onto those dreams as they mature. It is those dreams that the children's councils can manifest over generations. They can thereby help to guide caretaking of our nation's landscapes and natural resources. Such councils could provide insight to all levels of governmental organization (federal land management agencies, state agencies, the military, and private landowners) on matters of concern to children—matters that will surely be critical to future generations.

There is another important reason to form a children's council: to keep adults cognizant of the incessant march of time as our evolving social-environmental consciousness changes circumstances and social values.

First: We can be blessed by remembering, always, the child's ability to conjure up possibilities, hopes, and dreams. As heretofore stated, the young dwell in the present and the future, whereas adults, too often, are rooted in the past and cannot think beyond attempting to hold the illusion of the known past in the uncertain present.

Second: We forget how to ask questions that are truly relevant in the present for the present *and* the future. If we are going to ask relevant *and* wise questions about the future of the commons, such as public forests, national parks, oceans, and our place in the scheme of things, it is important to understand that most questions will deal with social values that are not amenable to scientific investigation. Nevertheless, scientific inquiry can help elucidate the outcome of decisions based on these values and must be so employed. In addition, we would be wise to accept the gift of Zen and approach the caretaking of our lands and communities with the mind of a child—a beginner's mind, one simply open to the wonders and mysteries of the universe, which includes us as inseparable components thereof.

A beginner sees what the answers might be and knows not what they should be, whereas an expert "knows" what the answers should be and can no longer see what they might be. Children, as beginners, are free to explore and to discover and so hear questions in a different context than do adults who think themselves experts. Children hear questions in the context of multiple realities and infinite possibilities. When determining what question to ask, therefore, it is critical to listen to what the children say because they represent that which is to come. To them, all things are possible until adults with narrow minds, who have forgotten how to dream, put fences around their imaginations.

We adults, on the other hand, too often think we know what the answers should be and can no longer see what they might be. To us—whose imaginations have been stifled by parents, schools, and the corporate-political instruction of what "reality" is—things have rigid limits of *impossibility*. We tend to become encrusted in a self-created prison of a single pet reality, with its self-limited possibilities. Because this adult reality is so limited in potential outcomes, it often turns into an obsession to be protected at any cost. We would do well, under this circumstance, to consider carefully what the children envision as possible in the future and what they want. The future, after all, is theirs.

That said, it is necessary to reiterate that, if we are going to ask truly wise questions about how best to strive for and protect social-environmental sustainability in all its myriad forms, we must be open to multiple hypotheses and explanations, and we must be willing to accept a challenge to our ideas in the spirit of learning, rather than as an invitation to combat. The greatest triumphs in science are not, after all, triumphs of facts but rather triumphs of new ways of seeing, thinking, perceiving, and asking questions. To nurture planet Earth as a biological living trust for all generations, we must learn to accept our ignorance and trust our intuition while doubting our knowledge. This reversal of our adult training requires the help of children.

Third: We, as adults and parents, think we know what is good for our children. But, we do so from our increasingly busy adult perspective—with all of its acquired baggage—from growing up and the current pressures of facing an ever-more-competitive world, which we seem committed to perpetuating. Instead of relegating children to the background of life, we would be wise to invite their participation in planning and evaluating the care we take of all they will inherit. In this way, real continuity might be better achieved as responsibility is passed from one generation of trustees to the next.

To form a children's council as fairly and representatively as possible, it would have to have an equal number of boys and girls, and all ethnic backgrounds would have to be represented as equally as possible. Children selected for the council would have to be from the inner city as well as rural and suburban areas. They would come from families in all walks of life and all regions of our nation, including towns along lake and ocean shores, the mountains, prairies, deserts, the snow-covered reaches of Alaska, and the Pacific isle of Hawaii.

To get the best possible representation of the children's collective voice, it would be necessary to visit different areas of our nation with the express purpose of interviewing children about what they want us, the trustees of their future, to leave for them. In addition, a new children's council would have to be appointed at intervals to keep their voices fresh and current; maximize their ability to be heard by adults who are often too busy and preoccupied to listen, no matter how sincerely, wisely, or loudly they speak; and to keep the vision of how we care for our home planet alive and vital.

With interviews in hand and the first children's councils formed, the members would meet with the respective government agencies and help to craft a collective vision of our home planet as a biological living trust of social-environmental sustainability for all generations.

SUMMARY

This book has three parts. Section I was nature's message to humanity, especially its leaders: Obey the inviolable biophysical principles that govern the universe and your place in it as the bedrock of leadership and decision making for human survival—or reap your self-imposed consequences, a legacy you bequeath all generations everlasting. Chapter 1 was a discussion of the inviolable biophysical principles that form the nonnegotiable limitations of social-environmental sustainability within which humans can shape the environment to their benefit. When those limits are ignored, environmental degradation ensues, along with a diminished productive capacity of those ecosystems on which we rely for a good quality of life—a situation that is not only progressively irreversible but also passed forward to all generations. Chapter 2 expanded the premise of Chapter 1 into the inviolable rules of leadership and decision making, which are necessary if social-environmental sustainability is to prevail.

Section II examined why a command-and-control form of leadership and decision making tends to become dysfunctional and ultimately fails and how to break the cycle of dysfunction. Chapter 3 examined the ways in which we carry our respective families with us throughout life—both the functional and dysfunctional aspects. Further, it pointed out that the governments, agencies, and various organizations are as functional or dysfunctional as we are as individuals. Therefore, we—individuals acting as a single, collective body—are society's driving force and thus enable or limit its ability to function within the purview of social-environmental sustainability or constrain it in a military-style command-and-control mode. Chapter 4 was a case study of the political duplicity within the sustainability programs at the Southern Nevada Water Authority. Chapter 5 illustrated the ways in which dysfunction and homeostasis are corrected for the benefit of all generations. These include understanding how job descriptions are used to control information and how positions of authority are used to control direction for political purposes. The correction culminates with the crafting of a shared vision, goals, and objectives, as well as understanding and implementing the seven steps in monitoring the vision's effectiveness.

Section III was a discussion of what it takes to be a psychologically mature, wise decision maker who actually serves as a leader with the good of all generations in mind. Here, it was pointed out that (other than natural disasters) all social-environmental problems begin with our thinking. Put differently, we create the vast majority of the social-environmental problems that we end up facing. We do it to ourselves. Understanding both this and nature's biophysical limitations, the discussion focused on how people must change their thinking through individual, personal growth if we, as a nation and world, are to have decision makers who actually lead from a position of psychological maturity for the social-environmental sake of all generations. Chapter 6 was an enumeration detailing the essential qualities of a psychologically mature decision maker and leader. Chapter 7 dealt with the responsibilities and pressures of social-environmental decision makers. These include such things as circumstances faced by leaders, being and disclosing oneself, avoiding the trap of perfectionism, inspiring performance, nurturing creativity, establishing realistic objectives, and how to maintain viability as a leader. Chapter 8 recognized the fact that children in a true democracy deserve a voice in determining the kind of future we adults leave for them. After all, they must live with the social-environmental consequences of our decisions and actions. In that, they have no choice—unless we consciously, purposefully give it to them, which means changing our adult thinking.

NOTES

1. Gaylord Nelson. http://en.wikiquote.org/wiki/Environment (accessed February 2, 2011).
2. Jonathan Rowe. The hidden commons. *Yes! A Journal of Positive Futures*, Summer (2001):12–17.
3. Edmund Burke. *Reflections on the French Revolution*, The Harvard Classics, Vol. 24, Part 3, Paragraph 62, Charles W. Eliot, ed. Collier, New York, 1909–1914.
4. Carolyn Raffensperger. The law of sharing: setting a policy and legal agenda for the commons. Science and Environmental Health Network: *The Networker*, 12(5), October 2007. http://www.sehn.org/Volume_12-5.html (accessed February 2, 2011).
5. Plato. http://www.iwise.com/D0N6A (accessed February 4, 2011).
6. Chris Maser. *The Perpetual Consequences of Fear and Violence: Rethinking the Future.* Maisonneuve Press, Washington, DC, 2004.
7. Hubert Humphrey. http://www.quoteland.com/author/Hubert-Humphrey-Quotes/451/ (accessed April 5, 2011).
8. Jack Ward Thomas, personal communication (2012).
9. Chris Maser. *The Perpetual Consequences.*
10. George Bernard Shaw. http://advancement.umd.edu/celebration/showStudent.php?main_id=1717 (accessed April 5, 2011).
11. Chris Maser. *The Perpetual Consequences.*
12. Gaylord Nelson.
13. Rachel Carson. *The Sense of Wonder.* Harper & Row, New York, 1965.
14. United Nations. *Universal Declaration of Human Rights*, abbreviated version. Minnesota Advocate for Human Rights and the Human Rights Resource Center, Minneapolis, MN, no date.
15. United Nations. *Convention on the Rights of the Child*, abbreviated version. Minnesota Advocate for Human Rights and the Human Rights Resource Center, Minneapolis, MN, no date.
16. David W. Orr. Beauty is the standard. *Resurgence*, 210 (2001):34–37.

17. The preceding three paragraphs are from Dean Button, who used the results of my work in Lakeview, OR, as a case study for his doctoral dissertation: Dean Button. Toward an environmental cosmology: the power of vision, values, and participation in planning for sustainable development. PhD dissertation, Antioch New England Graduate School, Keene, NH, 2002.
18. Johann von Goethe. http://www.quotationspage.com/quote/9806.html (accessed February 4, 2011).

Afterword

> During a fiery debate on financing education [in Rockford, IL, on December 9, 1997], the president of the school board ... chose a rather dramatic way to express his disagreement with a fellow board member. He grabbed him by the throat.
>
> The throat-grabbing incident is only the latest example of incivility in American civic life, as the work of town councils, county assemblies, and school boards becomes increasingly nettled by behavior that is boorish, menacing, even violent. The National League of Cities ... has made the problem of unruliness at local meetings its top focus during the past year, while *Governing Magazine*, a publication for state and local officials, recently fretted about a nationwide "epidemic of incivility."[1]

Such behavior all but precludes crafting a shared vision, let alone being a trustee for future generations with enough psychological maturity to implement any aspect of wise decision making for a sustainable future. It also exemplifies the abysmal psychological *immaturity* of many current "leaders" to whom we look for guidance. That raises a question

What are we teaching our children by example?

"History," wrote French historian Fernand Braudel, "may be divided into three movements: what moves rapidly, what moves slowly, and what appears not to move at all."[2] This statement contains the fears related to change—those for whom change comes too fast, those for whom change comes too slowly, and those for whom the world seems to stand still no matter how hard they try to institute change. In each case, those who want things to be different feel they have no control, which is universally frightening to many and thus isolating.

This pervading sense of isolation is precisely why human beings must remember that we are creatures for whom sharing is an *absolute necessity* to finding value in living. As Mother Teresa observed in a documentary about her life, "the worst kind of poverty is not being wanted."[3]

We cannot share in life's experiences and cannot offer the gift of our own experiences and talents if there is no one to accept our gift. Being unwanted and ignored is to shrivel inwardly like a dying leaf in a hot wind and, once soul dead, to blow listlessly about without purpose or hope.

As I travel around the world, I find over and over again that all we have of value, real value, as human beings is one another—someone with whom to share the experience of being human, with all its frailties, uncertainties, and magnificence. We *need* one another to care for if we are to be whole and healthy. To be alone is to have the essence of our humanity die within, while our physical bodies turn into fleshy robots.

If we, as adults, do nothing else, we can and must find the ethical courage to lead by example in how we treat one another and thereby help teach the children of today, tomorrow, and beyond *the absolute social necessity of common civility*. The future for humankind increasingly depends on it.

Why not choose peace and experience daily the beauty and wonder of life as it surrounds us? We are, after all, an inseparable part of nature, which is life itself. For the sake of all generations, let us choose life in all its splendor. Because the ideal is all that is worth writing about and striving toward, the purpose of this book is to illuminate a peaceful path toward social-environmental sustainability for all generations. Moreover, all decision making, and thus government, begins at home in one's community, where we teach children the fundamentals of social-environmental sustainability. From there, it reaches into the world, which makes local government the seedbed of world leadership. What and how we teach our children determines what kind of decision makers they will become. Why? The answer is simple. Everyone leads by example in daily life.

In closing, it would be good to remember that the well-being of nature and humans is served with grace, elegance, and sufficiency when humans revere the intrinsic values secreted in all parts of the life-support system that sustains humanity as an inseparable part of nature. In the words of Robert F. Kennedy:

> It is from numberless diverse acts of courage and belief that human history is shaped. Each time a man [or woman] stands up for an ideal, or acts to improve the lot of others, or strikes out against injustice, he [or she] sends forth a tiny ripple of hope; and crossing each other from a million different centers of energy and daring, those ripples build a current which can sweep down the mightiest walls of oppression and resistance.[4]

Today, we are a nation facing not only terrorism but also severe social-environmental bankruptcy with no "Chapter 11" protection. There is, however, a remedy for this situation: Let the young, innocent children of the elementary grades become our *spiritual* teachers even as we become their *material* teachers. In closing, I reiterate two lists of positive things you can do personally to strive for those relationships that create the ideal human web of life:

1. Develop a set of personal values that define for yourself who you really are and stick with them.
2. Listen to and obey your conscience because it is the guidance system of your behavior.
3. Learn to *accept* other people rather than merely tolerate them not only because acceptance allows others to be who, what, and how they are without residual rancor on your part but also because tolerance is really the repression of emotions that eventually leads to angry outbursts or acts of violence. We are, after all, responsible for what we do, regardless of how we feel at the moment.
4. Control your attitude, or it will control you.
5. Although the circumstances of your life have influenced who you are, you are responsible for who you become.
6. Do the work that is important to you not only because it will likely consume more than a third of your adult life but also because work that serves others creates a sense of fulfillment.
7. Compare yourself to the best *you* can do, not to the best others can do.

8. It is not what happens to you, but what you do about it that matters.
9. Develop both the right and left sides of your brain because that will lead to an improved ability to think and see things in greater perspective and help you to lead a balanced life.
10. Give unconditionally without any thought of compensation; this means giving for the sake of improving the person you are, not for recognition. Giving anonymously is an easy way to practice giving unconditionally.
11. Many people have helped you in great and small ways to create a fulfilling life; therefore, say "thank you" and mean it. Remember, people tend to be quick to criticize and slow to praise, so make it a point to acknowledge the efforts of others and reward every act of kindness with an act of kindness, even if it is only a heartfelt "thank you."
12. Although you cannot make someone love you, you can make yourself lovable.
13. Be humble because there is always someone who is cleverer than you are. If and when the urge for recognition comes over you, remember that you are where you are because of all the others who have gone before you and helped you along the way. Whatever praise comes to you therefore is partly theirs.
14. Enjoy conscious simplicity because the more content you are with simple things, the simpler and easier it is for you to experience joy.
15. Take time to reflect on your actions; action without reflection may well cause you to do something in an instant that will give you heartache for life. This just says that it is a lot easier to act or react than it is to reflect; for example, it takes years to build trust but only seconds to destroy it.
16. Think about big ideas because whatever you can imagine that will be helpful to someone else or to the health of Earth, you can begin to create and experience. We are, after all, what we think.
17. Finally, all we have of real value to give one another, including our children—*ever*—are choices and some things from which to choose. With every choice we pass forward, we give our children an unconditional gift of our love, trust, respect, and the benefit of our experience.[5]

Added to this list are suggestions by Professor David W. Orr for things adults can do for the sake of children:

1. Protect the purity of the air, water, soil, and food from pollution by artificial chemicals whose combined effects are not *and, perhaps, cannot* be known.
2. Connect children with the soils, forests, grasslands, waterways, plants, and animals that are so vital to their overall well-being.
3. Do everything humanly possible to protect the children from the risks of anthropogenically related climate change.
4. View renewable, natural resources as value-added products to be passed on to future generations.
5. Leave children a legacy of beauty and biophysical richness.
6. Protect prime agricultural lands.

7. Create and maintain societal priorities of establishing and maintaining educational opportunities for all citizens.
8. Protect neighborhoods and communities from ever-expanding transportation systems.
9. Protect good-quality open space in which children can safely play, explore, and discover both Nature and themselves.[6]

Ultimately, we adults of the world bequeath *all generations* a legacy of irreversible circumstances chosen from two behavioral alternatives:

1. We can continue on our present course of progressively degrading the current social-environmental conditions, *or*
2. We can strive for the ideal of social-environmental sustainability as psychologically mature trustees of planet Earth, which is the biological living trust in our care.

Because the legacy of social-environmental sustainability is achieved in the measure that each person is psychologically mature enough to strive for it in daily life, the question is: Which alternative will you choose?

NOTES

1. Dirk Johnson. Civility in politics: going, going, gone. *The New York Times*, December 10, 1997.
2. Fernand Braudel. http://www.indianchild.com/Quotes/history_quotes.htm (accessed February 8, 2011).
3. *Mother Teresa* [documentary film]. Today Home Entertainment. 1986.
4. Robert F. Kennedy. http://quotationsbook.com/quote/21029/#axzz1IhTy5ML7 (accessed February 8, 2011).
5. Chris Maser. *The Perpetual Consequences of Fear and Violence: Rethinking the Future*. Maisonneuve Press, Washington, DC, 2004.
6. David W. Orr. Beauty is the standard. *Resurgence*, 210 (2001):34–37.

Appendix: Common and Scientific Names of Plants and Animals

ALGAE

Algae	Protista

GRASSES

Switchgrass	*Panicum virgatum*

TREES AND SHRUBS

Bigleaf maple	*Acer macrophyllum*
Douglas-fir	*Pseudotsuga menziesii*
Juniper	*Juniperus* spp.
Lodgepole pine	*Pinus contorta*
Oregon ash	*Fraxinus latifolia*
Oregon white oak	*Quercus garryana*
Pacific madrone	*Arbutus menziesii*
Piñon pine	*Pinus edulis*
Ponderosa pine	*Pinus ponderosa*
Red alder	*Alnus rubra*
Spruce	*Picea* spp.
Western hemlock	*Tsuga heterophylla*
Western redcedar	*Thuja plicata*
White-bark pine	*Pinus albicaulis*
Willows	*Salix* spp.

INVERTEBRATES

MOLLUSKS

Mollusks	Mollusca

CORALS

Corals	Anthozoa

Spiders

Black and yellow garden spider	*Argiope aurantia*
Spiders	Arachnida

INSECTS

Mountain pine beetle	*Dendroctonus ponderosae*
Pinyon engraver beetle	*Ips confusus*
Spruce bark beetle	*Dendroctonus rufipennis*

BIRDS

Cliff swallows	*Petrochelidon pyrrhonota*

MAMMALS

Grizzly bear	*Ursus arctos horribilis*
North American elk	*Cervus elaphus*
Puma (mountain lion)	*Felis concolor*
Whales	Cetacea
Wolf	*Canis lupus*

Index

A

Abess, Leonard, Jr., 107
Absolute freedom, myth of, 14, 49, 154
Acceptance
  in detachment, 176
  unconditional, 175
Accountability
  of leadership, 3
  mutual, 183
Action, overemphasis on, 51
Adaptability, simplicity and, 72–73
Adaptation, 219
Adult thinking
  changing, 243–248
  difficulty with positive thoughts, 242
Adults
  actions on behalf of children, 248
  as trustees of children's welfare, 245
Advice-giving, compulsive, 210
Agency outsiders, 97
Aggression, as coping mechanism, 110–111
Agricultural chemical deposition, 14
Agricultural land, growing scarcity, 221
Agricultural research, 221
Air
  as breath of life and death, 38–39
  as foundation of global commons, 44
Air pollution, 43
  global transport of, 38
Air quality, 38–39
  as foundation of environmental quality, 43–44
  importance of, 5
Airborne contaminants, 38
Alchemy, 216
Algae, scientific names, 261
Algae blooms, due to corn-based ethanol, 63
Amberization, 10
American Petroleum Institute, 51
Anger
  as coping mechanism, 110–111
  as extreme fear, 110
Animals
  peace and, 239
  wish for well-being, 241
Anthropocene epoch, 5
Anxiety
  coping with, 192–193
  while working towards change, 207
Appraisal, as coping mechanism, 111–112
Appreciation, need for, 165
Approximation, 29
Asian monsoon cycle
  effects of climate change on, 20–21
  human effects on, 21
Aspiring leaders, 192
Attachment, 177
Attitudes, 147
Authenticity, 174, 200
  in leaders, 167–170, 169
  perceived lack of leaders', 167
  personal, 45
  in potential leaders, 178
  response to, 181
  through truth-telling, 171
Authority
  spiritual, 181
  willingness to delegate, 184–186
Autogenic succession, 29, 30
Avoidance, by displacement, 114

B

Balance of nature, myth of, 29
Bark beetle epidemic, 16
Beginner's mind, 224, 253
Behavioral triggers, identifying, 211
Bintuni Bay, 61
Biodiversity, 5
  as loser in SNWA project, 136
  in poorer world areas, 223
  in soil, 40
  as variety of life, 40–41
Biofuel production, cautions, 63
Biological backups, 41
Biological capital, 11
Biological threshold, 41
Biophysical costs, hidden in economic accounting, 148
Biophysical cycles, 27
Biophysical functions, 22–24
Biophysical integrity, 4
Biophysical resources, finiteness of, 4
Biophysical sustainability
  inviolable biophysical principles, 9–30
  law of cosmic unification and, 6–9
  waterbed principle and, 4–6
Biophysical sustainability principles, 3
  all-inclusive relationships, 10

change as eternal becoming process, 19–22
cumulative effects, 24–26
cyclical systems, 27–28
dynamic disequilibrium, 29–30
inviolability of, 9
irreversibility of relationships, 22
lag periods, 24–26
relationship outcomes, 10
relationships and energy transfer, 13–14
relationships as self-reinforcing feedback loops, 14–17
sunlight energy as sole true investment, 12–13
system basis in composition, structure, function, 22–24
systemic change and self-organized criticality, 28–29
systems and functions, 12–13
thresholds of tolerance, 24–26
trade-offs in relationships, 17–19
ubiquitous relationships, 9–10

Bioregions, 221
Birds, scientific names, 262
Blame, 194
effectiveness reduction through, 185
externalization of, 116
Border Patrol, attempted circumvention of environmental laws, 229–230
Boundaries, establishing, 205–206
Brahmaputra River, 202
Brothers, Kay, 131
Building discoloration, 17–18
Bureau of Land Management, 88, 93, 101, 104, 110, 120
dehumanizing isolation in, 97–98
dysfunctionality in, 95
Burke, Edmund, 147
Burke, James E., 179
Burnout, 225
avoiding, 225–226
causes, 226
preventing, 227–228
remedies, 226–227
Bush administration, 143
Butterfly effect, 70
Byzantine Empire, effects of climate change on, 19–20

C

California, growth control issues, 65
California Central Valley, 58
California floods of 1996-97, 57–58
California low-carbon fuels law, 103–104
Candor, 180
Capitalism, basis in dissatisfaction, 72
Carbon dioxide, 43
dependence of feedback loops on, 16–17
ocean content, 38
sequestration by groundwater, 63
Carbon emissions
failure to assign economic value to, 221
methods for conveying value to decision makers, 222
Carbon sequestration
human disruptions to, 20
impairment of ocean's, 10
increased by perennial plants, 60
straining of oceans' capacity, 64
Carbonic acid, ecosystem consequences, 64
Careerist subculture, 94
managers and, 144
Caring, 175
Carrying capacity, biological, 23
Case studies
recording, 208
sustainability programs, Southern Nevada Water Authority, 125–137
Cathedral of Learning, 17–19
Cattle driving analogy, 182
Causes, attachment to, 177
Censorship, 133
Center for Responsive Politics, 95
Center for Sustainable Resources, 73
Change
on brink of disaster, 3
crab shell analogy, 168
fear of, 100, 257
frightening aspect of, 168
fundamental, 50
as historical continuum, 20–22
immediate, 19
institutionalized resistance to, 87
irreversibility of, 66
as irreversible process of eternal becoming, 66–69
leader's need to, 176
as process of eternal becoming, 19–22
resistance to, 112, 119
risking, 101
systemic, 28–29, 69–70
as threat to business as usual, 120
time needed for reversal, 69
understanding historical, 19–20
Channel destabilization, downstream effects, 57
Checking behavior, as obstacle to decision making, 111, 112
Cheerfulness, in followers, 180
Cherished beliefs, 113, 119
clinging to, 150
Children
adult actions on behalf of, 248, 259–260
conflict resolution with, 250–251

creating community vision with, 249–250
desires for future, 237–238, 250, 251–252
essays by, 238–242
giving voice to, 233–236, 248–254
healing the land, 251
informing citizens about effects on, 163
innocence, 243
introducing to adult problems too early, 242
intuitive wisdom, 234
lack of voice in future, 236–243
nurturing, 246
participation in conflict resolution, 250
participation in planning, 253
rigid mind-sets, 242
shared decision making with, 233–236
as silent parties in environmental decisions, 66
teachers learning from, 242–243
Children's advisory councils, 251–254, 252, 253
Chinese history, dependence on Asian monsoon cycles, 21
Chlorofluorocarbons (CFCs), 43
Chlorophyll, 12
Choice, 71
of individual behavior, 143
myth of lack of, 100
people as products of, 59
power of, 87
as root of human relationships, 175
Churchill, Winston, 45
Citizenship, impoverished political views of, 150
Civil unrest
and climate change, 21
global warming and, 21
Civility, 46, 182, 257
Clean air, children's desire for, 237, 241
Clean water, children's desire for, 237, 241
Clear-cut logging, 15
economic justifications, 41
Climate
as global arbiter, 43
mineral aerosols' effect on, 14
Climate change
and civil unrest, 21
effect on Las Vegas water supply, 125
effects on Asian monsoon cycles, 20–21
effects on historical empires, 19–20
forced adaptation due to, 20
raising awareness for, 130
Closed-loop systems, myth of, 10
Coal, as stored energy from sunlight, 12
Coiled spring analogy, 27–28
Collective decision making, fostering through leadership, 166
Collective thinking, 89
Collective understanding, modeling, 156–157
Collective vision, 230, 254
Colorado River, finite water allocations, 125
Comfort zones, 169
Command-and-control decision making, 88–89, 93
dehumanization through, 105
homeostatic principle, 96–99
at Southern Nevada Water Authority, 128–129
versus true leadership, 164
Comment periods, manipulation of, 115
Commons, 30, 233. *See also* Global commons
law of, 234
question framing about, 252
sharing, 235
Community, building, maintaining, nurturing, 152
Community goals, 229
Community partners, 129
engaging, 128
Community worldview, 49
Competing values, 15, 108
Competition, 70
as basis of resource exploitation, 73
discarding view of, 248
overemphasis on, 46
social ethos of, 66
Complementarity, in followers, 180
Complex systems, 28
ecosystems as, 156
government agencies as, 93–94
Complexity, as disadvantage for cultural longevity, 73
Composition, 22, 61–62
Conduct guidelines, 213
Conduits, 13
floodplain example, 55–56
Conflict, defusing potential, 212
Conflict resolution, with children's participation, 250–251
Confucian ethics, 145
Confusion, resistance as, 120
Congress, average net worth, 96
Consensus building
among Shinto priests, 66
in response to criticism, 198–199
Consequences, nature's impartiality regarding, 15
Constant values, myth of, 13, 37
Constraints, 15, 59
on destructive human appetites, 147
guiding citizens to live within, 219
habitat-related, 15–16
internal, 147
need for social, 147
negotiability of, 153–154
self-imposed, 58
Contentment, simplicity and, 72–73

Control
 falsely equating with progress, 57
 feeling out of, 111
 myth of human, 4
 myth of lack of, 100–101
 need for, 165
 over response to circumstances, 191
 valuation in material world view, 147
Cooked meat, and facilitation of brain capacity, 6–7
Cooking, and transformation, 216
Coping mechanisms
 addiction to dysfunctional, 142
 anger and aggression, 110–111
 for anxiety, 192–193
 appraisal, 111–112
 defensiveness, 112
 denial, 112–114
 displacement, 114–115
 end result, 120–121
 engaging, 108–110
 filters, 116
 influence on effectiveness of decision making, 110
 projection, 116–117
 rationalization, 117
 repression, 118
 resistance, 119–120
Coral skeletons, environmental issues, 64
Corals, scientific names, 261
Corn-based ethanol, environmental consequences, 63
Corporate/political elite
 commodity production to satisfy, 143
 government agencies as tools of, 98
 status quo maintenance by, 235
 takeover by, 90
Cosmic unification, law of, 6–9
Counsel
 versus advice, 223
 asking for, 208
Cover, as habitat requirement, 15, 16
Crab shell analogy, 168
Creativity
 encouraging by recognition, 215
 encouraging in teams, 186
 in followers, 180
 imagination as source of, 150
 nurturing in others, 215–216
 risk as requirement of, 215
Credit sharing, 186, 216
Critical state evolution, 28
Criticism
 acknowledgment of, 198
 agreeing with truthful, 199
 avoiding personal interpretation, 199
 burnout due to, 226
 consensus building in response to, 198–199
 coping with, 194–196
 counterattacks, 197
 denial response, 197
 direct answers to, 200
 elevating discussion levels with, 198
 as form of projection, 194–196
 friendliness in response to, 198
 ignoring, 197
 listening to, 198, 199
 nondefensive response, 197
 paraphrasing, 199
 positive symbols, 199
 and public image, 196–197
 rationalization response, 197
 responding to negative, 197–200
 tactful refutation, 199
 truth telling, 198
 withdrawal from, 197
Cultural conflicts, creativity inherent in, 146
Cultural evolution, 168
 crab shell analogy, 168
 and fall of civilizations, 169
Cultural morality, decision basis in, 101
Cultural values, 48, 145
Cultural wisdom, 73–75
Culture, as person, 146
Cumulative effects, 14, 24–26, 65–66
 inability to recognize, 24, 25
 incubation period, 26
 sudden visibility of, 25
Curitiba, Brazil, 150, 151
Cyclical systems, 27–28
 frequencies, 27
Cynicism, 141–142

D

Dams
 failure to curb periodic flooding, 60
 false sense of security with, 59
 past support for, 147
 removal to recapture biophysical integrity, 148
DDT, waterborne contamination, 39
Dead zones
 due to corn-based ethanol production, 63
 oceanic, 55
Deception, 126
 results of modeling, 170
Decision antitheses, 37–38
 air quality, 38–39
 biodiversity, 40–41
 climate, 43
 human population-related, 41–42
 human survival and air quality, 43–44
 soil quality, 40

sunlight energy, 42–43
water quality, 39–40
Decision makers. *See also* Leaders
acceptance of slow/delayed results, 206–207
anxiety coping mechanisms, 192–193
avoiding burnout, 225–228
avoiding self-deception, 209
boundary issues, 205–206
characteristics of effective, 165–187
circumstances faced by, 191–192
communicating urgency, 222–223
coping with criticism, 194–196
coping with disagreeable people, 210–212
counsel *versus* advice, 223
dealing with uncommitted persons, 206
essence of responsible, 163
imaginary role playing, 212
inspiring performance, 213–214
maintaining viability as leaders, 225–228
making do, 216–219
Mulroy, Pat, 136
nature's urgent message to, 1
nurturing creativity in others, 215–216
overinvestment in followers, 209–210
personal limitations, 207–209
psychological maturity, 161
question framing, 223–225
realistic objectives, 219–220
remembering role as, 211
revolving, 183
teaching of leisure by, 66
use of power, 193–194
uses of humor, 210
value of listening, 204–205
Decision making
based on emotions, justified by facts, 172
breaking dysfunctional cycles, 141–143
and categories of value, 145–146
command-and-control, 88–89
coping mechanisms' influence on ineffective, 110–121
coping with responsibilities and pressures, 191
democratic revolution in, 248–254
dysfunctional, 93–96
engaging coping mechanisms, 108–110
fear of, 200–210
firewood analogy, 7–8
floodplain examples, 58–59
homeostatic command-and-control, 96–99
imagination as component of, 149–152
importance of vision, 146–152
job descriptions as limiting factor, 143–144
by leaders *versus* managers, 144–145
myth of abnegating personal responsibility, 99–105
myth of committing future leaders, 101
myth of facing reality, 103–104
myth of impossibility, 104–105
myth of informed denial, 102–103
myth of job description, 101
myth of lack of definitive data, 101–102
and nature's rules of engagement, 37
within organizations, 228–230
perception as element of, 148–149
perfectionism and, 200–210
power of, 87
roots at home, 258
self-disclosure and, 200
self-serving, 85
sharing with children, 233–236
too soon *versus* too late, 102
trust requirement, 165
uses of silence, 203–204
Decision-making rules, 45
basis of democratic society, 72
change and eternal becoming, 66–69
conscious limitation of wants, 72
cumulative effects, 65–66
dynamic disequilibrium, 70–71
energy transfers, 58–61
event interpretation, 71–72
human-environmental relationship, 48–50
intergenerational relationships, 50
interpersonal relationships, 46–48
intra-personal relationships, 45–46
lag periods, 65–66
present moment, 75–76
primacy of nature, environmental/cultural wisdom, and human well-being, 73–75
relationship outcomes, 51–54
relationships
simplicity and survival, 72–73
system basis in composition, structure, function, 61–62
systemic change and self-organized criticality, 69–70
thresholds of visibility, 65–66
trade-offs, 62–65
true investment, 55
voting rights, 75
Defensiveness, as coping mechanism, 112
Definitive data
avoiding decisions for lack of, 101–102
myth of lack of, 109–110, 113, 168
Dehumanization, 121
Delays, coping with, 206–207
Delegates, choice of, 184
Delegation
of authority, 184–186
of efficiency *versus* effectiveness, 182
steps in, 184
Delta growth, threshold of visibility, 26
Demanding people, coping with, 205–206

Democracy
    delegated compromise in, 109
    role of information, 72
Democratic process, 249
Democratic skills, cultivation of, 183
Denial. *See also* Informed denial
    as coping mechanism, 112–114
    of criticism, 197
    informed, 25, 26
Desert, groundwater pumping initiatives, 125
Detachment
    in effective leaders, 176–177
    equanimity as kernel of peace in, 177
Development
    encroachment on floodplains, 59
    water grab as stimulus package for, 130
Dictatorship, 164
Difference, as human construct, 235
Direct communication, 193
Disagreeable people
    coping with, 210
    facilitation with, 212
    identifying triggers, 211
    persistence, 211–212
    private discussions with, 211
    sandwich technique, 211
    timing issues, 211
Disconnect, science *versus* policy values, 110
Disequilibrium, 29–30
Disillusionment, 226
Displacement, as coping mechanism, 114–115
Dissonance, 126, 135
    within duplicity, 125–137
Distorted perceptions, 149
Disturbance regimes, 22, 28, 29
    biodiversity as buffer, 41
    resilience despite, 29
Diversity, 229
Dogen, 217, 219
Duplicity, 126
    dissonance within, 125–137
    job loss as evidence of, 135
    in water wars, 137
Dust storms, 16
Dynamic disequilibrium, 70–71
    in systems, 29–30
Dynamic relationships, 9
Dysfunctional agencies, 89, 93, 97
    breaking cycle of decision making, 141–143
    coping mechanisms in, 108
Dysfunctional decision making, 93–96
Dysfunctional families, 97

E

Earth flow, 28
Eccrine-based sweating system, 67
Ecological knowledge, deliberately ignoring, 106
Ecologically uninformed decisions, 98
Economic constants, reductionism in, 5
Economy-*versus*-ecology dilemma, 147
Ecosystem complexity, 156
Ecosystem integrity, 3
Ecosystem management, 95
Ecosystem simplification, 16, 229
Ecosystems, as dissipative structures, 28
Education
    role in democratic society, 249
    role in replacing fear with love, 151
Effectiveness
    confusing with efficiency, 182
    in decision makers, 165–187
    in followers, 180
    in leaders, 165
    monitoring for, 157–158
Efficiency
    confusing effectiveness with, 182
    in followers, 180
Ego, checking at the door, 176
Empathy, 204
Employee education, 128, 129
Endangered Species Act of 1973, 105
Energy-juggling act, 13
Energy transfer, 55–61
    as result of relationships, 13–14
Enmeshment, 209–210
Entropy
    law of maximum, 6
    maximum, 7
Environmental agencies
    adolescent/growth stage, 92
    breaking dysfunctional decision making cycles in, 141–143
    as collection of individuals, 90
    developmental stages, 90–92
    inception phase, 89–90, 90–91
    maturity stage, 92
    senescence/revitalization stage, 92
    stages, 87–88
Environmental change, symptomatic adaptation to, 219
Environmental laws, unpredictability of, 105–108
Environmental management, 87
Environmental Protection Agency, 88, 93, 110, 141, 143
    as collection of people, 97
    on greenhouse gas emissions, 112–113
    lack of ethics, 100
Environmental regulations, deliberate circumvention, 131
Eolian dust, 14
Equal Access to Justice Reform Act of 2005, 106
Equality, between adults and, 249

Equanimity
in detachment, 177
in effective leaders, 176–177
Eternal becoming, 19
change as process of, 66–69
Eternal moment, 75
Eternal now, 175, 209
Ethics
deficiencies in EPA, 100
trading for job safety, 99
Europe, superior land use policies, 106, 107
European Spatial Development Perspective of 1999, 107
Evidence-based policy, 221
Exploitive capitalism, long-term consequences, 95
Externalities
biophysical consequences, 37
myth of, 5
Extinction, 13
Eye contact, 200

F

Facilitators, for disagreeable people, 212
Failure
basis in human valuation, 71–72, 174, 199, 209, 224
with creativity, 215
Familial dysfunction, 246
Familial relationships, guidelines for improving, 246–248
Family
agencies as extensions of, 87, 90
hiding of dysfunction, 96
individual as extension of, 87
self as extension of, 142
Fear
alleviation, 142
versus love, 142–143
of making mistakes, 111
Fear of knowing, 102
Fears, helping others transcend, 181
Feedback
acceptance of, 186–187
from colleagues, 208
Feedback loops
overwhelm due to human activities, 17
relationships as self-reinforcing, 14–17
Fertilization, myth of investment, 11
Fessenden, Senator William Pitt, 195
Filters, avoiding decision making with, 116
Finite resources, 4, 41, 55
First Amendment rights, children and, 251
First law of thermodynamics, 6
Fish ladders, 155–156
Five Dynasties and Ten Kingdoms period, 21
Flood losses
as acts of hubris, 59
increases due to global warming, 202
reducing risk of, 61
U.S. statistics, 60
Flooding, periodic, 60
Floodplains
historic treatment of, 55–61
long-term risks of building in, 59
reconnecting, 60, 61
rethinking, 60
Focus Property Group, 127
Followers
characteristics of good, 179–180
honesty with, 201–203
leaders as good, 177–180
overinvestment in, 209–210
task matching, 185
Food sufficiency, children's desire for, 240
Foreboding, 222
Forest life cycle, 29–30
Forest simplification, 16
Free ecosystem services, 148
Free speech, 236
Freedom. *See also* Absolute freedom
links to morality, 154
relativity of, 14
Friendliness, in response to criticism, 198
Fuller, Richard Buckminster, 44, 121
Function, 61–62
defining systems by, 12–13
Functionary behavior, 99
judgment of, 104
Future
children's desires for, 237–238
children's lack of voice in, 236–243
giving children voice in, 248–254
Future generations, sacrifice for, 233, 243
Future leaders, committing to course of action, 101

G

Gandhi, Mahatma, 152, 176
life as message, 173
Ganges River, 202
Gender equality, 37, 41–42
as solution to population problems, 5, 41–42
Genuineness, 174. *See also* Authenticity
Glacier melting, 202
Global commons, 30, 233
air as foundation of, 44
Global ecosystem, sole true investment, 11–12
Global warming
bark beetle epidemic and, 16
and civil unrest, 21
effects on poor populations, 69

glacier melting due to, 202
greenhouse gases and, 43
and increased wet-bulb temperature, 68
informed denial of, 102, 112
intensified uncertainties due to, 27
in Kenyan highlands, 113
overfishing and, 55
and steady-state myth, 25
Goals, 153
crafting, 156
and vision, 152–153
Going within, 193
Government agencies, 88
as collections of people, 99
as extensions of family, 87
jurisdiction over public lands, 229
outliving of original functions, 93
policy *versus* heresy in, 96
as self-perpetuating machines, 92
subcultures, 94
three subcultures of, 93–94
Government employees, repressed feeling effects, 118
Grain crops, reduced global production, 27
Grasses, scientific names, 261
Gravity, water as captive of, 39–40
Great Plains, groundwater consequences, 63
Greatest good, 91
Greek tragedy, of modern society, 44
Green cities, 151
Green plants, 11, 12
Greenhouse gases, 44, 51
and global warming, 43
with groundwater pumping, 63
power plant emissions, 112–113
Groundwater pumping, Mojave Desert, 125
Growing up, as optional, 252
Growth, myth of continual, 66
Growth control, in Southern California, 65
Growth/no growth tug-of-war, 147
Growth rings
in stalactites, 19–20
in stalagmites, 20
Guilt, externalization of, 116

H

Habitat alteration, by timber industry, 15
Habitat corridors, 23, 24
Habitat requirements, 15
Hatfield, Senator Mark, 115
Healing relationships, creating through leadership, 165
Heat islands, 49
Hell on earth, 38
Helpfulness, in followers, 180
Heresy, 96
Hidden agendas, 126, 167
Hidden economy, commons in, 233
Hiding cover, 16
Hindsight, 148
Historical change
and cultural evolution, 169
initiation of collapse from within, 69–70
understanding, 19–20
Holocene period, 4
Home ranges, 205
Homeostasis, 108
in command-and-control decision making, 96–99
dehumanization through, 105
need to threaten, 143
preventing information leaks using, 98
at Southern Nevada Water Authority, 134
transformation into command-and-control, 98
Honesty
in followers, 180
with followers, 201–203
in leaders, 163
Honor, 171
in effective leaders, 170–171
Hospitality, 46
Hovercraft, children's desire for, 237
Human activity
effects on Asian monsoon cycle, 21
effects on survival variables, 43
Human desires, 15
Human dignity, 143
at heart of sustainability, 167
Human-environmental relationships, 48–50
Human equality, 235
Human overheating, biological strategies to prevent, 67–68
Human relationships, increasing commercialization, 74
Human skin temperature, 67–69
Human survival, air as foundation, 44
Human valuation, 71
narrowness of, 10
need to go beyond, 3
role of education, 151
success/failure basis in, 71–72, 199, 209
Human well-being, primacy of, 73–75
Humility, 174, 178, 183
authentic, 167
in leaders, 163
Humor, 210
Hyperthermia, risks with global warming, 68

I

Ignorance
as foundation of knowledge, 201–202
lifting, 142

Ignoring
as defense mechanism, 103
of ecological knowledge, 106
as response to criticism, 197
Imagination
children's, 248
as component of vision, 149–152
Impossibility, 253
myth of, 104–105
rationalizations about, 117
Impotence, pleading from position of authority, 104
In-State Groundwater Resources Project, 129
critical opinions of, 133
project approval by state engineer, 135
radioactive dust consequences, 136
Inclusivity, 183, 211, 252
Incongruent behavior, results of modeling, 170
Incubation period, for cumulative effects, 26
Independent variables, myth of, 13
Individual values, 145, 146
Individuality, loss of, in government agencies, 98, 99
Indus River, 202
Industrial Revolution, 20
Industrialized nations, competitive destruction of environments, 70
Influence issues, 192
Information distortion, 99, 118
Information leaks, homeostatic mechanisms for preventing, 98
Informed denial, 25, 26, 37, 43, 66, 113. *See also* Denial
avoiding decisions due to, 102–103
and floodplain approaches, 56
by tobacco industry, 114
Inner soul quality, 176
Inner space, conquering, 70
Innocence, 243
Insects, scientific names, 262
Insightfulness, in followers, 180
Integrity, 94, 118, 195
blacklisting of analysts with, 132
conflicts with SNWA, 126
in leaders, 170
maintaining personal and professional, 135
of personal space, 205
struggle to retain, 141
Interactive systems, 13, 28
Interconnectedness, 5. *See also* Waterbed principle
and myth of absolute freedom, 154
in sacred worldview, 147
Interdependent relationships, 13, 151
leaders' sense of, 165
Intergenerational relationships, 50, 249
Interpersonal relationships, 46–48
Interpretation, 224. *See also* Human valuation
Intrapersonal relationships, 45–46
Introspection, 207
Intuition, leading from, 193
Invertebrates, scientific names, 261
Investment
myth of fertilization as, 11
sunlight as sole true, 55
Inviolability
of biophysical principles, 9, 30
decision making rules, 45–76
Invisible present, 24, 25, 28
Irrawaddy River, 202
Irreversibility, 69
of eternal becoming, 66–69
of relationships, 22
Isolation, 257
dehumanizing, in government agencies, 98
pervasive sense of, 257
in power, 194

J

Japanese Soto Zen tradition, 217
Job creation, as socioeconomic priority, 51–52
Job descriptions
controlling employees through, 143
dissonance of reality with, 133
as isolating mechanisms, 99
as limitations on professional decision making, 143–144
myth of abnegating responsibility through, 101
narrow and rigid definitions, 97
sustainability programs, 127
Job loss
fear of, 117
as price of integrity, 126, 134
Job safety, trading dignity and ethics for, 99
Jung, Carl, 217

K

Kenyan highlands, global warming effects, 113
Keystone XL pipeline, 51, 52, 53
Kim Il Sung, 93
Kindness, 150, 185, 212
children's desire for, 239
treating children with, 245
Knowledge
ignorance as foundation of, 201–202
quest for, 203
as some version of truth, 114
Krakatoa, 38

L

Lag periods, 24–26, 65–66
Law of cosmic unification, 6–9
Law of maximum entropy, 6
Law of the Commons of the Natural World, 234
Leaders
  ability to make emotional connections, 171
  analogy to psychotherapy, 173
  characteristics of effective, 165–187
  conscious self-awareness in, 164
  decision making by, 144–145
  disadvantage as managers, 182
  duty to challenge others' values, 163, 164, 175
  emotionally present, 175
  essence of responsible, 163
  lack of public trust in, 167, 168
  maintaining viability as, 225–226
  modeling abilities, 225
  personal needs caveat, 164
  personal values, 163–165
  philosophy of life, 163–165
  power to motivate, 172
  sensitivity to unmet personal needs, 165
  as trustees of social-environmental sustainability, 144
  unfinished emotional business, 165
Leadership
  cattle driving analogy, 182
  versus command-and-control, 164
  encouraging in others, 186–187
  versus managerialship, 181
Leadership accountability, 3
Leadership characteristics, 165–166
  acceptance of ambiguity, 174
  admitting mistakes, 175
  authenticity, 167–170
  caring, 175
  detachment, 176–177
  encouraging leadership in others, 186–187
  equanimity, 176–177
  feedback acceptance, 186–187
  following ability, 177–180
  humility, 163, 174
  masculine-feminine balance, 171–172
  other centeredness, 167–170
  personal honor, 170–171
  personal power, 174
  rewarding effort, 186
  risk taking, 174
  self-awareness, 174
  self-disclosure, 163, 170
  self-respect, 174
  servant leadership, 180–183
  shared leadership, 183–184
  therapeutic effectiveness, 173–176
  willingness to delegate authority, 184–186
Leading by example, 76, 187, 207, 223, 229
  children and, 257
  from heart and intuition, 193
Leading from the heart, 144
LEED Certification, 129
  Molasky Corporate Center, 127
  Springs Preserve, 128
Left-brain thinking, as masculine trait, 171
Legal battles
  at SNWA, 131
  SNWA water grab, 130
Leisure, 66
  as virtue *versus* luxury, 66
Lerner, Jamie, 150, 151
Lessons learned, 208
Letter of the law, 101
Levant, historical climate changes, 20
Levee failures, 57–58
Levees, false sense of security with, 59
Liberal arts, role in democratic society, 249
Liberty, expensiveness of, 166
Limestone façades, effects of pollution, 17–18
Linear thinking, 49
Listening ability, 180, 204–205
  as half of communication, 204
  importance with children, 235
  in responding to criticism, 198
  validation through, 236
Local requirements, 48
Long-term trends, 48
Los Angeles Metropolitan Water District, 64–65
Loss, fear of potential, 103
Love
  ability to give and receive, 175
  as unconditional acceptance, 175
Loyalty, in good followers, 179

M

Machine mentality, 141
Making do, 216–219
Malaria, in Kenyan highlands, 113
Mammals, scientific names, 262
Managers
  decision making by, 144–145
  disadvantage as leaders, 182
  versus leaders, 181
Mangrove swamps, 61
Masculine-feminine balance, 181
  in effective leaders, 171–172
Material riches, false valuation of, 73
Material worldview, 147
Maximum entropy, 6, 7
Megacities, 106
Mekong River, 202
Mercury, waterborne contamination, 39
Mindless obedience, 89

Mineral aerosols, 14
Mining operations
  Colorado River degradation due to, 53
  controversy, 52
Missions, re-envisioning for future, 144
Mistakes
  ability to admit, 175
  admitting, 200
  learning from one's, 178, 179
Modeling, 156–157
Mojave Desert, 137
  SNWA groundwater pumping initiatives, 125
Molasky Corporate Center, 127, 128
Mollusks
  effects of oceanic carbonic acid, 64
  scientific names, 261
Money, *versus* wealth, 73–75, 106
Money chase, 106
Monitoring
  effective question framing, 155
  implementation, 157
  modeling collective understanding through, 156=157
  for outcome validation, 158
  outcome validation through, 158
  of plan effectiveness, 157–158
  plan preparation, 157
  preliminary, 156
  seven steps of, 155–158
  team progress, 185
  testing vision's effectiveness through, 154–158
  visioncrafting, 156
Motivation, for team members, 214
Motives, 167
Mulroy, Patricia, 126, 136
Multidimensional learning, 51
Myss, Caroline, 191, 193

N

National commons, 91
National Environmental Policy Act of 1969, 105
National forests, 91
National Park Service, 88
National parks, air pollution in, 38, 39
Native American proverbs, 40
Natural resources, myth of constant values, 37
Nature
  disengagement from, 220
  impartiality with respect to consequences, 15
  neutrality in, 47
  rules of engagement, 37
  urgent message to world's decision makers, 1
Necessity, as mother of invention, 149
Need to know, 102
Negative thinking
  in children, 237
  teaching children, 236
Neutrality, of nature, 47
New questions, 224
NIH factor, 103
Nitrogen seep, 63
Nitrous oxide, 43
Northern Resources Project, 125
Northwest Forest Plan, 105
Not invented here factor, 103

O

Obedience, mindless, 89
Objectives, 153, 157
  crafting, 156
  establishing realistic, 219–222
  quantitative nature of, 157
  vision and, 152–153
Ocean warming, effects on wind patterns, 16
Oceans, carbon sequestration by, 64
Off-grid experience, in Vermont, 135
Oil, as partly stored energy from sunlight, 12
Old-growth forests
  author's early preaching about, 176
  charade about allowable cuts, 120
  unstated liquidation policy, 98
Old questions, 224
Open systems, 10
Opportunity
  disguised, 179
  seizing, 178
Optimism, and psychological maturity, 246
Options, for future generations, 3
Organic pollutants, effects on buildings, 17–18
Organizational culture
  as root of dysfunction, 141
  at SNWA, 129
Other centeredness, 192, 194
  adult difficulties with, 242
  in leaders, 167–170
  in potential leaders, 178
Outcome validation, 158
Outcomes, 51–54
  detachment from, 176
  of silence, 204
  unpredictability of, 207
  validating through monitoring, 158
Outsider role, 126
Overfishing, 10, 55
  of predatory fishes, 16
Oxygen-18, 21
Ozone hole, 19
Ozone shield
  disappearance of, 43
  repair of, 44

P

Paradigm shift, 168
Paraphrasing, in response to criticism, 199
Participation, 183
  fostering, 213
Participatory democracy, including children in, 251
Peace, 246
  acting from, 177
  and animals, 239
  children's desire for, 237, 238
  choosing, 258
  and relationship, 239
  roles of equanimity and detachment, 177
  and safety, 240
People as solution, 141
Perception, as component of vision, 148–149
Perfectionism, 200–201
  and honesty with followers, 201–203
Performance, inspiring, 213–214
Persistence
  in followers, 180
  in potential leaders, 178
Personal integrity, 94, 126
  maintaining, 135
Personal limitations, 207–209
  admitting, 201
Personal needs
  enmeshment due to, 210
  for power, 194
  sensitivity to unmet, 165
Personal power, abdication from, 141
Personal responsibility, 46, 88, 143, 227
  accepting, 145
  avoiding through projection, 194–196
  avoiding through prostitution of beliefs, 100
  myth of lack of control, 100–101
  myths of abnegating, 99–105
  relationship and, 238–239
  as remedy for burnout, 226
Personal values
  of leaders, 163–165
  revealing, 163
Philosophy of life, of leaders, 163–165
Photosynthesis, 11, 12
Phytoplankton depletion, 16
Pinchot, Gifford, 90, 91
Plant communities, 23
Planting zones, updating due to global warming, 25
Points of view, as negotiable differences, 47, 48
Policy
  basis in values and preferences *versus* science, 110
  in government agencies, 96
  rationalizing, 117
  winners and losers, 109
Policy makers, communication with scientists, 222
Political correctness, 169, 225
Political infighting, 92
Political polarization, 108
Politician subculture, 94
Politicians
  average net worth, 96
  out-of-touch, 96
Population explosion, 41
  demands on biophysical resources, 4
  gender equality and, 41–42
  historical examples, 69–70
  solution to, 5
Positive framing, 153, 258
Positive projection, 196
Positive thoughts
  adult difficulties with, 242
  in potential leaders, 178
  teaching children, 236, 237, 238
Potable water, 4
Potential leaders, encouraging, 178
Poverty, and suffering due to global warming, 69
Power
  constructive uses, 166
  inappropriate use of, 194
  surrender of personal, 142
  use of, 193–194
  uses of, 193–194
Power debt, 193
Power sharing, 47, 194
Practicality, in followers, 180
Predictability, in followers, 180
Prediction, myth of, 4
Present moment, 75–76
Preservation, 9
  in nature, 10
Principle, 165
Privilege, fear of inadequacy disguised as, 72
Productive thinkers, 215
Products, as translations of information, 108
Professional impotence, 141
Professional integrity, 94, 126
  maintaining, 135
Professional subculture, 94
Professionalism, abuse in government agencies, 94
Profit, role in material worldview, 147
Projection
  of anger, 110
  to avoid responsibility, 194–196
  as coping mechanism, 116–117
  criticism as, 194–196
  negative and positive forms, 196
Propaganda, at SNWA, 130
Psychological maturity, 175, 176, 252
  children's beneficial effects on adults', 251

in decision makers, 161
in leaders, 170
and peace/optimism, 246
Psychotherapists, leaders as, 163
Public forests, 105
Public image, 196–197
Public lands, federal agency jurisdiction over, 229
Pyroclastic flow, 29

Q

Question framing, 155, 220–221, 222, 223–225, 253
about future of commons, 252
Questioning orders, 99
Questions
ecological, 220
listening to and repeating, 199
Quick-fix mentality, 52, 53, 55
air conditioning and, 69
L.A. Water District, 64–65
Quitting, 211–212

R

Rabin, Yitzhak, 171
Ranching boom, and dust deposition, 14
Rational Economic Man, 147
Rationalization
as coping mechanism, 117
as response to criticism, 197
Re-allocation of energy, 55
Realistic objectives, 219–222, 221
Reality, myth of facing, 103–104
Reciprocity, 47
Recognition, encouraging creativity by, 215
Recycling, 237
of sunlight energy, 11
Reductionist-mechanical thinking, 28
Reflection, neglect of, 51
Rejection, 217
Relationship outcomes, 51–54
Relationships
all-inclusive nature of, 10
children's essays on, 238
dynamic nature of, 9
human-environmental, 48–50
improving familial, 246–248
increasing commercialization of human, 74
intergenerational, 50
interpersonal, 46–48
intrapersonal, 45–46
irreversibility of, 22
outcomes of, 10
peace and, 239
and personal responsibility, 238–239
putting before achievement, 181
with self, 191
as self-reinforcing feedback loops, 14–17
trade-offs in, 17–19
transfer of energy in, 13–14
ubiquitousness of, 9–10, 45
Remorse, pretense of, 167
Renewable resources, 4
sunlight as, 55
Revolving decision makers, 183
Repression, 142
as coping mechanism, 118
Resilience, 29
increasing by reconnecting floodplains, 61
Resistance
as coping mechanism, 119–120
engaging, 206–207
Resource sharing, 47
Respect, 46
need for, 165
Right-brain thinking, as feminine trait, 171
Rights, 4
as blocks to personal evolution, 169
Universal Declaration of Human, 244
Riparian forests, dependence on floodplains, 60
Rising sea levels, 27
aquifer pollution by, 39
denial of, 26
Risk, 100
as requirement for creativity, 215
Risk-taking ability
in leaders, 174
in potential leaders, 179
Rodale, Robert, 53
Roman Empire, effects of climate change on, 19
Roosevelt, Theodore, 92, 195
Rules of engagement, nature's, 37
Rural communities, as losers in water grab, 137

S

Sacred worldview, 147
Sacrifice, for future generations, 233, 243
Safe past, 119
Safety, children's desire for, 240
Salmon migration, fish ladders and, 155–156
Salt Lake City, re-emergence as downwind radioactive community, 136
Sandwich technique, 211
Scapegoating, 116, 194, 218
School board, 257
engaging, 249
Science, 109
as inappropriate surrogate for value debates, 109
as non-basis of policy differences, 110
Scientific names, 261–262

Sea animals, carbonic acid and survival issues, 64
Second law of thermodynamics, 6, 8
Selective hearing/seeing, 116
Self-actualization, 184
Self-control, 228
Self-deception, avoiding, 209
Self-disclosure, 200
  in leaders, 163, 170
Self-empowerment, 194
Self-evaluation, 208, 228, 246
Self-image, 196
Self-knowledge, 191
Self-mastery, 228
Self-nourishment, 227
Self-organized criticality, 28–29, 69–70
  as butterfly effect, 70
Self-reinforcing feedback loops, 41, 62, 71
  behavioral, 46
  of reciprocity, 49
  relationships as, 14–17
Self-respect, 174
Self-serving decision making, 85
Senescence, in government agencies, 92
Servant leadership, 166, 180–183, 228
Service ideals, at agency inception phase, 90
Sexually inappropriate behavior, 210
Shadow, 217–218
Shared leadership, 183–184
Shared vision, 146, 213, 257
  and degree of negotiability, 154
  Gandhi's, 152
  of the heart, 152
  including children in, 251
Shelter, habitat requirements, 15–16
Shinto priests, consensus building process, 66
Short-term focus, 148
Short-term trends, 48
Shrubs, scientific names, 261
Silence, understanding, 203–204
Simplicity, as key to contentment, 72
Skepticism, as cop-out, 102
Snow melt, 39
Social capital, loss of, 75
Social collapses, initiation from within, 69
Social-environmental bankruptcy, 246, 258
Social-environmental decision making. *See also* Decision making
  responsibilities and pressures, 191
Social-environmental justice, 3, 235
Social-environmental leadership, 47, 54
  by example, 76
Social-environmental policy problems, 108–109
Social-environmental progress, falsely equating engineering and control with, 57
Social-environmental sustainability, 37, 48
  collective vision, 230
  dignity at heart of, 167
  as indicator of economic health, 74
  leaders as trustees of, 144
  monitoring in context of, 155
  as moving target, 29
  SNWA decisions regarding, 127, 137
Social impoverishment, 73
Soil, as great placenta, 40
Soil quality, 40
Sole causes, impossibility of, 24
Southern Nevada Water Authority
  conflict of interest with oil/gas companies, 131
  In-State Groundwater Resources Project, 129
  lack of due process for protesters, 131
  live web meeting broadcasts, 133
  Mulroy, Patricia, 126
  Northern Resources Project, 125
  outsider role at, 132
  as perceived community leader for sustainability, 129
  pipeline project opponents, 132
  pressure on state engineer, 136
  project approval from state engineer, 135
  project opponents, 136
  proposed groundwater pipeline, 125
  social-environmental project consequences, 136
  status quo maintenance, 134
  sustainability programs case study, 125–137
  "Teflon," 129, 131
  valuation of sustainability, 130
  withdraw plan requirements, 135
Special-interest groups, 95
Spiders, scientific names, 262
Spiritual authority, 181
Spotted owl controversy, displacement mechanisms, 114
Springs Preserve, 128, 133
Stability, myth of environmental, 4, 5
Stalactites
  evidence of climate changes, 19–20
  historical value, 20
Stalagmites, historical value, 20
Status quo
  agency maintenance of, 134
  maintenance by corporate/political elite, 235
  reversion to, 141
Steady state
  fallacy of, 71
  myth of, 25, 29
Stress-related illnesses, 70
Structure, 22, 61–62
Success, basis in human valuation, 71–72, 174, 199, 209, 224
Sunlight energy, 42–43
  as sole investment in global ecosystem, 11–12

Survival, simplicity and, 72–73
Sustainability
in agency employees' personal lives, 127
agency lip service to, 128
censored public face for, 133
constraints of, 146
corporate lip service to, 133
corporate ridicule of, 133
corporate valuation of, 128, 132
Las Vegas' desperate need for, 127
as moving target, 29
propaganda promotion at SNWA, 130
Sustainability plan, 129
drafting at SNWA, 128
hijacking of, 131
Sustainability programs
façade of, 126, 134
at SNWA, 126
Sustainability theme park, 128
Sustainable community development, 50
inclusivity as prerequisite, 228–229
Sustainable processes, 22
SWAT team mentality, 105, 106
Sweat glands, 67
Symbols, positive, 199
Symptomatic fixes, 52, 53, 109
failure of, 141
Systemic change, basis in self-organized criticality, 28–29, 69–70
Systemic solutions, 53–54
growth control in Southern California, 65
lack of patience for, 53
Systems
basis in composition, structure, function, 22–24, 61–62
cumulative effects, 24–26, 65–66
cyclical nature of, 27–28
defined by function, 12–13
as dissipative structures, 7
dynamic disequilibrium in, 70–71
interactivity of, 13
lag periods, 24–26, 65–66
thresholds of tolerance, 24–26
thresholds of visibility, 65–66
universality of dynamic disequilibrium, 29–30
Systems approach, 13
valuing symptomatic thinker contributions in, 150
Systems-oriented approach, 219

T

Tact, 199
Task definition, 185
Task matching, 185
Teachers, 217
children as spiritual, 246
disagreeable people as, 212
interviewing new, 249
learning from children, 242–243
as midwives of knowledge, 203
Teamwork, 48, 146
Technology, and disengagement from nature, 220
Temperature tolerance, human, 68
Terrible known, 101
Territory marking, 205
Terrorism, 246, 258
Therapeutic ability, in effective leaders, 173–176
Thermal cover, 16
Third Pole, 202
Threshold of visibility, 26, 65–66
Thresholds of tolerance, 24–26, 25
Tibetan Plateau, 202
Timber management, 15
lack of definitive data excuse, 102
Time
manipulation as moving target, 115
not enough, 114
Time is money attitude
consequences, 65
displacing uncomfortable situations onto, 114–115
Time management, 213
Timescales, 26
Tobacco industry, informed denial by, 114
Too soon, *versus* too late, 102
Toynbee, Arnold, 169
Trade-offs
Cathedral of Learning example, 18–19
dignity and professional ethics for job safety, 99
positive and negative consequences, 71
in relationships, 17–19, 62–65
Training, performance and, 213
Transformation, cooking and, 216
Transportation networks, reducing fragmentation by, 107
Tree-lined streets, as habitat corridors, 23
Tree planting, children's desire for, 238
Trees
children's desire for, 241–242
children's desire for more, 237
scientific names, 261
Truman, Harry, 178
Trust
dependence on healthy boundaries, 206
as requirement in decision makers, 165
Truth
agreeing with, 199
identifying with coping mechanisms, 121
knowledge as some version of, 114
risking revelation of, 170

telling as response to criticism, 198
telling in public, 171
Tyranny, remote, 37

U

Uncertainty, intensification with global warming, 27
Uncommitted persons, dealing with, 206
Understanding, 180
Unfinished business, 209
United Nations Convention on the Rights of the Child, 245, 246
United States, loss of community and mythology, 146
Universal Declaration of Human Rights, 244, 246
Universal values, 145
Unlimited resources, myth of, 4
Unpopular decisions, risk of making, 103
Urbanization, Curitiba innovations, 151
Urgency, communicating need for, 222–223
U.S. Army Corps of Engineers, 56–60
U.S. Fish and Wildlife Service, 88, 141
U.S. Forest Service, 88, 93, 141
contemporary goals, 95
dysfunctionality in, 94–95
U.S. National Park Service, 38
U.S. population, urban *versus* rural, 106

V

Validation, 158
through listening, 236
Value, categories of, 145–146
Value judgments, 109
Values, role in social-environmental policy, 110
Vermont, off-grid experience in, 135
Vested interests, 93, 176
Violence, not listening as act of, 235
Vision, 145
goals and, 152–153
imagination as component of, 149–152
importance of, 146
monitoring activities, 154–158
and negotiability of constraints, 153–154
objectives and, 152–153
perception as component of, 148–149
understanding, 152
worldview as component of, 147–148
Vision statement, 153
Visioncrafting, 156, 230, 257
with children's participation, 249–250
Visioning process, 154, 187, 250
Voters, as agency constituents, 90
Voting rights, 75

W

Wants, conscious limitation of, 72
Washington, George, step-down from power, 165–166
Water
as captive of gravity, 39–40
role in Las Vegas history, 125
Water conservation, *versus* urban development, 137
Water empress of Vegas, 126
Water grab, 125, 130
cover-up at SNWA, 131
Water quality, 39–40
Water rights, SNWA case study, 129–130
Water shortages, due to global warming, 202
Water supply, failing to account for long-term, 49
Water war, 130, 137
Waterbed principle, 4–6, 37, 51
Wealth, *versus* money, 73, 106
Weathering, 17
Western Airborne Containments Assessment Project, 38
Wet-bulb temperature, limits for human tolerance, 68
Wholeness, 200
Wind patterns, effects of ocean warming on, 16
Wisdom, 203
Withdrawal, from negative criticism, 197
Work, as labor of love, 175
Working styles, differences in, 186
Worldview, as component of vision, 147–148

Y

Yangtze River, 202
Yom Kippur, scapegoat tradition, 116

Z

Zen cook, 216, 217

For Product Safety Concerns and Information please contact our EU representative GPSR@taylorandfrancis.com
Taylor & Francis Verlag GmbH, Kaufingerstraße 24, 80331 München, Germany

www.ingramcontent.com/pod-product-compliance
Ingram Content Group UK Ltd.
Pitfield, Milton Keynes, MK11 3LW, UK
UKHW022013180926
475701UK00003B/342
9780367867256